Library of
Davidson College

LLOYD'S OF LONDON PRESS LIMITED
Sheepen Place Colchester Essex CO3 3LP. Tel. (0206) 69222

YARROW

Flotilla Leader "Dubrovnik"

Designed and constructed by Yarrow & Co. Ltd.
for the
Royal Yugoslav Navy, 1932
this Vessel is the largest and most powerful of its class ever built in Great Britain

YARROW — GLASGOW

DESIGN AND CONSTRUCT
TORPEDO BOAT DESTROYERS   SLOOPS
GUN BOATS   ARMED LAUNCHES ETC.

GLASGOW

Front-line technology

The unique and proven V/STOL combat capability of the Harrier has revolutionised deployment of front-line air power at sea and on land. Harriers are in service with the Royal Air Force, the US Marine Corps and the Spanish Navy: Sea Harriers are operated by the Royal Navy and the Indian Navy. Harrier II, the advanced version developed and built jointly with McDonnell Douglas, has been chosen to meet US Marine Corps and RAF requirements totalling over 400 aircraft.

In the South Atlantic campaign, during 1982, 8,000 miles from home bases, Sea Harriers gave dramatic proof of their operational effectiveness, dominating supersonic adversaries in air combat and generating up to six missions per aircraft per day in the most hostile of weather and sea state conditions. The unrivalled experience gained with Harrier operational V/STOL, afloat and ashore, firmly establishes British Aerospace in the front line of combat aircraft technology for decades to come.

British Aerospace
defence programmes
also include
(top to bottom, left to right):
Tornado F2 *Air Defence Variant*
(with MBB and Aeritalia)
Goshawk *US Navy jet trainer*
(with McDonnell Douglas)
EAP *Experimental Aircraft Programme*
Rapier *low-level missile systems*
ASRAAM *Advanced Short-Range Air-to-Air Missile*
(with Bodenseewerk Geratetechnik)
Sky Flash *medium-range air-to-air missile*
Sea Eagle *sea-skimming anti-ship missile*

BRITISH AEROSPACE
...up where we belong

British Aerospace plc, 100 Pall Mall, London.

C³I SUPREMACY

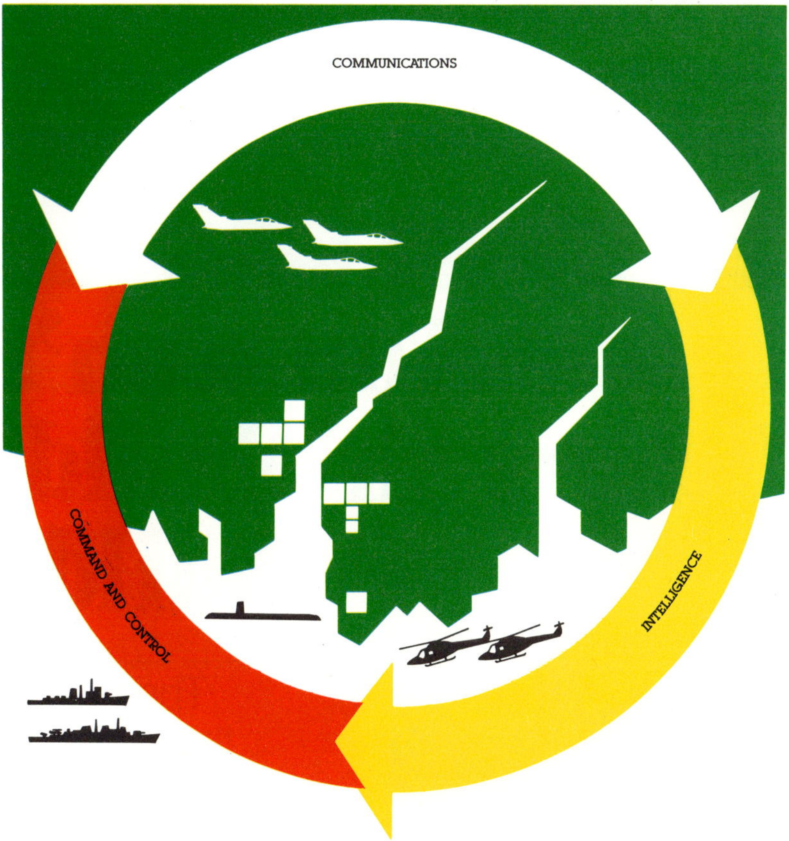

The Marconi Companies are collectively the most powerful defence electronics manufacturers in Europe, providing total system solutions for electronic defence projects on land, on and under the seas, in the air and in space.

Where the combined skills of several Marconi Companies are required, as in modern Command, Control, Communications and Intelligence (C³I) systems, Marconi Projects Limited provides a co-ordinated approach and 'single voice' management of all activities from concept to implementation and full life-cycle support.

Marconi

The Marconi Company Limited, The Grove, Warren Lane, Stanmore, Middlesex HA7 4LY, England. Tel: 01-954 2311. Telex: 22616.

1st BRITISH POLARIS MISSILE CARRYING SUBMARINE

1st TYPE 42 DESTROYER FOR THE ROYAL NAVY

You won't be
Vickers Shipbuilding
build you

1st BRITISH NUCLEAR SUBMARINE

1st EVER SUBMARINE FOR THE ROYAL NAVY

These 'milestone
the '1st's' that VSE
both in shipbuildi
manufacture of arn
And we are fi
because that's whe

Vickers
& Engine
A S
Barrow-in-Furness C
Telephone 0229

We bui
Because that's whe

1st BRITISH ANTI-SUBMARINE CARRIER

1st 'STRETCHED' TYPE 42 DESTROYER

he first to ask
nd Engineering to
"1st"

present some of
ve produced,
d the design and
ents over the years.
oice for so many,
e put their needs.

ing
ed
builders
1AF England
65411 VSEL G

sts".
 are in our field.

1st SECOND GENERATION PRESSURISED WATER REACTOR IN EUROPE

1st 2400 CLASS SUBMARINE FOR THE ROYAL NAVY WHICH ENTERS SERVICE IN 1987

A MODERN SHIPYARD COMMITTED TO TOMORROWS NEEDS

Cammell Laird maintains high standards both technically and in building of ships. We are pioneering new techniques and developing our facilities for the new generation of specialist ships. We have led the industry in Advanced Building Methods and Outfitting in the largest undercover warshipbuilding facility in Britain. This is being supported by the latest Computer Aided Design and Manufacturing (CADAM) techniques. We have an enviable record for building commercial vessels, and this high standard has been complementary to a continuous programme for the Royal Navy including submarines, destroyers, frigates, aircraft carriers as well as support ships for the Royal Fleet Auxiliary. The skills and flexibility of the management and workforce is testimony to this excellent record.

With every modern facility at our command, Cammell Laird can undertake the build of any size of surface ship or submarine for the navies of the world, and can convert merchant ships for special fleet roles. We also have the capability and experience to refit ships, submarines and auxiliaries to exact specifications. The recently announced amalgamation with Vickers Shipbuilding & Engineering Company marks another major landmark in our long history and makes us part of one of the largest shipbuilding organisations in the world.

This year, Cammell Laird will lay the keel of a Type 22 Batch III Frigate of the Broadsword Class under contract from the Ministry of Defence (N). This carries on Cammell Laird's programme of builds for the British Navy starting in 1846 with the 1,400 ton Man of War Birkenhead.

HMS EDINBURGH

RFA BRAMBLELEAF

HMS EDINBURGH — Advanced outfitting under cover

HMS REVENGE

CAMMELL LAIRD SHIPBUILDERS LIMITED
New Chester Road Birkenhead L41 9BP England
Telephone 051-647 7080 Telex 629463 Facsimile No. 051-647 7727

Defence Systems Consultants

- Command, Control, Communications
- Information Handling
- Avionics
- Consultancy
- Advanced Software Tools
- Signal Processing
- Electronic Warfare
- Project Management Services
- Operational Research

Contractors to:
UK Ministry of Defence
US Department of Defense
European Space Agency
NATO
PANAVIA
CERN

Systems Designers plc, Pembroke House
Pembroke Broadway, Camberley, Surrey GU15 3XD
Telephone: 0276 686200
Telex: 858280. Telecopier: 0276 683511

Systems Designers

"SINGAPORE" ALL-METAL FLYING-BOAT

The first all-metal flying-boat to be acquired by the Royal Air Force is the "Short-Singapore," constructed of light metal alloys; and built by the pioneers of British all-metal aeroplanes and seaplanes.

BROTHERS

(Rochester & Bedford), LIMITED,

WHITEHALL HOUSE, 29-30, CHARING CROSS, LONDON, S.W.1, ENGLAND.

Defending tomorrow's world

Short Brothers—leaders in the aerospace industry since the turn of the century—are today supplying high-technology defence products to peace-keeping forces throughout the world and are engaged in wide-ranging research, design and development programmes covering new generation products which will provide pace-setters in their fields through the present decade and far beyond.

JAVELIN SHERPA

TUCANO SEACAT

SHORTS
WINNERS OF FOURTEEN QUEEN'S AWARDS TO INDUSTRY

SHORT BROTHERS PLC Missile Systems Division Castlereagh Belfast BT6 9HN Northern Ireland Telephone 0232 703503 Telex 747087

Micro to Maxi-technology

Royal Ordnance designs and produces a comprehensive range of land, naval and air defence systems, sub-systems and components for the home and overseas markets.

From a micro miniaturised electronic time fuze for artillery ammunition to the 60 tonne Challenger main battle tank, Royal Ordnance deploys its special research, design and development skills. Skills which are fully integrated with some of the world's most advanced manufacturing facilities. Four divisions cover the fields of Weapons & Fighting Vehicles, Small Arms, Ammunition and Explosives—including propellants and rocket motors. Their combined technical expertise creates a unique structure which brings enormous customer benefits.

Royal Ordnance has been the foremost supplier of defence equipment to the British forces for over four centuries. Today our products and components are in service worldwide, and we are committed to continued investment in high technology. Proof indeed of our dedication to defence.

Royal Ordnance Marketing Services,
Griffin House,
PO Box 288, The Strand, London
WC2N 5BB, United Kingdom.
Telephone 01-930 4355. Telex 919661.

ROYAL ORDNANCE
Defence systems, sub-systems and components

"It is upon the navy under the Providence of God that the safety, honour and welfare of this realm do chiefly attend." Charles II

King Charles acted upon his conviction of the Navy's importance by doing much to make it more effective.

In something of the same spirit Ferranti today devotes the most advanced technology and resources to meeting the Royal Navy's current needs.

It is only 25 years since the first digital computer went to sea. Now practically every ship of the Royal Navy carries Ferranti equipment.

We think His Late Majesty would have approved.

FERRANTI Computer Systems

Ferranti Computer Systems Limited, Western Road, Bracknell, Berkshire, RG12 1RA.

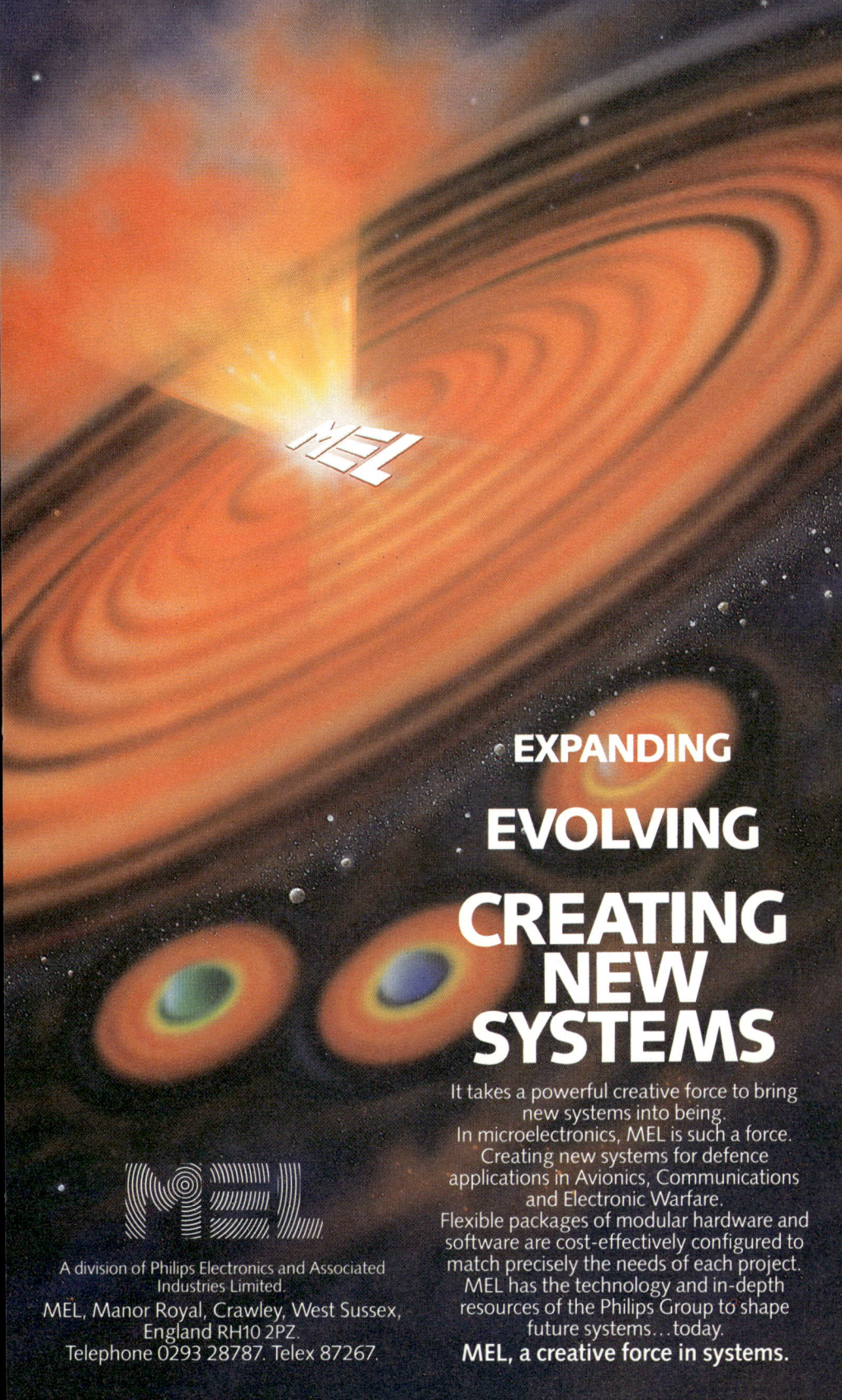

IRONCLAD to TRIDENT

100 Years of Defence Commentary

BRASSEY'S, 1886–1986

**BRASSEY'S DEFENCE PUBLISHERS
EDITORIAL ADVISORY BOARD**

Group-Captain D. Bolton
Director
Royal United Services Institute for Defence Studies

Lieutenant-General Sir Robin Carnegie KCB OBE

Armande Cohen CBE
International Civil Servant and formerly with the Assembly
of Western European Union

Professor L. D. Freedman
Department of War Studies, King's College, London

John Keegan
Author and Senior Lecturer, Royal Military Academy Sandhurst

The Rt Hon Lord Mulley

Henry Stanhope
The Times Foreign Policy Correspondent

John Terraine
Author

Some titles available from Brassey's

COKER
A Nation in Retreat: Britain's Defence Commitment

LAFFIN
Brassey's Battles: 3,500 Years of Conflict,
Campaigns and Wars from A to Z

LIDDLE
Home Fires and Foreign Fields: British Social and
Military Experience in the First World War

RUSI/BRASSEY'S Defence Yearbook 1986

SIMPKIN
Race to the Swift: Thoughts on Twenty-First Century Warfare

WINDASS
Avoiding Nuclear War: Common Security as a
Strategy for the Defence of the West

IRONCLAD to

100 Years of Defence Commentary
BRASSEY'S 1886–1986

Centenary Volume of Brassey's Naval Annual (Incorporating Brassey's Naval and Shipping Annual, Brassey's Armed Forces Yearbook and RUSI/Brassey's Defence Yearbook)

Edited by

BRYAN RANFT

BRASSEY'S DEFENCE PUBLISHERS
a Member of the Pergamon Group

LONDON · OXFORD · WASHINGTON D.C.
NEW YORK · TORONTO · SYDNEY · FRANKFURT

U.K. (Editorial)	Brassey's Defence Publishers Ltd., Maxwell House, 74 Worship Street, London EC2A 2EN
(Orders)	Brassey's Defence Publishers Ltd., Headington Hill Hall, Oxford OX3 0BW, England
U.S.A. (Editorial)	Pergamon-Brassey's International Defense Publishers, 1340 Old Chain Bridge Road, McLean, Virginia 22101, U.S.A.
(Orders)	Pergamon Press Inc., Maxwell House, Fairview Park, Elmsford, New York 10523, U.S.A.
CANADA	Pergamon Press Canada Ltd., Suite 104, 150 Consumers Road, Willowdale, Ontario M2J 1P9, Canada
AUSTRALIA	Pergamon Press (Aust.) Pty. Ltd., P.O. Box 544, Potts Point, N.S.W. 2011, Australia
FEDERAL REPUBLIC OF GERMANY	Pergamon Press GmbH, Hammerweg 6, D-6242 Kronberg, Federal Republic of Germany
JAPAN	Pergamon Press Ltd., 8th Floor, Matsuoka Central Building, 1-7-1 Nishishinjuku, Shinjuku-ku, Tokyo 160, Japan
BRAZIL	Pergamon Editora Ltda., Rua Eça de Queiros, 346, CEP 04011, São Paulo, Brazil
PEOPLE'S REPUBLIC OF CHINA	Pergamon Press, Qianmen Hotel, Beijing, People's Republic of China

Copyright © 1986 Brassey's Defence Publishers

All Rights Reserved. No part of this publication may be reproduced, stored in a retrieval system or transmitted in any form or by any means: electronic, electrostatic, magnetic tape, mechanical, photocopying, recording or otherwise, without permission in writing from the publishers.

First edition 1986

British Library Cataloguing in Publication Data
Ironclad to Trident: 100 years of defence
commentary, Brassey's, 1886–1986: centenary
volume of Brassey's naval annual (incorporating
Brassey's naval and shipping annual, Brassey's
armed forces yearbook and RUSI/Brassey's defence
yearbook).
1. Armed Forces—History
I. Ranft, Bryan
355'.009   U39
ISBN 0-08-031191-1

Printed in Great Britain by A. Wheaton & Co. Ltd., Exeter

Editors of Brassey's Annuals 1886–1986

NAVAL ANNUAL

1886–1891	Lord Brassey
1892–1899	T. A. Brassey
1900–1901	J. Leyland
1902–1905	T. A. Brassey
1906	J. Leyland and T. A. Brassey
1907–1913	T. A. Brassey (Viscount Hythe from 1911)
1914	Viscount Hythe and J. Leyland
1915–1916	J. Leyland
1917–1918	Not published
1919	2nd Earl Brassey and J. Leyland

NAVAL AND SHIPPING ANNUAL

1920–1928	A. Richardson and A. Hurd
1929	C. N. Robinson
1930–1935	C. N. Robinson and H. M. Ross

NAVAL ANNUAL

1936	C. N. Robinson
1937–1949	H. G. Thursfield

BRASSEY'S ANNUAL: THE ARMED FORCES YEARBOOK

1950–1963	H. G. Thursfield
1964–1973	J. L. Moulton

ROYAL UNITED SERVICES INSTITUTE AND BRASSEY'S DEFENCE YEARBOOK

1974–1975	Editorial Board: S. W. B. Menaul, R. G. S. Bidwell, R. H. F. Cox

EDITORS OF BRASSEY'S ANNUALS 1886–1986

1976	Editorial Board: S. W. B. Menaul, A. E. Younger, R. H. F. Cox
1977–1979	Editorial Board: A. E. Younger, E. F. Gueritz, R. H. F. Cox
1980–1982	Editorial Board: E. F. Gueritz, H. Stanhope, Jennifer Shaw
1983–1984	Editorial Board: D. Bolton, H. Stanhope, Jennifer Shaw
1985	Editorial Board: D. Bolton, H. Stanhope, Jennifer Shaw, A. J. Trythall
	Editor: B. H. Reid

Preface

The main problems of editing this celebratory volume have been finding methods of selecting material from 100 years of publication to fit within a reasonably sized book which would both present a comprehensive survey of Brassey's achievement and be more than a collection of unconnected pieces.

The solutions have been sought in the nature of the Annual itself. Since its foundation as the Naval Annual in 1886 to its present form as the Defence Yearbook, its aim has been to provide accurate information and provoke intelligent discussion on defence matters. To achieve this it has summarised and analysed the developments in international relations, strategic thinking and military technology which have affected the nature of war.

The method of selection, therefore, has been to divide the century into chronological periods which have some thematic unity and within each to select those articles which best depict the most significant developments, both as they were seen at the time and for their relevance today. Considerations of space have necessitated further narrowings of selection. The articles finally chosen concentrate largely on politico-strategic matters and their interaction with technological change. Highly important contributions on personnel matters and detailed engineering subjects have had to be excluded. Many important articles have been omitted, especially in the most recent volumes, because of their length. Others have been cut (indicated by * * * * * in the text) and apologies are due to authors for this. The aim has been to produce a volume which is both a centenary celebration and which can be read in its own right as a historical survey of modern war and an introduction to contemporary strategic dilemmas.

Acknowledgements

My thanks are due to Tony Trythall and Jenny Shaw of Brassey's Defence Publishers who originally thought of this book and who, together with their colleagues, have given continuous help in its production. The bulk of the research was done at the National Maritime Museum Greenwich, where the Library and Photocopying staff were unfailingly helpful, as was Sylvia Smither with her typing.

May 1985   BRYAN RANFT

Contents

General Introduction   xv

List of Illustrations   xxi

Advertisements as History—following page 55

Part I: 1886–1904
Global Rivalry and Technological Change   1

1. **1886**
   Recent Naval Administration   5
   LORD BRASSEY
2. **1887**
   Shipbuilding Policy: Protection of Commerce   11
   ANON
3. **1888–89**
   The New Shipbuilding Programme   15
   LORD GEORGE HAMILTON
4. **1893**
   Naval Manoeuvres in 1892   18
   J. R. THURSFIELD
5. **1894**
   British Naval Manoeuvres in 1893   23
   J. R. THURSFIELD
6. **1896**
   The Value of Torpedo Boats in Wartime   24
   COMMANDER R. H. BACON
7. **1897**
   The Progress of Foreign Navies   28
   E. WEYL
8. Relative Strength   30
   T. A. BRASSEY
9. Principles of Imperial Defence   32
   T. A. BRASSEY
10. **1900**
    Progress of the British Navy   38
    C. N. ROBINSON
11. The Progress of Foreign Navies   41
    J. LEYLAND
12. Naval Manoeuvres in 1899   43
    J. R. THURSFIELD
13. **1901**
    The Transport Operations to South Africa   45
    J. LEYLAND

x CONTENTS

14. 1904
    Submarines and Torpedo Warfare   49
    ANON

Part II: 1905–1914
Anglo-German Rivalry and Continuing Technical Advance   57

15. 1906
    Steam Engineering—The Turbine   60
    G. R. DUNNELL
16. 1907
    The Tactical Qualities of the Dreadnought Type of Battleship   62
    LIEUTENANT-COMMANDER W. S. SIMS, USN
17. 1909
    The Naval Expansion of Germany   67
    J. LEYLAND
18. Speech on the Naval Estimates   72
    R. McKENNA, FIRST LORD OF THE ADMIRALTY
19. 1912
    Naval War Staffs   76
    J. LEYLAND
20. Recent Changes in Warship Design   80
    SIR WILLIAM WHITE
21. 1913
    The Progress of Naval Aeronautics   84
    ANON
22. 1914
    Armour and Ordnance   89
    C. N. ROBINSON

Part III: 1915–1919
The Great War   93

23. 1915
    The World War. Narrative of Naval Events and Incidents   96
    C. N. ROBINSON
24. First Lord's Statement in the House of Commons, November 27, 1914   100
    WINSTON CHURCHILL
25. German Declaration—War Area   102
    ADMIRALTY STAFF MEMORANDUM, FEBRUARY 2, 1915
26. 1919
    The Enemy Navies   104
    J. LEYLAND
27. The Submarine War on Merchant Shipping   107
    COMMANDER L. H. HORDEN RN

Part IV: 1920–1935
Disarmament and the Air Power Controversy   113

28. 1920–1921
    Foreign Navies: United States and Japan   117
    J. LEYLAND
29. 1921–1922
    The Capital Ship   121
    REAR-ADMIRAL SIR REGINALD BACON

30.	The Possibilities of the Torpedo REAR-ADMIRAL S. S. HALL	126
31.	Air Power and Sea Power MAJOR-GENERAL SIR S. W. BRANCKER	132
32.	**1934** Foreign Navies CAPTAIN E. ALTHAM RN	139
33.	**1935** Japan and her Navy COMMANDER S. TAKAGI IJN	144
34.	The Fleet Air Arm 'VOLAGE'	153

Part V: 1936–1939
The Approach to War — 159

35.	**1936** Relative Naval Strength G. H. HURFORD	163
36.	United States Naval Aviation REAR-ADMIRAL E. J. KING USN	166
37.	**1937** The Merchant Navy as a Factor in Imperial Defence E. H. WATTS	170
38.	**1938** British Naval Air Progress: The Fleet Air Arm Decision H. G. THURSFIELD	174
39.	**1939** The Interdependence of Home Defence and Commerce Protection 'SECURUS'	178

Part VI: 1940–1949
World War II: Interpreting the Lessons — 183

40.	**1940** The Merchant Navy in War and Peace ARCHIBALD HURD	186
41.	**1945** The Organisation of Fighting Forces H. G. THURSFIELD	190
42.	**1946** The Air War at Sea OLIVER STEWART	199
43.	**1948** The Future Employment of Naval Forces FLEET ADMIRAL CHESTER W. NIMITZ USN	204

Part VII: 1950–1956
Infantry Wars in a Nuclear Age — 209

44.	**1951** Western Defence COLONEL E. H. WYNDHAM	212
45.	**1954** The Role of Modern Infantry in Battle MAJOR-GENERAL SIR HAROLD E. FRANKLAND	217

xii CONTENTS

46. 1955
   The Strategic Air Command US Air Force ... 221
   AIR VICE-MARSHAL W. M. YOOL
47. Operations by Regular Troops Against a Guerrilla Enemy ... 226
   MAJOR-GENERAL E. K. G. SIXSMITH

Part VIII: 1957–1965
Strains within NATO: The Intercontinental Ballistic Missile ... 233

48. 1957
   Defence Policy. A New Approach? ... 237
   H. G. THURSFIELD
49. The Royal Air Force in a Time of Growth and Change ... 242
   D. M. DESOUTTER
50. 1958
   Those Against the H-Bomb ... 251
   RICHARD GOOLD-ADAMS
51. 1963
   West European Defence ... 258
   AIR VICE-MARSHAL W. M. YOOL
52. 1965
   NATO ... 263
   ALASTAIR BUCHAN
53. The Future of the Aircraft Carrier ... 272
   VICE-ADMIRAL SIR PETER GRETTON

Part IX: 1966–1981
Defence Reviews: NATO and the Increasing Soviet Threat ... 279

54. 1966
   The 1966 Defence White Paper and Debate ... 283
   MAJOR-GENERAL J. L. MOULTON
55. Review of the Military Situation in Europe ... 287
   MAJOR-GENERAL E. K. G. SIXSMITH
56. 1967
   People's Revolutionary War ... 293
   SIR ROBERT THOMPSON
57. 1969
   NATO's Northern Flank ... 299
   CLAUS G. M. KOREN
58. 1970
   American Defence Policy and Strategy for the 1970s ... 306
   COLONEL R. D. HEINL JR
59. 1971
   NATO from a SACLANT Viewpoint ... 312
   ADMIRAL E. P. HOLMES USN (RET)
60. 1972
   The Transformation of Strategy ... 318
   MICHAEL HOWARD
61. The Army in Northern Ireland ... 326
   MICHAEL BANKS
62. 1974
   The Defence of France and the Defence of Europe ... 333
   PIERRE DABEZIES

63. 1976–1977
    The Soviet Military Effort in the 1970s   338
    JOHN ERICKSON
64. 1978–1979
    Perspectives of NATO Defence   342
    J. M. A. H. LUNS
65. 1980
    Intelligence—The Handmaiden of Policy   347
    LIEUTENANT-GENERAL SIR DAVID WILLISON
66. 1981
    Civil Defence—A View for 1981   354
    C. N. DONNELLY

Part X: 1982–1986
Full Circle   361

67. 1982
    Trident: A Candidate for Cancellation   365
    IAN BELLANY
68. Merchant Shipping and the Maritime Threat   373
    H. G. DAVY
69. 1983
    Strategic Weapons   380
    LAWRENCE FREEDMAN
70. 1985
    Space: The Military Applications Today and Tomorrow   388
    GROUP CAPTAIN T. GARDEN
71. 1986
    British Defence Issues   396
    ADMIRAL SIR JOHN FIELDHOUSE

Index   **405**

General Introduction

The Founder

Thomas Brassey, founder and first Editor of the Annual, was born in 1836. His father, also Thomas (1805–1870), achieved prominence and wealth as a civil engineer. He was associated with George Stephenson and, carried on the wave of railway building which characterised the nineteenth century, became a large-scale international contractor, especially in India and South America.

Like so many self-made Victorian entrepreneurs, he gave his son the education and access to social status he himself lacked. The younger Thomas went to Rugby and University College Oxford where he read law and modern history. He was called to the Parliamentary Bar in 1866, but soon abandoned it for politics. Already in 1861 he had stood unsuccessfully as a Liberal at Birkenhead, where his father's career had begun. In June 1865 he was successful at the naval town of Devonport, but was defeated at the General Election a few weeks later before he had taken his seat. It was not until 1868 that he finally entered the House as member for Hastings, another seaside constituency, which he represented until 1886, the year he founded the Annual.

As an M.P., Brassey sought a reputation as a specialist expert rather than as a political activist. Following his father's interests, he studied industrial labour problems and published two books, *Work and Wages* (1872) and *Foreign Work and English Wages* (1879). But already his maritime interests had emerged with the publication of *British Seamen* in 1877, to be followed in 1882–1883 by the encyclopaedic *The British Navy*, the five volumes of which laid the foundations of the Annual. These publications and his parliamentary speeches were complemented by a flow of letters to *The Times* and other journalism on maritime matters. He was particularly concerned with the impact of technological change on both the Royal and Merchant Navies, naval pay, the administration of dockyards, ship design and, a favourite hobby horse, the provision of naval reserves—all of which figured largely in his editorship of the Annual.

Brassey's naval expertise led to his appointment as Civil Lord of the Admiralty in Gladstone's 1880 Administration and as Parliamentary Secretary in 1884. As Civil Lord his direct responsibilities were confined to buildings and works and the administration of Greenwich Hospital. But his passion for naval affairs led him to bombard his colleagues with detailed memoranda which, in the critical comment of Vincent Baddeley, an Admiralty civil servant, although appreciated for their information, "seldom led to any concrete conclusion and the effect of them was rather to ventilate the subject than to produce any tangible result". In his article on Brassey in the *Dictionary of National Biography* (1912–1921) Baddeley adds that during his short term as Parliamentary Secretary, acting as Admiralty spokesman in the House of Commons, he "did not add to his reputation, for he was no parliamentary debater, not quick at taking up points made against his department in the House of Commons".

Although he was given a peerage as Baron Bulkeley and held minor office in Gladstone's government of 1893, his domestic political career was over, and his energies and most long-lived achievement centred on the Annual. His appointment as Governor of Victoria (1895–1900), although it took him away from direct control of the Annual, did not break his connection with naval affairs. He was closely concerned with the federation of the Australian states which he saw as a forerunner to the country becoming a member of a future Imperial Federation of all the self-governing colonies, each of them contributing to the Imperial Navy, to diminish the burden of increasing naval expenditure on the Mother Country. This theme was continuously taken up in his later speeches in the House of Lords and in his contributions to the Annual, and to the proceedings of the Institute of Naval Architects, of which he became President in 1893.

He was created Earl Brassey in 1911 and died in 1918. He was succeeded in the earldom and in the control of the Annual by his son Thomas, who had already acted as editor since 1892. He died without issue in the following year.

Changing Format: Sustained Aim

From its beginning Brassey's has been a compilation depending for its accuracy and comprehensiveness on the contributions of a wide range of specialists working under editors remarkable for their balanced views on defence matters. Although the format and organisation of the Annual have had several changes, the still prevailing pattern was set in the first few editions. The content was divided between discussion and information. The former dealt with international events and domestic politics as far as they affected defence,

GENERAL INTRODUCTION   xvii

significant changes in British and foreign naval policy, especially as they were influenced by technological change, and matters of recruitment, training, and conditions of service. Active service operations and analyses of the significance of British and foreign manoeuvres provided some of the livelier articles. The information section consisted of detailed statistics of all the world's navies, supported by a large number of illustrations, silhouettes and plans of warships and diagrams of guns, engines and other equipment. To these two main sections were added appendices based on official documents and speeches. The Annual was unique in the completeness of its coverage and, although edited by Britons, made frequent use of foreign experts who were given full freedom to put their countries' point of view. As long as it remained purely a naval annual, it naturally put the case for the maintenance of Britain's maritime power and resources, but never stridently. It may have been complacent in its treatment of Britain's imperial and commercial power, but rarely jingoistic. For instance, it showed understanding of Germany's claim to maritime strength before 1914, just as in another context it was to give fair treatment to all sides of the bitter argument about the control of maritime airpower in the years between the wars.

Naturally it criticised British governments of all parties who seemed to be sacrificing Britain's security to short-term political expediency, but never intemperately. This was largely due to the intellectual quality and sense of responsibility of a line of distinguished editors, military and civilian. In this avoidance of partisanship the Annual has followed the precepts of its founder who, in his 1897 Preface, stressed both that his material was drawn from sources open to the public and not provided by government favour and that, while Brassey's would criticise past and present policies in drawing lessons for the future, such criticism would not be "barren and would be free from party political partisanship". Wider international events did occasionally impinge on production. In 1899 E. Weyl, a long-standing French contributor, withdrew in the aftermath of the Fashoda incident and in 1901 the younger Brassey had to relinquish the editorship while serving in the South African War. A further complication came in 1911 when the strains of mounting naval rivalry made governments more cautious in publishing details of naval equipment.

It was at the height of the Anglo-German naval race in 1912 that the younger Brassey, now Viscount Hythe, considered discontinuing the Annual because its basic purpose of providing information for naval officers was now being more adequately fulfilled by the Admiralty, and the task of informing the general public by the Navy League. Fortunately, after consultations with naval officers and others, he decided to continue. The Annual was far from being in a rut and in 1914 demonstrated its concern with new developments by beginning a systematic examination of aeronautics and a listing of the air forces of the major powers. This last pre-war volume also marked a domestic change: it was the first to be printed by William Clowes, replacing J. Griffin of Portsmouth, the original printers. It was also the first volume to contain illustrated advertisements, which are themselves of considerable historical interest, as will be seen from those included in this book.

The war inevitably brought difficulties. The volumes produced in 1915 and 1916 were much shorter than their predecessors. This was largely due to the omission for security reasons of detailed information on the British and allied navies. They did include as much as was available on enemy forces and narratives of naval events in the war drawn from official publications. These deliberately avoided criticisms of operations and evaluations of tactics. These were left to future historians free from security restrictions. No volumes were published in 1917 and 1918 and it was not until 1919 and after that critical assessments of war experiences appeared.

The 1920 volume recorded the beginning of a new era. On the international scene it was marked by the ending of Britain's naval predominance and by widespread demands for naval disarmament. For Brassey's it meant the end of the family editorship and the succession of two established contributors, Alexander Richardson MP and Archibald Hurd, an expert on the Mercantile Marine and author of its official war history. It was presumably Hurd's influence, and his belief in the vital part merchant shipping had played in the war, which brought about the change of title to *Brassey's Naval and Shipping Annual*, and the division of subsequent editions into a Naval and a Merchant Shipping section. A change which the 1921 Preface records had been widely welcomed and accepted as providing "the only publication of its wide scope dealing with naval and shipping affairs in the English speaking world". This team continued until Hurd resigned owing to illness in 1927 and Richardson died in the following year. Control passed to his son James, a marine engineer, who died soon after. He had arranged for the editorship to pass to Commander C. N. Robinson, who had contributed on British naval matters since 1901. Robinson was to continue until 1936, assisted from 1930 by H. M. Ross. The acceptance of the growing importance of air power was made apparent in 1935 with the appearance of an enlarged Air Section, dealing with civil as well as military maritime aviation. "Thus", as the Preface proclaimed, "in one volume are now brought together the three great branches of that sea power which is so vitally concerned in the security of the Empire—the Royal Navy, the Merchant Navy and the Air Force."

In the Jubilee volume of 1936, the last Robinson was to edit, perhaps wishing to put naval rearmament at the centre of his compilation, he restored the original title of *Naval Annual*. He was succeeded by Rear-Admiral H. G. Thursfield, naval correspondent of *The Times*. He was to continue as Editor until 1963, contributing many perceptive articles as well as presiding over the consequences of the change in title to *Brassey's Annual: The Armed Forces Yearbook* in 1950. He saw this as a natural development justified by the increasing acceptance of the unity of defence. He described the Annual's aim as being to serve all three services as it had the navy since 1886, by reviewing the events, developments and opinions of the past year and also by emphasising the importance of defence to a wider public. He also re-emphasised the Annual's independence: "it has never been up till now and will not be in the future in any way official, in the sense of being officially controlled or even officially inspired. Opinions will be freely expressed and are the responsibility of the contributor and the editor." This was followed by a qualification echoing the tones of the founder:

it should however be emphasised that independence does not connote irresponsibility. Every care will continue to be taken that only authoritative information will appear. Gossip or speculation on matters which are officially secret is tightly barred from our pages.

The Annual's wider coverage was emphasised by the appointment of two assistant editors, Brigadier C. N. Barclay and Air Vice-Marshal H. M. Yool, who made significant regular contributions for many years. In 1956, at a time of increasing defence controversy, especially on the relative weight to be given to nuclear and conventional forces and growing inter-service conflict between the Navy and the Air Force, Thursfield again thought it necessary to define the Annual's position:

Articles appearing in Brassey's Annual which contain theories or opinions upon strategical subjects, upon Defence Policy and upon administration in the Services, are included because, in the opinion of the Editors, they constitute valuable contributions to the discussion of such matters. But the theories of opinions expressed are those of the authors alone and may or may not accord with official views. The fact that they appear in Brassey's Annual must not be taken as indicating any official approval or inspiration.

This firm declaration, absolutely necessary in a publication which to be effective must be independent and yet maintain friendly relations with officialdom as well as opening its pages to serving officers and officials, has continued to be echoed until today.

Thursfield continued as Editor until his death in 1963 and was succeeded by Major-General J. L. Moulton who had retired from the Royal Marines in 1961 as Chief of Amphibious Warfare. He was already a contributor and well qualified to maintain the Annual's joint service emphasis. He was to continue until 1973. In the following year the Annual appeared in its present guise, the close association with the

Royal United Services Institute for Defence Studies, bringing together the two major British organisations solely concerned with defence studies. The editorial control passed to a Board usually headed by the Director of the RUSI. The final change came in 1980 with the Annual's absorption into the Pergamon publishing group, leaving the editorial control unchanged.

List of Illustrations

1924	1. Earl Brassey—following page xxii
1898	2. The Spithead Review 1897

The Man of War in Transition—following page 91

1886	3. H.M.S. *Imperieuse*
1892	4. H.M.S. *Royal Sovereign*
1907	5. H.M.S. *Dreadnought*
1895	6. H.M.S. *Daring*

The Great War—Victors and Vanquished—following page 111

1913	7. Torpedo tube
1901	8. H.M.S. *Cressy*
1909	9. German Armoured Cruiser *Scharnhorst*
1912	10. German Battle Cruiser *Von Der Tann*
1913	11. H.M.S. *Queen Mary*

The Industrial Base—following page 158

1923	12. Boring machine for heavy guns
1934	13. Turbine erecting pits

Danger in the Far East—following page 158

1921/22	14. Japanese Battleship *Nagato*
1927	15. H.M. Submarine *Oberon*

The Coming of the Aircraft Carrier—following page 181

1966	16. Two *Ark Royals*
1924	17. H.M.S. *Hermes*
1935	18. Hawker *Nimrod* and *Osprey* aircraft
1926	19. U.S.S. *Saratoga*

xxii   LIST OF ILLUSTRATIONS

World War II—Men in Action—following page 207

1940	20. *Sunderland* Flying Boat-control cabin
1943	21. Life in a British Submarine
1945	22. With an Arctic Convoy
1944	23. Salerno Landing
1945	24. First Men ashore in Normandy

World War II—Merchant Ships at War—following page 207

1924	25. The *Athenia*
1925	26. The *Rawalpindi*
1935	27. The *Queen Mary*

Dawn of the Nuclear Age—following page 232

1954	28. First atomic shell
1950	29. Boeing B 47 *Stratojet* Bomber
1953	30. Avro *Vulcan* Bomber
1962	31. *Polaris* Missile
1965	32. U.S. *Enterprise*

Aerial Flexibility—following page 277

1951	33. Piasecki Helicopter
1963	34. Hawker P. 1127

The Lives of a Soldier—following page 360

1969	35. The British Army in Germany—The Royal Artillery
1954	36. Malaya-Casualty evacuation
1966	37. Southern Arabia
1971	38. The Northern Flank
1970	39. Belfast 1970

Russia Goes to Sea—following page 360

1968	40. "Z" Class Submarines
1971	41. Confrontation in the Air and at Sea
1976/77	42. The Soviet Warship *Kiev*

EARL BRASSEY
(1836-1918)

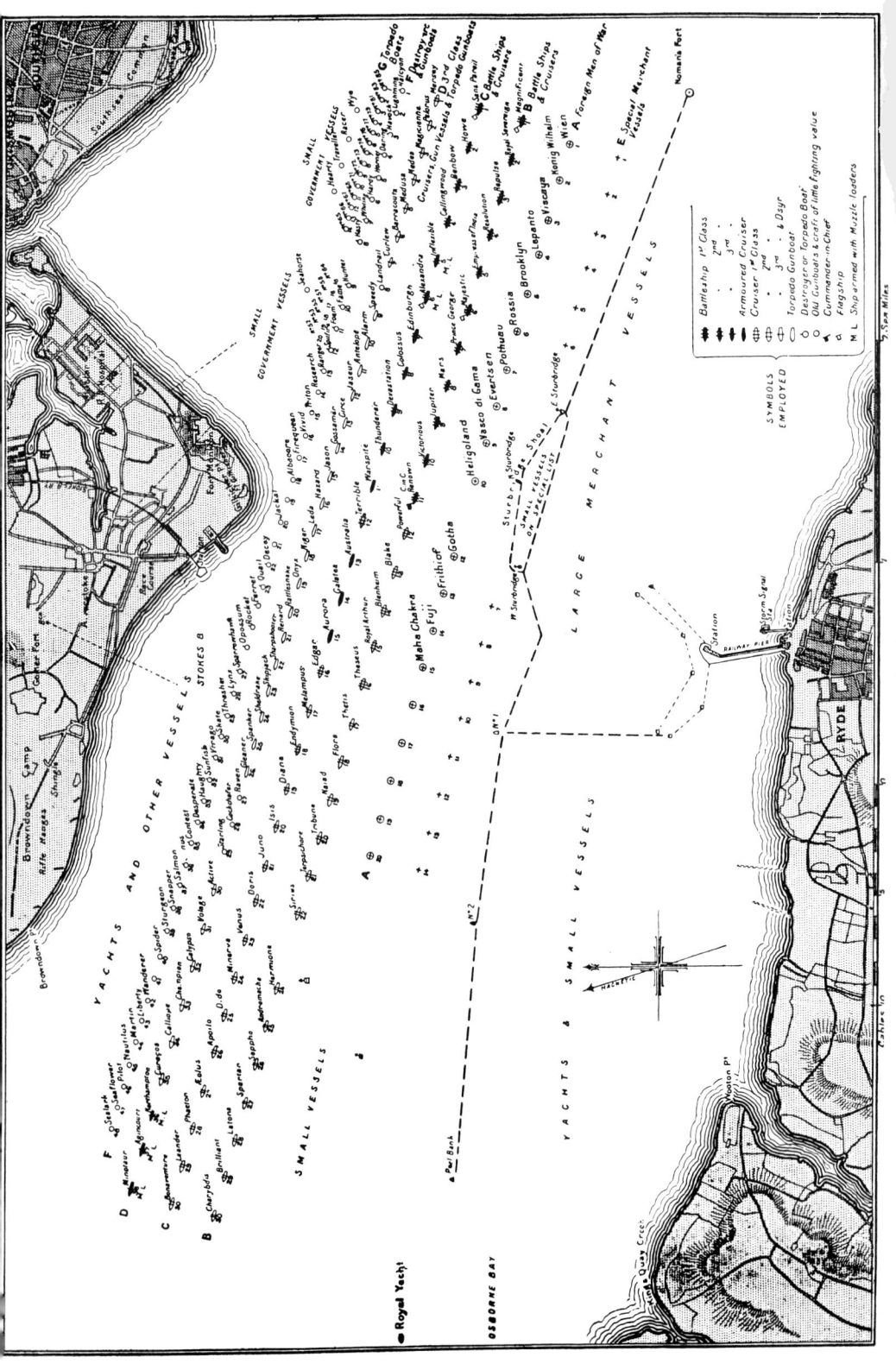

PART I: 1886-1904
Global Rivalry and Technological Change

Introduction	3
1. 1886 Naval Administration: General Policy LORD BRASSEY	5
2. 1887 Shipbuilding Policy: Protection of Commerce ANON	11
3. 1888-89 The New Shipbuilding Programme LORD GEORGE HAMILTON	15
4. 1893 Naval Manoeuvres in 1892 J. R. THURSFIELD	18
5. 1894 British Naval Manoeuvres in 1893 J. R. THURSFIELD	23
6. 1896 The Value of Torpedo Boats in Wartime COMMANDER R. H. BACON	24
7. 1897 The Progress of Foreign Navies E. WEYL	28
8. 1897 Relative Strength T. A. BRASSEY	30

GLOBAL RIVALRY AND TECHNOLOGICAL CHANGE

9. **1897**
Principles of Imperial Defence 31
T. A. BRASSEY

10. **1900**
Progress of the British Navy 38
C. N. ROBINSON

11. **1900**
The Progress of Foreign Navies 41
J. LEYLAND

12. **1900**
Naval Manoeuvres in 1899 43
J. R. THURSFIELD

13. **1901**
The Transport Operations to South Africa 45
J. LEYLAND

14. **1904**
Submarines and Torpedo Warfare 49
ANON

Introduction

After the Franco-Prussian War of 1870 rivalry between the European powers—Britain, France, Germany and Russia—increasingly took the form of competition for colonies and overseas markets, which in turn led to naval expansion; and the rise of Japan was eyed apprehensively by both Russia and the United States. It was also in the United States that the most substantial arguments linking national greatness and prosperity with sea power originated in the globally influential writings of A. T. Mahan.

In Britain it was universally accepted that her imperial and trading position depended on the world-wide dominance of her navy. Therefore the rise of foreign navies was watched with apprehension combined with a determination to match it, whatever the cost. The most pressing danger was seen as the capability of France and Russia, especially in alliance, to launch crippling attacks on Britain's merchant fleet, then larger than those of all the rest of the world combined. This fear produced the massive shipbuilding programme of the Naval Defence Act of 1889. This was the real beginning of the naval race commonly associated with Britain's response to Germany's naval expansion after 1898.

If this was the international scene in which Brassey's emerged, the 1880s also marked the consolidation of the technological innovations which had revolutionised naval matériel since 1815 and cast clouds of uncertainty over the nature of future maritime war. Increasingly efficient boilers and engines hastened the abandonment of sail even as an auxiliary to steam and increased the significance of speed in combat. Iron and steel similarly displaced wood in hull construction and the growing destructiveness of guns and projectiles competed with heavier protective armour. By the end of the decade the principal problems in designing warships to embody all these advances had been identified, if not completely solved. Huge ironclads became symbols of national prestige as well as evidence of the technical triumphs of the age. Private shipbuilders entered with gusto into competition to provide any government with the necessary cash with an up to date fleet.

As our own age witnesses, technology never stands still. Well before the battleship reached its most powerful form, new underwater weapons, especially the Whitehead torpedo, were seen by some to mark their obsolescence as they became vulnerable to attacks by small and inexpensive torpedo boats. This extreme view was accepted by few professionals. The majority thought that quick-firing guns and torpedo boat destroyers would provide an effective defence. The emergence of the submarine at the turn of the century was to revive these doubts, as was the advent of aircraft.

Recommended Reading

Arthur J. Marder, *The Anatomy of British Sea Power 1880–1905* (1964).

1. 1886

Recent Naval Administration: General Policy
Lord Brassey

* * * * *

For the early years of the late Liberal Administration* no more authoritative statement of their shipbuilding policy can be quoted than that supplied by the following observations taken from the speech delivered by Mr. Trevelyan in the debate raised by Lord Henry Lennox in April, 1882. Mr. Trevelyan, on that ocasion, spoke as follows:

"When Lord Northbrook became First Lord of the Admiralty, it became their business to examine the state of their own and foreign Navies; and the conclusion at which they arrived was, that the time had come when they ought to increase the rate of iron shipbuilding; and so, without making any noise about it, they began building the ships which the right hon. gentleman opposite had commenced, at a faster rate. In 1880-1, instead of 7,231 tons, Lord Northbrook actually built 9,235, and in 1881-2, 10,816 tons; and this latter estimate was adhered to within 100, or, he might say, 40 tons. ... That was the quiet and unsensational, and not ineffectual way, in which they had dealt with the situation."—(3 *Hansard* [268] 1057.)

A general statement of the policy of the Board in relation to armoured shipbuilding is contained in the following extracts from speeches delivered by the Earl of Northbrook in the House of Lords.

On the 12th April, 1883, replying to Viscount Sidmouth, he spoke as follows:

"I do not wish in the least to say anything offensive to our neighbours across the Channel. They have in their discussions, and in the reports of their Committees, especially in the report of a Committee presided over by no less a man than M. Gambetta, frankly admitted that England stood unrivalled in regard to her Navy. I hold that no Government of this country—and certainly it is not the intention of

---

*Gladstone's government of 1882-5 (*Ed.*).

the present Government—would for a moment allow any nation to take a position of equality with England at sea. Such a policy would be discountenanced on whatever side or party. There can be no doubt, as the noble viscount pointed out, that for the last six or seven years great activity has prevailed in France in the construction of ships. But this activity is due to a policy, which has been openly declared, and the reason for which I would desire to state, because it has been stated in authentic documents which have been laid before the French Chamber. During the Franco-German war—as your lordships may well understand—the French Government were unable to devote the average attention and expenditure to their Navy, added to which they have taken a different course to ourselves in respect to the construction of their ironclad ships. Almost all the French ironclads up to a recent date have been built of wood, whereas we have adopted iron and steel from a very early date in our construction. Those of your lordships who have a knowledge of the subject are aware that wooden ships do not bear the same length of existence as iron and steel vessels, and, therefore, it was the opinion of a Commission of the French Chamber in the year 1878 that in a certain number of years a considerable number of wooden armour-plated ships in France would become unserviceable, and that it would be necessary to replace them with new ships. The programme laid down in accordance with that has been gradually carried out by the French Government. No one who has considered the question will think for a moment that the gradual fulfilment of such a programme should for a moment give rise to any susceptiblity on our part.

"In comparing the two navies, any officer or gentleman who wishes to make it as bad as possible for us, omits a lot of ships from a list of our Navy, and adds a lot of ships to the French or other navy which are perfectly useless. I can assure the noble viscount I have carefully considered the question, and I do not believe in his apprehensions. I am bound to say that I think it is desirable to increase very materially the amount of construction, but I may explain to your lordships that five years ago new dimensions were adopted, and it was discovered that greater power was produced in a gun by means of the use of slower-burning powder, which very much revolutionized the construction of ships by making it necessary to place very much thicker armour upon the ships. Then, again, there had been important improvements in engines and other technical matters with which I need not trouble your lordships; all of which changes have delayed construction. And I have no doubt they had a very material effect upon the policy of the Board of Admiralty which we succeeded in the matter of construction. Taking into consideration all these questions, the present Board of Admiralty have found that they can increase the amount of armour-plated tonnage constructed, not at an extravagant rate, but at a rate more in accordance with the rate laid down in former years."

In the Session of 1884 Lord Northbrook was again called upon from his place in the House of Lords to defend the policy of the Admiralty. On that occasion, he said:

"We decided, gradually but substantially, to increase the provision made for the construction of armour-plated ships. In 1880, although, as I have before remarked, we found the Navy Estimates already introduced, we were able to make an arrangement by which the progress of the armour-clad ships on the stocks was accelerated; and in the years 1881, 1882, and 1883 we very largely increased the provision for the construction of such ships. Probably, the best way of showing this is to compare the expenditure on the hulls and machinery of such ships year by year. In 1879–80 it was £631,724; in 1880–1 it was £698,798; in 1881–2 it was £949,313; in 1882–3 it was £990,710; in 1883–4 it was 1,260,137; and in 1884–5 £1,232,900; or, in round numbers, the sum appropriated to armour-plated construction had been doubled. I will now give the names of the ships laid down during the last four years. In 1880, in England, the Collingwood was laid down; in France none. In 1881 we laid down the Impérieuse and Warspite; France the Hoche. In 1882 we laid down four—the Howe, Rodney, Camperdown, and Benbow; France two, the Neptune and Marceau. In 1883 we laid down the Anson, and France the Magenta. In the four years we have laid down eight ships, and France four."

Subsequently to the date of the speech from which this extract is borrowed, Lord Northbrook's Board laid down the Hero, Sans Pareil, and Renown. In France two ironclads have been laid down, but the work of construction has been suspended.

The principles enunciated by Lord Northbrook in his succesive speeches as First Lord of the Admiralty were approved by the most competent critics of naval administration. In an article published on the 17th July, 1885, the able correspondent of the *Times* at Portsmouth expressed himself as follows: "This policy was not a matter of haphazard, but was founded upon certain definite principles which in course of years, and when the urgent necessity for an irresistible navy had become less pressing, might have given the country an efficient fleet of armour-clads. The first principle was not to ask the House of Commons for a large vote to secure an immediate increase in armour-plated tonnage, but to prefer the course which his lordship considered the wiser one, of making the necessary increase 'gradually and steadily, and not at one time by a very large expenditure of public money.' Another principle followed was to avoid the example of other Powers in laying down monster fighting ships, and to confine the armour-clads built to the moderate displacement of 10,000 tons, and costing not very far from three-quarters of a million each. A third principle was to take sufficient money to press forward as quickly as possible the ships under construction; a fourth was to avoid the multiplication of types; while the fifth and last enumerated was to attach special importance to the element of speed. It is not necessary to describe the satisfactory results to which the observance of these vital principles conduced."

* * * * *

Viewing the policy of the Board of Admiralty, of which the compiler was a member, in the light of experience, it cannot with justice be said that undue hesitation was shown in the development of the shipbuilding operations of the Navy. From the first the expenditure and the construction had been steadily, and indeed very considerably, increased until the year 1884–5. The mistake consisted in yielding to financial considerations which, owing to the falling revenue, were doubtless of a pressing nature, and keep the estimates for 1884–5 at the same figure as in the preceding year.

The deficiency in the construction of torpedo boats, and the necessity for doing more in the defence of coaling stations, and in providing graving-dock accommodation abroad, especially in India, were prominent topics in a speech delivered by the present writer at Portsmouth on the 12th August, 1884. That speech was answered by Lord Henry Lennox, and public attention was shortly afterwards awakened by a series of able articles in the *Pall Mall Gazette* on the condition of the Navy. It was not contended that the Government had spent money badly, but that they had not spent enough. The increased risks at stake were insisted upon, as evidenced by the statistics of the growth of our imports, our exports and our shipping. The views taken by the writers of those articles were unduly disparaging to the British Navy. Our superiority in ironclads was not sufficiently acknowledged. It was pushing the argument unfairly to compare the war fleet despatched by the French to the China seas with the numerical strength of our ordinary peace squadron on that station. The value of the reserve of cruisers in the mercantile marine was not taken into the account of our means of defending our commerce, but when every allowance has been made for the exaggeration of our deficiencies, the fact remains that the demands were fully justified for increased estimates. It had become urgently necessary to strengthen the Navy in fast cruisers, and in the torpedo flotilla. The long delay in the adoption of breech-loading guns could only be compensated for by increased activity in construction. The coaling stations abroad were not in a proper state of defence.

It is not a condemnation of the Board of Admiralty to acknowledge their obligations to the Press. In an article on "Politics as a Profession" in the *Quarterly Review* of January, 1869, it was truly said:—

"It is impossible for anyone conversant with the motive agencies of the time not to recognise the extent to which the Press now performs—some might say usurps—many of the functions of Parliament and the Executive Government. The Administration may not be so much to blame for this seeming abnegation of its functions as at first appears, for it is busy, feeble, and hampered."

Under a Liberal constitution, such as that so happily established in this country, the people are more powerful than their rulers. The force which governs is not the bureaucracy, but public opinion.

The demands originally formulated in the columns of the *Pall Mall Gazette* were unanimously supported in the Press and by public opinion. The permanent strength of the Navy must ever depend on the will of the nation; and no movement such as that which has lately taken place could proceed from the action of a particular party or the influence of an individual Minister. "It is," said Lord Beaconsfield, "quite a wild idea that a body of men, though they may be Ministers, can meet in a room and suddenly alter the establishments of a country. The establishments of a country are adapted to the policy which the country pursues."

Under Lord Northbrook's Administration the building votes were increased from £3,082,000 in 1880–1 to £5,047,000 in 1885–6. The truth is that at the bidding of the nation, we entered upon a new policy.

The national anxiety, as expressed in the columns of the public journals, had been awakened by the ambitious policy of our most powerful naval rival. The remarkable growth of the French Navy is due to a comparatively recent effort. This is clearly explained in an article contributed by Sir Nathaniel Barnaby to the last edition of the *Encyclopaedia Britannica*. In the period 1873–7 France expended in building and completing armoured ships for sea not more than half the sum which was spent in the corresponding period for our own Navy. Since 1877, the French expenditure on building had always equalled, and generally exceeded that of England.

This statement is confirmed by figures taken from the tables published in the work of M. Gougeard.

Expenditure on armoured Shipbuilding—Hulls only

		£
1874–78 4 years	England.	3,978,712
	France.	1,852,000
1878–85 7 years	England.	5,247,342
	France.	5,990,375

A comparison between the shipbuilding votes (including repairs) in

France and England, pointing to the same conclusion, is given to the next Table, taken from the Navy Extimates presented to Parliament in the two countries:

Date	England (Votes 6, 10)	Date	France (Votes 11, 12, 15, 16, 17, 18, 19, 20, and Budget Extraordinaire)
	£		£
1875–76	3,636,538	1875	2,050,000
1876–77	3,983,605	1876	2,038,880
1877–78	5,179,601	1877	2,391,910
1878–79	3,861,317	1878	2,398,758
1879–80	3,162,622	1879	2,840,860
1880–81	3,147,585	1880	2,898,112
1881–82	3,376,484	1881	3,120,899
1882–83	3,516,216	1882	3,254,569
1883–84	3,830,633	1883	3,383,346
1884–85	3,810,105	1884	3,175,357

2. 1887

Shipbuilding Policy: Protection of Commerce

Anon

* * * * *

In any comparison of our forces with those of other nations, it should be borne in mind that the whole of the external trade of the United Kingdom is carried by sea. The external trade of other maritime nations is to a large extent carried overland. In the year 1883 the percentage of imports and exports by land and by sea was as follows:

	Imports		Exports	
	By Land	By Sea	By Land	By Sea
France	34	65	31	68
Germany	65	34	63	37
Russia (1881)	47	52	30	70
Italy	42	57	51	48

That these nations have means of obtaining supplies and of carrying on their external trade independently of that which in time of war might be interrupted or wholly destroyed by the cruisers of an enemy, is apparent, and any necessary supplies could probably be obtained in quantities sufficient at least for immediate requirements. With the United Kingdom, the case is otherwise. With our teeming population we are dependent on importation for the necessaries of life, and an interruption of supplies by an enemy even temporarily successful by sea might lead to irretrievable disaster. In the circumstances the extent of our preparations for the defence of our trade must ever be to us a matter of supreme interest and concern.

The following Table shows the extent of the Mercantile Marine of the principal European nations in 1883, and the value of their seaborne trade:

	Merchant Vessels		Sea-borne Trade	
	Number	Tonnage	Imports	Exports
			£	£
France	15,222	1,003,679	153,054,200	124,078,240
Germany	4,315	1,269,477	85,873,800	92,855,200
Russia (1881)	4,790	500,554	42,624,920	55,428,800
Italy	7,471	973,333	31,336,890	24,381,120
Total	31,798	3,747,043	312,889,810	294,743,360
United Kingdom	24,675	7,242,216	426,892,000	305,437,000

It will be seen that the tonnage of the British shipping was nearly double, and that the imports and exports of this country were largely in excess of those of France, Germany, Russia, and Italy taken together. The corresponding figures for the present time would show a similar result.

Having shown the relative amount of the tonnage and trade of France and England, we may proceed to compare the extent and importance of the colonial possessions of the two countries, possessions which, together with their commerce, demand protection in time of war.

	Area	Population	Imports and Exports
	Square Miles		£
French Colonies	931,000	31,000,000	30,180,000
British Colonies	8,958,000	283,676,000	405,579,000

From this statement it would appear that in extent and population our colonial possessions are more than nine times as great as those of France, while in respect of commerce our Colonies surpass those of France more than twelve times.

The mutual trade of the United Kingdom with its Colonies is estimated at the value of £250,000,000 to £300,000,000 per annum, and the total trade of the United Kingdom and its Colonies at not less than £1000,000,000 per annum. The shipping engaged is valued at £200,000,000. Further, it has been computed that there are constantly at sea, the property of the United Kingdom and its Colonies, goods to the value of £150,000,000, together with shipping to the value of £130,000,000. If we take our normal expenditure on the Navy at

£12,000,000, this amount represent a premium of but 1 per cent. on the value of the trade whose safety it is intended to insure.

In view of these statistics, the demand of Lord Charles Beresford for further additions to our unarmoured vessels of war is amply justified. We can indeed improvise in time of war armed cruisers from among the faster vessels of the Mercantile Marine, and we can do this to a greater extent than any other nation; but at the best, such vessels are makeshifts, and not equally matched with the regularly constructed and appointed ships of the enemy. The protection afforded by such vessels in time of war would nevertheless be of the highest value to our scattered merchantman, and the measures taken by the Admiralty in the course of the present year to ensure that some of the fastest vessels about to be built for the American mail service shall be adapted for conversion into armed cruisers, and be available for that purpose should they be required, merit a warm tribute of praise.

To realise the havoc that cruisers of an enemy can work when ably commanded, we have but to recall the damage inflicted on our commerce by the French in the last century, under commanders of energy and determination, such as were Surcouf and Lemême. It is stated that in the years 1793 and 1794, no less than 788 British vessels were captured by the French cruisers.

The question of providing adequate protection for the Mercantile Marine cannot be too earnestly studied by those who are responsible. No question more urgently demands the attention of the Government and of the people of this country. In providing this protection the Colonies are deeply interested, and it is not unreasonable to expect that, the facts being fairly represented, the Colonies would be prepared to contribute a share of the cost proportionate to the interests involved.

The exponents of the public opinion of our maritime rivals are open in their declaration that they would assail us on the side where they know we are most vulnerable. Their efforts would be concentrated on the cutting up of our seaborne trade. To quote from M. Gabriel Charmes:—"There are a few trade routes, along which passes the wealth of the world, and on which is developed the very life of that immense world-wide British Empire. These routes, to some extent, correspond to the arteries. There are five or six, at the most ten. It would be an easy matter for us to scour them continually. True it is that formidable fortresses, such as Aden, Malta, and Gibraltar, afford some protection, but it is not under the guns of these fortresses that our cruisers would perform their exploits. They should avoid every known or avoidable risk. A warfare directed against the commerce of an enemy has its rules, and one must have the courage to formulate them exactly. The rule we would lay down should be this: Mercilessly attack the weak; without shame fly before the strong."

This language may be contrasted with the more generous and lion-hearted counsels of Lord Nelson, as given in the fourth volume of his collected despatches:—"Respecting privateers, I own I am decidedly of opinion that with very few exceptions they are a disgrace to our country, and it would be truly honourable never to permit one after this war."

The line of policy in shipbuilding indicated in the writings of M. Gabriel Charmes was practically taken under the administration of Admiral Aube. The expenditure on ironclads was reduced to a sum of £510,000 distributed on the building and completing, necessarily at a leisurely rate of progress, the ships previously laid down. While the ironclads were comparatively neglected, the French have actually under construction the cruisers shown in the following list:

		Tons
20-knot ships	Davoust	3,027
	Suchet	3,027
19-knot ships	Tage	7,045
	Cecille	5,766
	Isly	4,160
	Jean Bart	4,160
	Alger	4,160
	Coëtlogon	1,850
	Cosmao	1,850
	Surcouf	1,850
	Forbin	1,850
	Lalande	1,877
	Tronde	1,877
Collective Tonnage		43,499

We have here a total of fourteen ships, for the most part cruisers of the first class, and all are 19-knot ships at least.

* * * * *

3. 1888-89

The New Shipbuilding Programme
Lord George Hamilton*

* * * * *

Now I wish to say a word on the subject of designs. We have a large number of ships which are coming on, and we wish to impose a test more effective than that of the measured mile and the ordinary experimental cruise, because we have formed certain opinions regarding their merits and defects and are conscious that they should be thoroughly tested. We have had the advantage of the naval manœuvres, and that has fully confirmed the opinion we have formed, and all the vessels which we propose to build will be characterised by a high freeboard, more length, more engine room, and we shall avoid cramping too much. (Hear, hear.) We have designed a new ship of war which shall not only be capable of fighting, but shall also give decent accommodation for the officers and men on board and in all weathers. (Hear, hear.) So far as the question of design, construction, or armament is concerned, we can give every assurance that any increase of expenditure which may be sanctioned shall be given effect to in a prompt and business-like manner. (Hear, hear.) I now turn to the exact number of ships which we propose to build, and the expenditure associated with them; but I do not propose to give the individual cost of each vessel, following the precedent set by Lord Northbrook in 1885, because, if we are to derive the benefit of contractors, it is not advisable to give them too close a line as to their tenders. (Laughter.) The number of ships which we consider should be added to Her Majesty's Navy in our new shipbuilding programme is 70 (Ministerial cheers), and their estimated cost, including armament, is £21,500,000. (Ministerial cheers and counter-cheers.) We propose to build eight first-class battle-ships, with a displacement of 14,000 tons; two second-class battle-ships, with a displacement of 9000 tons; nine first-class cruisers, with a displacement of 7300 tons; 29 smaller cruisers, of the Medea class, with a displacement of 3400 tons; four smaller cruisers, of the Pandora class, with a displacement of 2600 tons; 18 torpedo gunboats,

*1st Lord of the Admiralty (*Ed.*).

of the Sharpshooter type, with a displacement of 735 tons—making a total aggregate tonnage of 318,000 tons. As to the time each of these vessels will take to complete, the first-class battle-ships will be finished in from three-and-a-half to four years; the second-class battle-ships, in three years; the first-class cruisers, in two-and-a-half years; the smaller cruisers, in two years or somewhat less; and the gun vessels in one-and-a-half-years. Of the aggregate cost of £21,500,000 for these 70 vessels, the sum of £16,150,000 will be for engines and hulls, and £5,350,000 for the armament. Dividing the sum of £21,500,000 into two portions, one of £11,500,000 and another of £10,000,000, the sum of £10,000,000 represents the work which we propose to put out to contract, and the £11,500,000 is the amount of work which we propose to assign to the dockyards. The first sum of £10,000,000 will enable us to build, arm, and equip the following vessels:—4 battle-ships, 5 first-class cruisers, 17 second-class cruisers, and 6 torpedo gunboats—making a total of 32 vessels; and I propose to put the whole of these 32 contracts out in the course of the present financial year. (Ministerial cheers.) Funds for this expenditure will be raised by methods and under conditions which the Chancellor of the Exchequer will subsequently explain. If I find there is any tendency to run up prices against the Admiralty we may have to alter our decision on that point, but with that one reservation we intend to put this amount of work out. For the purpose of building and completing the remaining 38 vessels and their armament there is £11,500,000. This sum, which is to be included in the ordinary estimates, can be divided into two heads—£8,650,000 for construction of engines and hulls, and £2,850,000 for armament. We propose to lay down in the dockyards this year 20 vessels, as follows:—4 battle-ships of the first class, 1 of the second class, 3 first-class cruisers, 6 second-class cruiser, and 6 torpedo gunboats. Adding these 20 to the 32 put out to contract we have 52, leaving 18 out of the 70 still to be accounted for. A second-class battle-ship will be laid down early in the financial year 1891, and the other vessels will take their places on the slips in the dockyards as soon as they become vacant by the launch of the vessels laid down in the first year. The whole of the programme, including both dockyard and contract work, is to be finished in four-and-a-half years from the date of the commencement of the first vessel. (Cheers.) There is a naval and an administrative advantage in this distribution of work to which I should like to call attention. Taking the programme as a whole it consists of 70 ships, 10 of which are battle-ships and 60 cruisers of different types. A battle-ship takes very much longer to build than a cruiser—three-and-a-half or four years—and if they are to be finished within four-and-a-half years they must all be commenced in the first year. But a battle-ship when completed is not efficient unless she has certain small vessels attached to her as scouts (hear, hear), and we consider that out of the 70 vessels 20 are the satellites of the battle-ships. The remaining 40 cruisers will be effective whether used in squadrons or individually. We propose to commence the construction of the whole of those this year, and thus make an immediate addition to the Navy of what it most requires. Later on, when an increase is made to our battle-ships, each battle-ship will be accompanied by two smaller vessels, and thus there will be no drain upon our force of independent cruisers. This enables us to carry out effectively the whole of our shipbuilding programme, and it also enables us to give continuous employment in the dockyards, thus avoiding the great evils of a sudden expansion of business. (Cheers.) In order to absorb the amount reprsented by this sum of £11,500,000 we raise the shipbuilding vote in the ordinary estimates by £615,000, and reduce the ordnance vote by £400,000, because we intend to purchase all the ordnance for the contract ships out of the £10,000,000 to which I have referred. If the shipbuilding vote we kept at the level at which we propose to put it, and the ordnance vote at the same level, there will be sufficient to provide for the whole of this work as well as for all the work we have in hand, and to complete in the period of four-and-a-half years. (Hear, hear.) The Committee will observe the extreme rapidity with which we propose to build these ships. In France a battle-ship takes nearly ten years to complete. In this country, previous to 1885, ironclads took six years and large cruisers four years. We propose to lay down 70 vessels, including 10 battleships, in four-and-a-half years. It is in the interests of economy that we make this proposal.

* * * * *

4. 1893

Naval Manoeuvres in 1892
J. R. Thursfield

"A Torpedo-boat is essentially the weapon of the weaker, and of the assailant rather than the defender of commerce." So wrote the late First Lord of the Admiralty in his Statement explanatory of the Navy Estimates for 1892–3. Some exception may, perhaps, be taken to the latter half of the proposition. The primary object of the assailant of maritime commerce is not destruction, but capture. For this latter purpose the torpedo-boat is manifestly a very inefficient instrument. But no one who realises the indefeasible conditions of naval warfare will dispute the proposition that the torpedo-boat is essentially the weapon of the weaker naval combatant. Its range of action is limited to a certain distance from a port of shelter, the *maximum* extent of which is capable of almost exact calculation. Outside that range the command of the sea belongs *ex vi termini* to the superior naval Power, and even if the command of the sea were in dispute, as it might be at the outbreak of hostilities between two naval Powers of the first rank, the relative strength of two maritime belligerents contending for it could not be affected by a weapon which both would be incapable of using in such circumstances. Inside the range of action accessible to a torpedo-boat, however, the conditions are widely different. Here the torpedo-boat may, in certain circumstances, be held to place the weaker Power more or less on a footing of equality with its stronger antagonist. Its actual offensive power is tremendous, while its menace is still more formidable. Hence it is not at all surprising, that since the invention and development of the so-called automobile torpedo, the torpedo-boat armed with it, and specially designed to give effect to its offensive capability, has exercised an immense influence over the naval imagination, and has always occupied a prominent place in the naval manœuvres of the principal maritime Powers. Last year, in particular, it may be said to have been the prime factor in the only manœuvres of which we have detailed information, those, namely, of England, France, and Russia. It will be convenient, therefore, to examine the history and results of these manœuvres mainly from this particular point of view. It will be seen from this comparative method of treatment, that each Power has virtually framed its manœuvres on the assumption to which the late First Lord of the Admiralty gave expression when he wrote that "a torpedo-boat is essentially the weapon of the weaker".

Both France and Russia might have to engage in a naval conflict in which their naval power was inferior to that of their antagonist, and hence they are both compelled, by the circumstances of their position, to regard the torpedo-boat as a potent weapon of defence, both active and passive. England, on the other hand, has rather to regard the torpedo-boat as an offensive weapon in the hands of her antagonists, and to study the dispositions, both strategical and tactical, which are best calculated to neutralise its efficiency. We shall see how these governing considerations affected the manœuvres of each Power in turn.

In the case of our own manœuvres, the operations of last year may be regarded as the logical outcome of the experience of previous years. In 1890 a successful attack on a fleet lying at Plymouth was made by torpedo-boats having their base in the Channel Islands; and, as the Official Report on the Manœuvres of that year stated, "it was proved by actual experience that there are officers who can navigate their boats for hours together across a crowded route, can reach their objective punctually at the pre-arranged time, and can then manœuvre at very high speed at night in an anchorage so filled with shipping that manœuvring in it when fresh and in broad daylight would require much care and attention. Lieutenant Wells, commanding No. 87, upon one occasion made a continuous run of 420 miles, during which he examined fifty-three vessels. A result of the 1890 manœuvres is that opinions on the effective radius of torpedo-boat action will have to be reconsidered." Taking Alderney as the base, it may be said to have been shown in 1890 that the effective radius has its extremity somewhere between Falmouth and Scilly. Falmouth was found to be readily accessible; but all attempts at Commander Barry's torpedo-boats to reach Sir George Tryon's colliers stationed at Scilly were failures. This means that every port in the Channel, and *à fortiori* every portion of its waters, is accessible to hostile torpedo-boats having their base on either coast. Hence it follows that fleets lying in any of the ports of the Channel must, in the event of war, be prepared to resist the attacks of torpedo-boats, either by means of their own resources, offensive and defensive, or by means of a passive and local defence attached to the port. It means, also, that ships or fleets operating in any part of the Channel would also be liable to the attacks of an enemy's torpedo-boats.

It followed naturally from the remarkable and more or less unexpected results of the manœuvres of 1890, and in 1891 an attempt should be made to ascertain by experiment the most effective answer to the attack of the torpedo-boat as developed in 1890. The question still remains confined within very narrow limits, as compared with the necessary range of action of British fleets in time of war. The function of these fleets is to secure and maintain the command of the sea by the control of all the important strategic points involved; and it needs no argument to show that many of such points must in all circumstances lie altogether outside the utmost possible range of action that can be assigned to the torpedo-boat. But in order to be free to take and keep the sea, the British fleets much enjoy unmolested freedom of access to the ports and anchorages of the United Kingdom. Their local and passive defence while lying in those ports is a question which need not concern us here. A far more important question is their defence at sea within the effective range of torpedo-boat action. This again involves two questions, namely, the tactics adopted by the torpedo boats, and the tactics to be adopted in answer to them by the ships and fleets liable to attack.

These questions were officially propounded in the following terms, taken from the Report on the Manœuvres of 1891:

"The objects were—

"(a) To ascertain the tactics which would probably be adopted by flotillas of torpedo-boats stationed at several points on one shore of a channel in order to harass or destroy an enemy's ships in the channel, or lying at anchorages on the other shore.

"(b) To ascertain the measures which should be taken to give security against the attacks of those torpedo-boats."

* * * * *

The operations of 1891 were thus a direct duel between torpedo-boats and a sea-going squadron, the object of the torpedo-boats being to harass the squadron, to impede its operations, and to destroy it if possible, while the object of the squadron was, by clearing the sea of torpedo-boats, to recover its freedom of operation, and, in the meanwhile, to take effective measures for its defence against their attack. In this it was completely successful, materially aided as it was by the conditions of weather and daylight prevailing during the operations. The superiority of the active over the passive defence is inherent in the nature of the torpedo-boat, and in the conditions which determine its action. It is very formidable in attack, but very feeble in defence. In fact, its only secure defence in daylight lies either in a shelter inaccessible to large vessels or in a clean pair of heels. It is now generally admitted that a torpedo-boat caught at sea in daylight can be readily destroyed by the smallest modern cruiser afloat. The effective striking range of its characteristic weapon, the torpedo, does not exceed 500 yards; it is itself vulnerable to quick-firing guns up to a distance of at least 2000 yards. It can discharge only one missile, or two at the outside. Its missile is liable to many accidents and misadventures, and its chance of hitting its adversary in an end-on position, or in any other position than that of broadside to broadside, is quite inconsiderable. Visible in clear weather from the deck or masthead of a man-of-war at from seven to ten miles, and coming within range at 2000 yards, it must traverse nearly the whole of this distance under the fire of its adversary before it can deliver a single blow in return. It follows that in daylight the odds are so tremendously against it that its only safety is in flight. But here, again, it is subject to a characteristic weakness in defence. Its speed rapidly declines in a heavy sea, and a sea which would hardly affect the speed of a cruiser might reduce that of a torpedo-boat to a maximum of 12 knots. Thus it appears that, whereas the effective range of action of a torpedo-boat is limited in space, as we have seen, by a circle drawn from its port of shelter with a radius approximately equal to the distance between Alderney and the Land's End, it is limited in time by the period between dusk and dawn. Hence its range varies according to the season of the year, the time limit being smallest in the summer and largest in the winter, while the space limit varies irregularly in an inverse direction, being governed largely by conditions of weather and sea.

Thus does the active defence against the torpedo-boat, by its tremendous menace of capture and destruction, immensely limit the effective range of action of the latter. It also possesses other incidental but indefeasible advantages. The conditions most favourable to torpedo-boat operations are those in which an attack is to be made on a fleet at anchor in a known position. The conditions become far less favourable if the position of the fleet is not known to the assailants, and positively unfavourable if, before an attack can be made, the fleet has to be discovered at sea. It is by no means easy to effect this discovery in any circumstances, and in some circumstances it is practically impossible. On the other hand, it is by no means equally difficult for the cruisers and torpedo-gunboats of a sea-going fleet to discover the hostile torpedo-boats. The ports of shelter frequented by the latter are known and can be watched, and though the torpedo-boats may evade the blockade after nightfall, they must always run the risk of having to return to one port of shelter or another after daybreak, when their destruction or capture may be made almost a matter of certainty. In fact, whereas there is no single point within the whole sphere of operations where the torpedo-boat can be certain of finding the fleet, there is always a series of points—namely, the neighbourhood of the ports of shelter—where, sooner or later, the torpedo-boats must be found, and where, when found, they are always liable to be overpowered by a disposition of naval force adequate in itself and adapted to the local circumstances. It follows that an active defence adequately organised and skilfully disposed must in the end completely neutralise the offensive capacity of the torpedo-boat. The operation is a difficult one, full of hazards, surprises, and adventures, and intensely trying to the nerves and endurance of officers and crews engaged in it. But experience seems to show that it must in the end be successful, although the mischief wrought in the meantime by torpedo-boats daringly handled might be so great as materially to affect the issue of a campaign. In any case the tactics of a sea-going fleet will always be more or less determined by the menace of torpedo-boats in its neighbourhood.

* * * * *

5. 1894

British Naval Manoeuvres in 1893
J. R. Thursfield

* * * * *

Thus the continuous experience of three years' manœuvres, those of 1891, 1892, and 1893, would seem to show that the sea-going torpedo-boat is an overrated weapon of offence. In 1891 the late Admiral Long showed, as was pointed out in the *Naval Annual* of last year, "that an active defence adequately organized and skilfully disposed must in the end completely neutralize the offensive capacity of the torpedo-boat." This demonstration was reinforced by the manœuvres of 1892, which also showed further that the extinction of the torpedo-boat menace follows immediately on the destruction of the shelter provided for the hostile torpedo-boats and on the surrender of the sea-going squadron to which they are attached as auxiliaries. Lastly, the manœuvres of 1893 completed the demonstration by showing that, even in default of an active defence adequately organised and skilfully disposed, torpedo-boats are very apt to suppress themselves and to attain a very high rate of extinction in the normal course of their attacks on a powerful and vigilant sea-going adversary. The truth seems to be that a torpedo-boat ought properly to be regarded not as an independent sea-going unit of naval force, but as a peculiar and very destructive kind of projectile with a very extended range which varies according to circumstances, but is by no means unlimited in any circumstances, and with an intelligent power of altering its direction in the course of its flight, but also with a considerable liability to be destroyed or intercepted before it attains its mark. As such its menace is tremendous, and its influence on all strategical dispositions within its range is dominant and decisive so long as its menace is unabated. But experience, now repeatedly tested in our own and other navies under conditions as closely analogous to those of actual warfare as peace manœuvres can be made to afford, would seem to have shown that its strategic menace is far more formidable than its real offensive capacity, and that, regarded as a projectile, it is endowed with a really remarkable capacity for hitting wide of the mark and destroying itself before it has delivered its blow—to say nothing of its very awkward habit of occasionally mistaking a friend for an enemy.

6. 1896

The Value of Torpedo Boats in Wartime
Commander R. H. Bacon

The value of torpedo-boats in war time will be felt in two distinct ways—actively and potentially; actively in damage done to men-of-war and merchant vessels, potentially by their reputation and menace governing and checking dispositions and plans, and exercising an unending nightly strain on the officers and men of the opposing fleet. A torpedo-boat may be defined as "a fast small light-draught vessel, whose main armament is a torpedo, and whose function is to attack and destroy larger vessels either in harbour or on the open sea." The main requisite is speed, which is necessary so as to enable them to approach a ship at a rapid rate even when she is running away, to make fast journeys from a base to raiding grounds, or in daytime to flee from, or keep at a desired distance from, a hostile ship. Size is of the next importance. If of exaggerated dimensions they become unhandy, a target for the enemy's larger guns, which ordinarily would not be used against them, and also a target for the torpedoes of hostile boats. If too small they are unequal to the sea-going work required of them. All other considerations are of secondary importance. A modern torpedo-boat is merely a sea-going torpedo-tube of high speed, and as such it should be treated. It will be of advantage, before dealing with the functions of torpedo-boats in war and their capabilities of performing the duties desired of them, to examine the changes which have taken place in the design of boats and the causes which led to such alterations.

The evolution of torpedo-boats is instructive. The original idea comprised two classes of boats. The first class were the smallest boats that could keep the open sea, the second class were the largest boats which a ship could hoist in and carry on board. It is with the former class that I will first deal. In the early days the idea of the main functions of torpedo-boats was to attack ships at anchor in a harbour, consequently small size and good manœuvring powers were considered of great importance, and the size of the boats was reduced to a point which endangered their sea-going qualities. This led to the building of a larger class, the 125 ft. boat, with a sea-going speed of about 17 knots. These boats, first commissioned in 1887, were excellent in their day, and from repeated trials on subsequent manœuvres were found to be of great value both for the attack of ships in harbours and also at sea. But within a very short time two important changes in war material and defence took place, namely, the speed of ships was increased above that possessed by these boats, and a commencement was made in defending harbours with breakwaters. The former limited their value for attack at sea, the latter for attack in harbour. The breakwater is the only unanswerable argument to a torpedo-boat. Behind it ships are practically secure against boat attack. The only vulnerable point is the entrance or gateway, which is capable of such concentration of defence by gunfire, booms, and boat mines, as to make entrance by unsupported boats practically impossible. Thus these boats have been the cause of enormous outlay in harbour defence, and the threat of their existence compels vessels seeking shelter for coaling or repairs to take refuge in certain definite harbours where they can lie with moderate security.

The falling-off of the speed of these boats by becoming old, and the increase of speed in modern ships, has reduced their efficiency as sea-attacking boats, so that a new class of greater speed and size became a necessity if the boats were to keep their relative value as a weapon for the destruction of ships. But let it be remembered that it is with these boats, with all their disadvantages, that the successes of past yearly naval manœuvres have been gained, and that with great disadvantages against them they have succeeded in harassing and sinking ships of the fleets opposed to them. What must we expect, therefore, if boats are built of treble their efficiency? Such boats we now have. The 125 ft. boat has recently been surpassed by boats of larger build, culminating in the torpedo-boat destroyer, which attains a sea-going speed of 25 knots, has twin screws, and carries the latest improved torpedoes in addition to her guns. The object in view that originally led to the construction of these destroyers was the annihilation of the smaller patterns of torpedo-boats, but it is impossible to doubt that in addition to such work a portion of them must be used as torpedo-boats, and also that when performing their duties as destroyers they will have the opportunity of acting in their torpedo capacity. Briefly, the capabilities of such boats are to keep the sea for a week or less, the exact time being dependent on the speed and nature of the duty undertaken, to command a speed of about 24 or 25 knots, and to carry two torpedoes of great accuracy whose speed is 31 knots, and which have a charge of 150 to 200 lbs. of gun-cotton. Such a torpedo endows them with the power of sinking the largest ship afloat.

There are several general questions which closely affect the value of the torpedo-boat in war time, which it will be well to discuss. The actual chance that a single boat has of torpedoing a single ship must be the main factor in the problem of the value of boats, and it is one which the experience of war alone can solve with absolute certainty. Boats not only vary in speed, size, and nature of torpedo carried, but also both ship and boat vary in the skilfulness of their officers and men— officers in manœuvring and supervision, men in accuracy of aim and discipline. It is useless to appeal to the results of peace manœuvres to decide this question. Arbitrary time allowances to the boats, or number of aimed rounds from the ship, have to be resorted to. No direct experiments have or can be conveniently carried out to decide this question. The result is that we are absolutely in the dark. Time allowances in manœuvres have two important limits. If over a certain time be allowed, the boat would invariably approach the ship near enough to make certain of hitting her before being put out of action. If less than a certain time be allowed, the boat would always be out of action before having a chance of firing her torpedo. These limitations are quite independent of direct experiment with gun-fire at night, and can only be based on the speed of the approaching boat; this system teaches boats to regard gun-fire at close ranges as equal with that at 1,500 yards. That the system of time allowances is the best that at present is available during large manœuvres is shown by it being adopted in those of late years, but for the purpose of assessing the value of boats versus gun-fire in war, the data of all manœuvres is valueless. War will admit of no hypothesis, or compromise with convenience— facts will then be facts. Suddenness of attack, invisibility, rapidity of movement, as well as the large size of target that the ship presents, are points in favour of the boat; whereas detection at a distance, errors in adjustment of torpedoes and director would help the ship. But above everything let us beware of giving too false a feeling of security to the ship by assessing necessarily hurried and excited gun-firing too highly.

For the sake of argument we will assume while considering this question that over a large number of cases the chances of success would be equal, and that speaking generally a torpedo boat will, in half the cases of attack, disable an ironclad, and that conversely in half the number of cases the boat will be destroyed. With more than one boat against a single ship the chances will therefore lie with the boats. Nor will these chances be practically altered if more than one ship is in company, since all the boats will concentrate their attack on one or more ships which cannot receive much support from the gun-fire of the others. Again, the larger the number of ships the greater the chances of choice of an advantageous ship to attack. Ships must always be ignorant of the plan and direction of attack as well as of the number of boats attacking. It is hardly conceivable that with even only three boats, they would never be detected, separately fired at, and disabled by the gun-fire of a ship within the two or three moments available before the torpedoes were fired. If once a boat has got to close quarters at night it is an unpalatable though potent fact that owing to size of target the chances of success lie largely with the boat.

The inherent feature of torpedo-boats which governs their use in war time is that the boats are of small value compared with a battleship or cruiser. By value is meant small cost, that they can be rapidly replaced, and that the number of men they carry is small. So that for this reason they can be asked, risked, and if necessary lost, to obtain ends which may be of great moment without more than small relative loss to their own side. This is no age for squeamishness in using the weapons a country supplies. The tools we handle now are far keener than any hitherto employed, the ends to be obtained are of vital worth, and I venture to predict that the successful side will be that on which the admirals and captains do not hesitate to employ the ships and boats in the way that leads most surely to the disablement of their opponents without being unduly deterred by sentiment, or by risks which may be legitimately incurred. In history we have on record many cases when to obtain strategical advantages inferior squadrons did, or should have engaged superior ones to reduce their numbers, although themselves courting annihilation. How much the more may not such a necessity arise in the future, since with modern weapons a single ship may, if well handled, inflict much and lasting damage on two or more. How much more vulnerable have not our ships become at close quarters to the powerful weapons now in use than ever was the case in the early part of the century, when a possibility of a single shot totally disabling a ship, or of the smallest craft floating having the power of destroying a battleship, were factors absolutely unknown in warfare. Those vessels which can be risked are those of least worth, and which can be most easily replaced; so that putting sentiment on one side, the *raison d'être* of the power of a torpedo-boat lies in the fact that it can be used and risked without its loss being of great moment to the fleet.

* * * * *

7. 1897

The Progress of Foreign Navies
E. Weyl

* * * * *

Germany

The Emperor William wishes to have a powerful Navy, and the efforts of his Government are directed to this object. The commerce of the Germany Empire is increasing all over the world with a rapidity which causes her competitors some anxiety. German colonists are making their way everywhere, and, owing to their excellent qualities, are ousting their rivals. The Mercantile Marine is rapidly increasing, and if there is still no comparison between it and that of Great Britain, it has taken the first place amongst the Merchant Navies of the Continent. The Canal from the Baltic to the North Sea, from a commercial point of view, has not fulfilled the hopes of its promoters, but has quite done so from a military point of view. Last year, during the Naval manœuvres, the whole German Fleet passed through it in thirty hours. The Canal thus doubles the efficiency of the Imperial Navy by allowing it to concentrate rapidly either in the Baltic or in the North Sea—an immense strategical advantage. Nevertheless, the strength of her Naval position appears insufficient to satisfy the views at present held by the Empire, and the Imperial Government is asking the Reichstag for an important increase to the Navy Estimates.

The Navy Estimates for the year 1897-8 amount—including extraordinary expenditure—to a total of £6,467,977—the highest sum yet reached since the inauguration of the rapid Naval expansion by the Emperor William II. From 1874 to 1889-90 the Navy Estimates increased gradually from about £1,950,000 to about £2,750,000. Upon the accession of the present Emperor, General von Caprivi, who some years previously had been appointed Minister of Marine, gave way to a Minister more in harmony with the ideas of the young sovereign. As a result, in the first budget of the present reign—that of 1890-1—the Naval Estimates suddenly increased to nearly £3,600,000. In the following year they amounted to more than £4,750,000, and in 1892-3 to more than £4,500,000. For the last three years they have averaged about £4,150,000.

Early in March (that is, after this chapter was in print) the Secretary of State for the Imperial Navy, Admiral Hollman, electrified the Budget Committee by laying before them a memorandum intimating that for the years 1898–1901 an expenditure on new construction is proposed to £9,144,000, in addition to the sums provided in the Navy Estimates for 1897–8 for vessels already in hand to be laid down during the present year. Admiral Hollman's proposals include the construction of four first-class battleships, six first-class cruisers, besides numerous smaller vessels, principally torpedo-boats and torpedo-division boats.

In the Estimates for 1897–8 it is proposed to lay down one first-class battleship, two second-class cruisers, one despatch-boat, two gunboats, one torpedo-division boat, and eight torpedo-boats. The first instalments for all these vessels amount to about £300,000. The new battleship is to take the place of the König Wilhelm, which is twenty-nine years old; the two cruisers O and P will be modified Gefions; the despatch-boat is to take the place of the Falke, the two gunboats will replace the Hyane and the Iltis (lost in the China Seas on July 29th, 1896). The new battleship is to cost £756,000, and the cruisers £400,000 each.

* * * * *

8. 1897

Relative Strength

T. A. Brassey

No year since the *Naval Annual* was first published has been so full of anxiety for those at the helm of the British Empire, in no year has there been such imminent danger of war, as in the one just passed away. At the moment of writing, the action of the Greeks in Crete is giving fresh cause for anxiety. In former years we used to confine the comparisons made in this chapter to the fleets of France and Russia; but during the past year it has become more and more evident that we must take into consideration the Navies of all the principal Naval Powers. We have been in a position to hold our own against our probable enemies. Our Navy is practically equal in strength to the combination of any two foreign Navies. Is the standard of strength hitherto accepted sufficient?

There has been little change in the relative position since last year. Both in England and in France powerful battleships have been completed. In France and Russia the sums allotted to new construction in this coming year are slightly reduced. In Italy the drain on the resources of the country by the attempt to create an empire in Africa has seriously hindered the progress of her Navy. In Germany, on the other hand, there is a decided increase in shipbuilding activity. The arguments by which Admiral Hollman recommended the proposals for the increase of the German Navy, to which allusion has been made in the previous chapter, are even more worthy of attention than the proposals themselves, as they indicate a new departure in German policy. The gist of them was that Germany must be in a position to fight with strong forces on sea as well as on land, and that a position of power in the world can only be assumed by Germany if she has a powerful Navy. The people of the United States seem to be determined to take a position amongst the Naval Powers of the world. Though Congress did not accede to all the demands of the Secretary of the Navy, it authorised the construction of some powerful ships in addition to those already building. By the end of the century the United States will possess a fleet of some eight first-class battleships. Japan has started on her ambitious career somewhat later in the day, but by the close of the year she will possess two powerful battleships, and there is no indication that she will relax her efforts to create a powerful Navy. A contract has recently been made with the Thames Ironworks for the construction of a battleship of 14,850 tons. The South American Republics continue to obtain from Elswick some of the fastest cruisers in the world.

* * * * *

9. 1897

Principles of Imperial Defence
T. A. Brassey*

* * * * *

    The main principle which I wish to lay down at the outset is that the defence of the Empire rests absolutely on our power to retain the command of the sea—in other words, on sea-power. I do not wish to minimise the functions which the army will have to perform in case of war, but I do wish to insist very strongly that no army which it is conceivable we could raise and maintain would compensate for inferior Naval strength.

    In the year 1892–3 the gross cost to the British taxpayer of defending the Empire amounted to over £35,500,000, £20,500,000 of which was devoted to expenditure on the Army, and £15,000,000 on the Navy. To those who had grasped the principles of warfare which are applicable to a sea-power like Britain, it appeared that if the relative proportions of Naval and Military expenditure were reversed, the Empire would be better defended. The proportions of Naval and Military expenditure, though not reversed, have been entirely altered in the last four years. The Navy Estimtes for 1896 amount to £22,800,000 gross, or £21,800,000 net. The Army Estimates amount to £20,900,000 gross, £18,000,000 net. It is impossible to deny that the British Empire is better defended today than it was two years ago. We owe the change that has taken place to the fact that the principles of Imperial Defence are becoming better understood. The deepest gratitude of every Englishman is due to Captain Mahan, of the United States Navy, for so clearly setting forth those principles in his two admirable books.

    I will endeavour to illustrate the assertion that the defence of the Empire rests on sea-power by considering the forms of attack which we may have to meet in case of war with a first-class European power, or combination of European powers. We shall have to meet attacks on commerce, attacks on colonies and dependencies, and, possibly, invasion. I have put them in the order in which they are likely to occur.

    The *Jeune École* of French Naval officers has laid it down that in the

*Speech in Melbourne to the Imperial Federation League, 19 October 1896 (*Ed.*).

event of war with England the Naval force of France should be mainly directed to the destruction of British commerce.

* * * * *

The British Merchant Navy holds a higher position to-day than it has ever done before relatively to the Merchant Navies of other countries. The aggregate merchant tonnage of the British Empire amounts to 10,512,272 tons, made up as follows:

The United Kingdom	8,956,181
Canada	951,210
Australasia	359,614
British India	65,140
Other British Possessions	180,127
Total British Possessions	1,556,091

The aggregate tonnage of the Merchant Navies of all other countries amounts to 8,449,000; or, if we include vessels employed on lakes and rivers in the United States, to 10,305,000. Taking steamships alone, which are generally considered to possess three times the carrying efficiency of sailing ships, 6,377,000 tons sail under the British, 3,624,000 tons sail under foreign flags; or, including vessels employed on the lakes and rivers of the United States, 5,332,000 tons. Including only those vessels which ply upon the ocean, the British Empire possesses at the present time more than half the total merchant tonnage of the world, and nearly two-thirds of the tonnage of steamships. In any future war in which we may become involved, British commerce will undoubtedly suffer losses. Their number and extent will depend on the strength and efficiency of the British Navy. Judging from the experience of previous wars, the losses will almost certainly be more numerous, but they should represent a less percentage of the whole. If the command of the sea is lost, the ruin of British commerce is assured. It is idle for British merchants to talk of securing the safety of their ships under a neutral flag. No Power with which we may be at war would respect the neutral flag where ships were carrying food supplies absolutely vital to the existence of its enemy. One hundred years ago England was nearly, if not quite, self-supporting; to-day we are not provisioned for more than a few months.

Canada and India alone of British possessions are open to serious attack by land. British South Africa has a long land frontier, but no first-class Power could contemplate a serious attack except with troops transported over sea. The duty of repelling an attack on either Canada or India may depend primarily on the army, but our real power to defend them depends absolutely on the command of the sea. In event of war with Russia, we can put troops on the north-west frontier more easily than Russia can bring forward her invading forces. The contingency of war with the United States no Englishman likes to contemplate. Should Canada ever again be liable to invasion, our power to defend Canada depends, as in the case of India, on the power of transporting British troops by sea. Australasia, South Africa, Canada (except in the contingencies I have mentioned) are in a great measure secured from attack by their wide extent of territory and their numerous population. An army of 20,000 men would be required to conquer and hold any of these great colonies. Such an army cannot be collected and despatched across the ocean surreptitiously. To make the attempt while the command of the sea was in doubt would be madness. Attacks on commerce by cruisers keeping generally out of sight of land are the most probable form which operations of the enemy would take on the coast of India, South Africa, or Australia. Occasional raids on territory might be attempted by small expeditions, either with a view of obtaining supplies or inflicting damage. It is certain that few captains would waste ammunition on bombarding a seaport, with the chance of falling in with an enemy's cruiser before they could return to their base to obtain a fresh supply. Against such attacks the best defence is an active Naval defence by ships which are able to pursue and fight the cruisers of the enemy wherever they may be found.

In accepting the localisation of the vessels of the special Australian Auxiliary Squadron, we have acted on a principle universally condemned by students of Naval strategy, and seriously hampered their utility. I will endeavour to give an illustration to bring this home to the minds of everyone in this hall. You know that during the past fortnight British and Russian fleets have been watching one another through the Dardanelles. If the British Government had been influenced by the agitation raised in England, there is little doubt that we should have been at war with Russia, and possibly with France as well, at this moment. The Naval force, maintained by foreign Powers in waters in the neighbourhood of Australia, whether in the Pacific or Indian Ocean, is absolutely insignificant compared to our own. In China the Russian and French Squadrons are equal, if not slightly superior, to the British Squadron. They can oppose one battleship and five armoured cruisers to one battleship, three armoured cruisers, and a first-class protected cruiser. If the British China Squadron were to be defeated in battle, the command of the Pacific and neighbouring seas would be temporarily lost. British commerce would be interrupted, and Australia would be liable to invasion by Russian troops from Vladivostok or French troops from Saigon. The squadron now in Australian waters would be powerless to prevent it. I have no hesitation in saying that if the British China Squadron were immediately reinforced on the outbreak of war by the flagships here and in the Pacific, it would have a reasonable prospect of defeating, or at any rate of holding in check, the combined squadrons of France and Russia. There would most probably be a great popular outcry against any such action on the part of the Admiralty, but it is absolutely certain that the Orlando and Warspite would do more to defend the coasts of Canada and Australia in Chinese waters than they could ever do if they remained in Canadian or Australian waters. Against small raiding expeditions, accompanied by troops which are not likely to, but still might, escape our cruisers, you in Australia must be prepared to defend yourself by maintaining a military force, not necessarily numerous, but certainly efficient, and capable of taking the field against disciplined troops.

* * * * *

Having considered the three forms of attack to which we are exposed, we can form some opinion as to the ends to which our efforts should be directed in case of war. Our first and principal object must obviously be to defeat the enemy's main fleet in battle, and to completely checkmate his operations. An effective army, powerful fortifications, superiority in cruisers will not compensate for a deficiency in the line of battle. Battleships alone can give us that command of the sea which is indispensable alike to the safety of our commerce, our colonies, and the shores of the United Kingdom. Our second object must be to maintain a sufficient force of cruisers to deal with the hostile cruisers or privateers designed to prey upon our commerce, or with the expeditions intended to attack our colonies which might escape our principal fleets. We should always endeavour to deal with the latter at or near the point of departure rather than at their destination, for in this case the cruisers defend not only the point to be attacked, but the intervening ocean. Our third object should be to capture the coaling stations and colonies of the enemy, which are far more indispensble now than they were before the introduction of steam to depredations on commerce. During the Great War, French cruisers and privateers, issuing from Mauritius and the West Indian Islands, did us considerable harm. It was not till 1812 that all the colonies of France, Holland and Denmark had fallen before the British arms. Many millions of pounds would have been saved if we had seized Mauritius, Martinique, and Guadeloupe earlier in the war.

In view of the military forces now maintained by Continental Powers under conscription, the part which the British army can play in war with any first-class power, except Russia or the United States, is only a secondary one, but it is still important. With the assistance of the Navy it must lend its energies to the capture of the colonies and coaling stations of the enemy. The capture of St. Pierre or Réunion would not be great achievements for the British army, but the conquest of Algeria would test its powers to the uttermost. With Algeria hostile in time of war the trade route up the Mediterranean would never be absolutely secure and might have to be abandoned altogether.

It would be impossible to deal thoroughly with the question whether the Navy and army are sufficient for the duties imposed on them in our national defence. Not many months ago His Excellency gave an address on our Naval position in 1896. We are, I believe, just strong enough at sea to hold our own against a combination of any two other first-class Powers. Is our present standard of strength sufficient? We have to reckon with the fact that our very greatness, the splendid growth of our self-governing colonies under free institutions, the talent we have shown for the government of native races in Egypt and India, make us the most unpopular Power in the world.

Hitherto the burden of defending this great Empire has fallen almost exclusively on the inhabitants of the mother country. During the past two years we have added our £7,000,000 to our Navy Estimates alone, irrespective of £14,000,000 provided in the Naval Works Bill. In many of the colonies, certainly in the Australasian colonies, expenditure on defence has been cut down, and the tendency seems towards still further reduction. You have been passing through a period of severe depression. We in the old country have had a revival in material prosperity. The addition to the Naval expenditure has hardly been felt, certainly not by the general body of taxpayers. We have been able to hold our own well up till now against our probable enemies, but should these enemies become more numerous at a time when commerce and industry are not so prosperous as they are now, the British taxpayer may find the burden almost too heavy for his shoulders alone. Speaking as a representative of British working men, and putting it to you as purely an abstract question, is it just that we who live in the old country should contribute twenty times what you do to the common defence? Is it right that the sons and the brothers of British workmen should uphold the British flag in every corner of the world, while, if I am to judge from what I sometimes read in Australian newspapers, it is considered unreasonable to expect an Australian to serve anywhere except in defence of Australia? Though I am a member of the Imperial Federation Defence Committee; though I believe that it is well that we should turn these questions over in our minds, I certainly deprecate the tone sometimes adopted by members of the Committee in discussing the question. Believe me, Englishmen as a body recognise that Australians as well as Canadians have done much for the defence of the Empire in the past. We do not forget that Melbourne and Sydney have been well defended at colonial expense. We do not forget the presence of the New South Wales contingent in the Sudan, a great object lesson to European nations of the unity of sentiment which animates all who live under the Union Jack. A contribution of £135,000 a year does not loom very large in Navy Estimates which amount to £22,000,000, but it is valuable as the recognition of a principle and as an earnest of what our colonies may some day be prepared to do. We shall not repeat the mistakes of the past. We do not and we have no right to expect that you will make any serious money contribution to the defence of the Empire until we are prepared to give you a constitutional voice in the control of that expenditure. That is impossible under our present constitution.

* * * * *

I have had unrivalled opportunities of seeing the British Empire. Let me say in conclusion that it is the highest ambition of my life to help to bind the colonies and the mother country more closely together, and whatever may be my political career, I can undertake that my best energies will be devoted to that object. This is no more than could be expected from the son of your Governor, who, at a time of life when many men are looking to rest from their labours, left his home and his children, who were settled round him, to serve his Queen and his country for the sake of the cause which we both have so much at heart.

10. 1900

Progress of the British Navy
C. N. Robinson

In preparing this statement of the progress of British warship construction it seems natural and appropriate to preface the chronicle of work done at home by a glance around the naval horizon. Several reasons incline me to this course, and not the least that we have it on the best authority that what is occurring abroad in the field of naval exertion has a most important bearing upon the promise and fulfilment of the various programmes of shipbuilding projected from year to year by our own Admiralty. The British standard of strength by sea is admittedly relative to something which depends on the action, actual or proposed, of others; and although it may be modified or limited by factors over which those others have no direct control, these factors again, be they political or material, must be always of secondary importance. It is not so much the multiplicity or the value of the interests to be protected that determines the strength of the defensive force as the number and activity of possible depredators. It appears therefore that this thought should be in one's mind when reviewing our naval progress, that whether we find the aims and ends authoritatively set forth as essential to have been achieved, or if, on the contrary, we discover retardation and postponement, the result should be measured against what is going on elsewhere, and no kind of persuasion or argument based upon other conditions should content or satisfy us.

Regarding the naval horizon from this standpoint, it has to be confessed that the prospect is not altogether so clear as could be wished. There are signs to which we cannot shut our eyes, signs which may presage a coming storm, may herald passing squalls, or may be merely clouds which will disappear with the dawn. That there is unusual activity and movement in the naval establishments of several of the foreign Powers is obvious, but there is no reason to doubt the knowledge by our own naval authorities of all that is occurring in foreign dockyards and arsenals. The significance of the circumstances just referred to rests rather with the future than with the present. It is quite possible to agree with the Secretary of the Admiralty when he asserts that the Navy of this country is perfectly competent to deal with any attack upon us that could now be meditated by any possible combination of Powers, and yet to feel a certain amount of apprehension lest our preparations should not be altogether sufficient or adequate to enable us to continue in this entirely satisfactory condition. The indications which it appears to be most necessary to watch carefully are those which virtually denote the introduction of new naval elements in the direction of world policy.

In our naval survey it is Germany that looms largest because she exhibits a desire and a determination to raise herself from the position of a comparatively weak sea Power to that of one among the more potent. Elsewhere in the *Annual* will be found a translation of the material points in the far-reaching programme of shipbuilding for which the Emperor and his ministers are still seeking the authority of the Reichstag. That the necessary sanction will be obtained need not be questioned, and providing the work be carried out promptly, the extension of her fleet must become all that the most ardent naval enthusiasts among her people can wish. Of the Emperor's intentions with regard to his Navy there is no concealment. In his New Year's speech to his generals he is reported to have said:

"At the beginning of the century the army of Frederick the Great had fallen asleep on its laurels. It was guided by senile generals, and its officers were ruined by luxury and stupid arrogance. Our punishment was severe; the army was thrown into the dust. Frederick's glory faded; his banners were broken. During seven years of hard slavery God taught us to recover ourselves under the pressure of an overbearing conqueror. Our nation established general military service, which gained the greatest importance under my grandfather, who reorganised the army in spite of all stupid opposition. His spirit revived the army, and his confidence in God carried the army away to unforeseen victories. So he united the German nationalities. By our army Germany regained her position in the council of nations. You, my generals, must preserve and prove the old qualities in the new country. Simplicity and modesty in life and daily sacrifice to the royal service must be your rule. *As my grandfather did for the army so will I for the navy, carry out the work of reorganisation. The navy must be equal to the army. Then I shall be enabled to procure for Germany the place among foreign nations which she has not yet obtained.*'

Practically the result of the German programme, if fully carried out, will be to double the Fleet during the next sixteen years. We should not hastily deem such an increase to be a menace to our supremacy at sea, and individually it cannot be so. At the same time the conditions are conceivable in which if such a new force were thrown into the balance against us it might make our position one of great jeopardy, for assuming that the German Fleet remained intact during the continuance of a war in which this country was involved with other sea Powers it might be used to impose terms of peace, or partition, inimical to our interests and the continued maintenance of our Colonial Empire.

Two other Powers are adding or projecting additions to their naval strength in a manner which demands notice. They are Powers which happily may be considered as actuated by motives towards this country which are entirely friendly. Both the United States and Japan have recently been engaged in naval war, and in each case the truth of Captain Mahan's teachings as to the influence of sea power has been borne home to their peoples. It is not surprising therefore to find that they are making great efforts to be strong on the ocean. When her present programme of shipbuilding is completed—and this will be almost immediately—Japan will have quadrupled her naval strength as compared with what it was when she drove China from the sea. As a result, there will be forged in the Far East a weapon of immense power, and one that we may be sure will be used skilfully and ruthlessly to enforce the wishes of the enlightened rulers of the Land of the Rising Sun. We have to look back over a somewhat longer period to obtain the contrast, but the rehabilitation of the Navy of the United States presents a picture quite as striking as that of the Eastern Power. The natural and characteristic aspiration of our American cousins that their Fleet shall be home-grown seems to be alone responsible for the fact that its increase has not advanced so rapidly as has that of Japan. The progressive strides, however, have been continuous, and are consistent with a continuity of policy which exhibits no signs of abatement or change.

* * * * *

11. 1900

The Progress of Foreign Navies
J. Leyland

* * * * *

Germany

The opening of the year 1900 was made remarkable by the presentation to the Reichstag of a new shipbuilding programme of extended duration. The German Navy Law of April 10th, 1898, had been found insufficient, though when it was presented it was supposed to be invested with something of definitive character. The new scheme, which was adopted by the Federal Council, was laid before the Reichstag with an important explanatory memorandum (given almost in full in Part IV.), which begins by declaring that the protection of national interests and especially of foreign commerce is a vital question. "For this purpose the German Empire requires peace—not only upon land, but upon the sea—not, however, peace at any price but peace with honour." The memorandum expounds certain naval principles which are familiar in this country. A war touching commercial interests is likely to last long, and will last longer according to the object of the superior enemy. To that enemy such a war might cost little comparatively, but if it proved unfortunate for Germany it would result in the destruction of her maritime commerce and perhaps of her colonies, and a commerce destroyed requires long to recover. Again, the result of a naval engagement would be to disable many ships, but the stronger adversary would be possessed of other forces; and therefore, though the fleet now ready and in hand might render a blockade difficult, it would be powerless to prevent it.

It is unnecessary to describe the details of the measure here, because the essential portions and many details are given elsewhere.

The fleet necessary to Germany, we are told, must have the tactical formation of two double squadrons of efficient battleships, with the essential auxiliaries of cruisers and torpedo-boats, and the second double squadron or fleet is to have the same constitution as that adopted for the first under the law of 1898. The first fleet is to consist of the most modern vessels, in order that it may be a "school for tactical training," and be always ready for an outbreak of hostilities, while the second fleet may consist of the older vesels, not all of which will be kept continuously in commission. Although the fleet in contemplation is for employment in home waters, it is evidently intended to increase the number of ships abroad, and reference is made to the occupation of Kiao-chau and to the increase of foreign commerce. The last matter is enforced in an appendix to the programme, wherein the development of German maritime interests and affairs is expounded under the headings of the increase of population at home and in the colonies, the development of commerce at sea and of German shipping, the expansion of the shipbuilding industry, the enlargement and increase of harbours, the magnitude of the German fisheries, the protection of cables, and the growth and development of the colonies generally.

The German Government certainly take a long look ahead, for the financial provision extends to the year 1916, and the ships will not then have all been completed. When the Navy Law of 1898 was adopted, a plan was laid down for determining the obsolescence of warships, which was really at the base of the building programme, and under this scheme it is estimated that by the year 1917 seventeen battleships and coast-defence vessels, ten first and second-class cruisers, twenty-nine third-class cruisers and gunboats, and twelve divisions of various torpedo-boats will have become obsolete. In the list of ships thus to be condemned are the four ships of the Brandenburg class, the new armoured cruiser Fürst Bismarck, the vessels of the Hertha class, and some small cruisers which have not yet been built, and are indicated by letters only. The four ships of the Sachsen class, with the König Wilhelm, Kaiser, and Deutschland, will disappear from the active list in 1901, but after that date there will only be three condemnations up to the year 1914, when the now modern ships will begin to disappear from the list. As these vessels become obsolete others are to be laid down to take their places, so that the fleet should always remain of the strength designed.

* * * * *

12. 1900

Naval Manoeuvres in 1899
J. R. Thursfield

* * * * *

"The Marconi system of wireless telegraphy was tried in the Naval Manœuvres of 1899, and proved very successful so long as only one ship was signalling. Signals were taken in at a distance of sixty miles." Such is the judgment of the Admiralty, as recorded in the First Lord's Statement explanatory of the Naval Estimates. The qualification, "so long as only one ship was signalling," is significant. Wireless telegraphy has been shown to be trustworthy for the transmission of signals between two ships sixty miles apart, but it is evident that the Admiralty are not yet satisfied that the system is applicable to the ordinary signalling of a squadron, where from the nature of the case many signals from different ships must often be made simultaneously. On the other hand, the limitation indicated applies to all signals available during a fog—that is, to all audible signals. Visible signals made simultaneously do not interfere with each other. Audible signals do. But as a whole fleet can be addressed in a fog by audible signals made from the flagship, the ships answering successively in the order of their fleet numbers, so the Marconi signals can be made available in like manner for many manœuvring purposes both in fog and clear weather. There is, moreover, another point to be considered. At present a Marconi signal made by any ship can apparently be taken in by any other ship, provided with the necessary apparatus, within a circumference of sixty miles. Thus, a cruiser giving important information to her own flag in one direction may quite unconsciously give the same information to an enemy in another direction. Ciphers may be used, of course, but unless a cipher is frequently changed it will very soon be deciphered by an enemy who knows his business, and to change ciphers daily on a system preconcerted beforehand is not very easy in practice, and is cumbrous and dilatory in any case. It may hereafter be found possible to direct the vibrations which transmit the message only in a given direction, as the beam of a searchlight is directed. But this at once limits the power of signalling to two ships each of which knows accurately the position of the other. It is true that the beam of vibration—if the expression may be permitted—may be slowly swept through an arc of the horizon, as the beam of a searchlight can be swept, in the hope of picking up the friendly ship with which it is desired to communicate. But if the position of the friendly ship is not known, the beam may have to be swept through a large arc of the horizon, or even through the whole circumference, with the chance of missing the friendly ship after all, and with no certainty of not betraying the presence of the transmitting ship to enemies on the look out for her. This has often happened in manœuvres through the incautious or injudicious use of a searchlight, and would be quite as likely to happen through the use of the Marconi system in the manner indicated. There is another method of wireless telegraphy known as the "Syntonic" system. In this system, unless the receiver is attuned to the transmitter beforehand, no effect is produced on it. This system does not appear as yet to have passed beyond the laboratory stage of experiment; but its principle, if susceptible of practical development, seems to promise far-reaching results.

* * * * *

13. 1901

The Transport Operations to South Africa
J. Leyland

The transport of the troops to South Africa during the war has been by far the most considerable operation of the kind in which any nation has ever engaged, and the more one thinks of the embarkation and despatch of the troops, the more profoundly is one impressed with the significance of the great historic success. The distance at which the hostilities were waged, the vast numbers of men and animals employed, the huge aggregate of stores of every class to be conveyed, all demanded the resources of a merchant marine such as is possessed by no Power save our own. No other nation has ever put into the field an army of a quarter of a million men, with lines of communications covering 7,000 miles of sea and land, provided with horses, transport animals, field and siege guns, ammunition, waggons, vehicles, traction-engines, bridge-building, pontooning, and telegraph materials, and tents, tools, and equipments, as well as with food, forage, and hospitals, not to speak of the thousands of objects that are necessary for the efficiency and the operations of forces in the field.* These forces were to be employed in a country that may be described as almost destitute of military supplies, and where few requirements beyond waggons and draught oxen could be procured, and they were to continue on active service for a period which has not yet come to an end. From the strategic point of view the naval transport was simplified, owing to the fact that we were not at war with a sea power, and that beyond a general patrol of the route to South Africa no special precautions were called for. On the other hand, having regard to national considerations, it was necessary that the immense strain thrown upon our merchant marine should not dislocate commerce or industries, and should not disarrange the rates of freight. When we remember that this

---

*The siege train despatched in the Tantallon Castle in December, 1899, consisted of eight 6-in. howitzers on heavy mountings and carriages, and four 4·7-in. guns, with very cumbrous platforms, which weighed 36 cwt. each. The ship also carried twenty pontoon waggons, fourteen other waggons, 5,000 boxes of shell, 400 cases of 4·7-in. cartridges, and a great weight of other ammunition. She also conveyed 470 officers and men.

was accomplished with complete success, we recognise that the fact speaks volumes for the strength of our commercial fleet, and there need be no surprise that the achievement has aroused the admiration of the world, for it has been a revelation of resource, energy, organisation, and national spirit, for which there is perhaps no parallel, as also a triumph of good management and business-like capacity, reflecting the highest credit upon all concerned.

The magnitude of the operations undertaken by the Transport Department of the Admiralty, under the direction of Rear-Admiral Bouverie F. Clark, can only be appreciated by the evidence of statistics, though these can but imperfectly suggest the vastness of the task. Before the war broke out, and in the period from June to September, 1899, preliminary reinforcements from Home and the Mediterranean, to the number of 8,168, reached South Africa. The following table shows the total number of officers and men landed at the various South African ports from all sources during the succeeding period, from October 1st, 1899, to October 31st, 1900:

	From Home and Mediterranean	From India	From Mauritius	Colonial Contingents	Totals
1899					
October	1,632	6,004	450	..	8,086
November	42,642	..	..	1,922	44,564
December	23,548	572	..	689	24,809
1900					
January	25,134	1,007	..	..	26,141
February	19,259	391	..	2,625	22,275
March	37,144	315	..	953	38,412
April	21,425	19	..	2,750	24,194
May	11,554	..	..	1,734	13,288
June	9,795	..	..	918	10,713
July	7,675	..	..	..	7,675
August	1,765	..	..	..	1,765
September	4,143	..	..	..	4,143
October	3,267	..	..	..	3,267
Totals	208,983	8,308	450	11,591	
				Grand Total	229,332

The subsequent reinforcements to South Africa up to January 31st, 1901, were 8,072. We thus reach a grand total of 245,572 officers and men; but this aggregate has since been very largely increased, and the despatch of transports still goes on.

It must, however, be remembered that the transport service has not merely been for the conveyance of the forces from certain places to

South Africa. It was, for example, necessary to detain certain of the transports, to the number of about twenty, on the Cape station, always in readiness to move a division from Cape Town to Durban or intermediate ports, or *vice versâ*, according to the military requirements. Moreover, provision had to be made for conveyance to England, India, and the Colonies of returning troops and invalids, sick, and wounded. The total number in the last categories up to January 31st, 1901, was returned as 58,911.*

Scarcely less important than the men were the horses and other transport animals employed in the field. The total number of horses from Home, India, Austria, America, and Australia, including those with the Colonial contingents, embarked for South Africa up to October 19th, 1900, was 124,834. The mules were conveyed from North America, Spain, Italy, India, England, Cyprus, and Australia, and the total number landed in South Africa from October, 1899, to the end of October, 1900, was 62,690.

It is unnecessary to burden these pages with a classification of the military and other stores, vast in variety as in volume, that have been transported from various points, and chiefly from England, to South Africa. But it will illustrate the magnitude of the work if I take the month of December, 1899, promising that, in some of the early months of 1900, the aggregate was still higher. In the month selected the total bulk of stores conveyed to South Africa in transports and freight-ships may be estimated at 52,000 tons, while ships not taken up by the Admiralty conveyed, during the same period, nearly 17,000 tons of hay, about 7,000 tons of oats, over 600 tons of meat, and about 1,000 head of cattle, besides nearly 15,000 tons of coal.

The troops were principally embarked in transports hired by the Admiralty, of which there were 102, chartered at a rate estimated on their gross tonnage—perhaps I shall not be far wrong in saying that the mean rate was 20s. per ton per mensem—and these were under their own officers, and were manned by their own men, whose wages were paid by their owners; but the vessels were completely at the disposal of the Admiralty, who coaled them, superintended the order of embarkation, fixed the dates of departure, and, in a word, had complete control of the whole operation. The freight-ships were in a different category altogether, being hired to carry so many men, horses, or tons of goods, at an agreed rate for the voyage, the management remaining in the hands of the owners. For example, the great majority of the boats of the Union-Castle Line were so employed, and did splendid work. I am indebted to a table prepared by Dr. Benedict William Ginsburg,

---

*It was estimated that, up to the end of March, 1901, about 50,000 officers and men, sick and wounded, had been disembarked at Southampton, many of them requiring very special handling at the port.

showing the number of ships and the gross tonnage of each shipping company engaged in chartered transport work.

Ships	Companies	Gross Tonnage	Ships	Companies	Gross Tonnage
9	Elder, Dempster & Co.	52,164	2	Orient Line	10,953
6	Cunard S.S. Co.	38,414	2	Pacific S. N. Co.	10,355
4	Leyland Line	36,756	2	lamport and Holt	10,287
5	West India and Pacific S.S. Co.	33,275	2	Lund Line	10,156
4	White Star Line	33,270	2	Bibby Line	9,969
6	P. and O. Co.	32,103	2	Geo. Smith and Sons	8,516
6	B. I. S. N. Co. . . .13,739 B. I. Association . .16,959	30,698	1 2	J. Glynn Union Line	7,359 7,139
5	Allan Line	29,559	1	Rankin, Gilmour & Co.	6,900
4	Dominion Line	28,019	1	Welsford & Co.	6,215
4	Johnston Line	26,534	1	Asiatic S. N. Co.	5,660
5	Castle Line	25,831	1	Federal Line	5,464
3	Manchester Liners	16,682	2	Plate S.S. Co.	4,767
3	Anchor Line	16,559	1	Atlantic Transport Co.	4,212
2	British Shipowners	14,651	1	Lawther, Latta & Co.	4,009
2	Harrison Line	13,330	1	McGregor, Gow & Co.	3,455
2	National Line	13,162	1	Houlder Bros.	3,444
2	Wilson Line	12,382	1	Houston Line	3,261
2	Royal Mail S. P. Co.	11,491			
2	Brocklebank Line	11,353	102	Grand Total	598,354

In addition to these vessels, thirty-one ships of the British India Steam Navigation Company, with a gross tonnage of 109,730, were chartered for the transport of troops from India to South Africa. We thus arrive at a total of 133 ships, with an aggregate tonnage of 708,084. This immense fleet did not, however, represent anything like the full number of vessels engaged in transport operations. Troops were conveyed in ships engaged in the regular service to South Africa, and large numbers of vessels were employed in the transport of Colonial contingents, and others again in the conveyance of horses, mules, and stores. Indeed, a War Office return of "Embarkations in connection with the South African Campaign, 1899–1900 (up to October 19th, 1900)," which was issued on March 1st, 1901, indicates that 283 transport and freight ships had been employed in the operations, and, of course, many of these had made repeated passages to and fro.

* * * * *

14. 1904

Submarines and Torpedo Warfare
Anon

In the *Naval Annual* of 1902 Commander Robinson gave a succinct review of the progress and development of submarine vessels up to that year, dealing especially with such structural details as were known of the craft—then complete or building. In the present article the intention is to deal somewhat more broadly with submarine attack, not only the attack of submerged or nearly submerged small craft on larger ships, but also with torpedo warfare generally. That torpedo warfare has been and will be greatly influenced by the development of the submarine there is no doubt. But apart from the submarine, the introduction of the gyroscope has made a very great difference in the importance of under-water attack. It is usual to trace back the ancestry of the submarine to Bushnell's craft, invented towards the close of the eighteenth century; but Bushnell's vessel was a combination of the submarine and the mine, and this arrangement has never been and can never be of any great fighting value. It is the combination of the submarine and the torpedo that is so valuable; but for the Whitehead the submarine would remain as an interesting toy, and but little more.

The first vessel in which the Whitehead was used from a submerged tube was the British Vesuvius, which dates back to 1874. Like the present day submersibles, she aimed at getting within torpedo range of a big ship without being noticed. For this purpose she was very low in the water, and presented a very small target. But small as the target was it was large enough to be clearly visible at quite a long range, and a very few hits on the above water portion would suffice to sink her. The advent of the quick-firing gun rendered the Vesuvius (with her slow speed of 10 knots) quite unserviceable, and her place was taken by high-speed, light-draft torpedo craft, with not only the torpedo tubes but also the engines and boilers well above water. These vessels used their high speed both for getting within range quickly and for escaping after firing. And when it was endeavoured to build a type of vessel with high speed, but showing very little above the surface, and fitted with submerged torpedo tubes, the great cost of the first ship, the Polyphemus, prevented any similar craft being built. In order to get a number of invulnerable torpedo vessels for a moderate sum, it was considered desirable in the submarine or submersible to return to the small size and cost of the Vesuvius, with the same low speed, and to provide means for taking cover under water when attacked by the quick-firing gun. Thus it is the desire to evade the gun that has caused the evolution of the modern submersible, which escapes under water when hard pressed. It must be clearly understood that the submersible (a much more explicit designation than submarine) is steered to the best position for attack in precisely the same manner as the low freeboard Vesuvius or Polyphemus. The officer who commands a submersible on seeing his opponent steers to intercept her, and he cannot choose the right course to do so unless he has a clear view. A vague impression still prevails in some quarters that some instrument may be invented, or has actually been devised, by which the captain of a submersible may be enabled to keep his craft entirely under water, and yet adjust his course so as to intercept a moving enemy. This is quite impossible, as even 100 feet of water is quite as efficient an obstacle to vision as a brick wall. The first necessity for a submersible is that the officer in command should have a clear view of the enemy up to the time when he fires his torpedo. The torpedo once fired he can dive to escape, but when fairly under water the submersible is practically harmless.

The problem of how to obtain a good view without being seen by the ship that is desired to attack is the all-important one, and much of the success or otherwise of the submersible must depend on the arrangements made to enable her captain to see and not be seen. The simplest plan, and one that should be very efficacious when the weather is misty or there is a certain amount of sea on, is to have the top of the conning-tower above water whilst the hull is still immersed. This appears to be fully appreciated in France, and both the early submarines Gymnote and Zédé, together with the up-to-date submersible Triton have conning-towers rising some 5 feet above the hull of the boat. The Lake type of boat in the United States also has a high conning-tower, but the original Hollands have very low ones. This is, however, altered in the modified Holland boat A 1, recently run down when exercising at Portsmouth, for the conning-tower is about the same height as in the French craft. If the sea is very smooth a circular conning-tower makes a good deal of splash in going through the water, and it is therefore desirable to make it boat-shaped, as has been done in the Lake boat, and also in the French Triton and the latest Hollands. If the top of the conning-tower shows too much, then the boat must be sunk till only the the top of the periscope or optical tube remains above water. With the help of this instrument, objects can still be clearly seen, but the arc of vision is necessarily limited, so that it is very difficult to judge the course of a ship which it is intended to attack, and it is almost impossible to guess her distance. As the submersible is very slow, it is above all things necessary that some idea of the range of the enemy shall be obtained, otherwise there is a great liability to get left astern— a hopeless position for attack. Whether the approach to the point of attack is made with the top of the conning-tower above the surface, or with the periscope only showing, it is evidently absolutely essential that the boat shall keep at an even depth. This is the more necessary because the more evenly the boat goes along, the less tendency there is to splash, and thus give away the boat's position. Moreover, if the conning-tower bobs in and out of the water the view is spoilt, and this equally applies to the periscope. In the earlier pattern of boat a good deal was said about the difficulty of maintaining a uniform depth, but both in England, France and America this difficulty has been quite overcome. For example, the Lake boat was kept within 1 ft. of the depth ordered for a considerable time with a stranger at the helm who had never before steered a submersible. The Hollands have an equally good record, and, whatever the exact type of rudder or water-plane, it may be taken as certain that the problem of keeping a submarine at a uniform depth when running submerged has been practically solved.

The early French boats were all submarines pure and simple, driven by electric motor only. They proved serviceable but very slow, their radius of action is limited to 50 miles or thereabouts, after which the accumulators needed recharging. After some years of experiments the French have now decided to fit special engines for surface-running to all their boats, and it is this class of boat that they designate a submersible. The first engines used by the French submersibles for surface-running were ordinary steam-engines with boilers heated by liquid fuel, but there were grave difficulties in submerging quickly owing to the necessity of closing up the extensive ventilation arrangments. The heat was also very objectionable. They have now come into line with England and America in using gasolene motors for running on the surface and recharging the accumulators. All boats fitted for running under water, whether submarines or submersibles, use an electric motor actuated by accumulators when submerged. It is the invention of the accumulator that has rendered under-water navigation a practical matter. A motor actuated by an accumulator consumes no air, and there are no deleterious gases to be got rid of. The steam boilers and reservoirs used in the Nordenfelt boat and the early French submersibles made it so hot when under water as to be well-nigh unbearable. Compressed air is out of the question, as taking up too much space. All other motors consume large quantities of air, which cannot, of course, be procured when submerged. The great defect of the accumulator system is its enormous weight. Not more than 2 to

2½ I.H.P. can be obtained per ton of batteries and motor. A surface torpedo craft in which the total weight of engines, boilers, coal and water is the same as that of a submarine of 100 I.H.P. would develop some 3000 I.H.P., but such a craft burns a boatful of air in a few seconds, and this enormous supply of air is essential to all ordinary motors. Thus a very weighty installation is required to propel a boat 8 knots under water, and if an under-water speed of 10 knots was asked for, the displacement would have to be increased to 400 or 500 tons in order to admit of some 300 tons being devoted to the plant for electric propulsion.

When a boat is on the surface, a comparatively light air consuming motor will produce much more horse-power than the heavy electric plant, and can be used for recharging the accumulators. As to the type of motor for use on the surface, the French have tried both steam and petrol engines, the steam boiler being heated by liquid fuel. In England and America the petrol or gasolene engine is exclusively used, the power being generally about some three times that of the electric motor, and the resulting speed about 3 to 4 knots greater. Thus the smaller submersibles have speeds of about 6 knots submerged, 9 knots on surface, and the largest and latest about 8 and 12 knots respectively. Speeds of 15 knots on the surface are talked of, but the boat would have to be very large and very expensive, whilst, as pointed out above, 10 knots is a most unlikely speed under water. The cost of submersible boats running 12 knots on the surface and 8 knots submerged will probably be about £40,000. Thus the idea of a swarm of very cheap vessels stationed at every port or important strategic point does not seem likely to be realised. It is with a view to getting large numbers that the French are building a class of boat of only 70 tons displacement, costing some £18,000. They are confessedly for harbour defence only, their speed and radius of action is very small, and they are generally stigmatised as being unsatisfactory.

*   *   *   *   *

Summing up, there is no doubt that the submersible has a useful future before it. The problem of maintaining a regular depth has been completely solved, and both the surface and submerged motors are serviceable and satisfactory, though doubtless improvements will be made. Their speed must continue very low, especially when submerged, and the radius of action must be somewhat limited; but, still, in the narrow waters of the Channel, North Sea and Mediterranean they will exert a considerable influence. And generally any waters in which a number of submersibles are stationed will be well-nigh untenable for hostile ships, though by taking special precautions they may pass through unharmed. As yet no satisfactory means of attacking these craft has been developed, but it does not appear likely that they will long retain their present immunity, and it is rather to the torpedo than to the gun that we must look as the weapon to be used. Harbours can be defended against submarine attack by the use of mines, but breakwaters and booms with heavy and strong under-water parts are very much to be preferred. It must be remembered that, even when running on the surface, the submersible has a very deep draught, say 10 to 14 ft., and when submerged she requires at least 30 ft. depth of water. If a boom goes within, say 12 ft. of the bottom, the submersible can only run under it by smashing up her conning tower and periscope, which, if it did not sink her altogether, would render her harmless. But ships will no longer be able to lie stationary during the day in positions known to the enemy, unless they have taken adequate precautions by the use of nets or a ring of scouts to shelter themselves against attack by submersibles. It remains to be seen whether this will have to be demonstrated in the next war, as a similar truth with reference to surface torpedo craft and anchoring at night in an exposed position, was forcibly demonstrated to the Russians outside Port Arthur. And it behoves those who would not wish to act as warning beacons to the world, but only to prepare in peace time the necessary defences but also to continually exercise all concerned in their use.

It is at least four years since the gyroscope was so far perfected that it was adopted in both our own and foreign Navies. But for some time the number of torpedoes fitted was not very large, and as always happens when a new invention comes into use, it was some time before the enhanced power of the gyroscopically-guided torpedo was generally appreciated. Before the introduction of the gyroscope the limit of range of the torpedo was placed at 800 yds., for any inaccuracy being cumulative the torpedo turned aside more and more as it progressed, thus it was almost certain to be so deflected in, say a minute and a half that it would be 4 or 5 points off its proper course before it reached 1200 yds. The air-supply of the 18-in. torpedo, which has now been in use for ten years, was quite sufficient to propel it some 2000 yds. at a moderate speed, but long range not being required this quality remained in abeyance.

Now, however, it is fully realised that long range torpedo fire may be of the highest value, not so much that it is expected that it will be easy to hit a single ship at 2000 yds., but because it will be easy enough to make sure of hitting a line of ships at that range. If the ships in line are broadside on, the most natural position for a gun action, the intervals in the line are only twice as great as the spaces occupied by the ships. Thus a torpedo running through a line ahead has a chance of hitting of 1 in 3. If the ships turn till they are in line abreast, a torpedo run at right angles to the line has only a chance of about 1 in 16. But such a shot would be quite exceptional, and if the torpedo is run at a line abreast at an angle of only 1½ points from the direction in which the ships are steering, the chance of hitting is doubled, and amounts to 1 in 8. A line of 12 ships can fire simultaneously from 12 to 24 torpedoes on the broadside. If we take the lowest number, such a line would, with one discharge, be practically certain of disabling one ship of an opposing line, and with average good fortune they might expect to hit 3 or 4. Seeing that such decisive results may be expected by simply firing torpedoes through the enemy's line, it is perfectly obvious that the fleet which has the longest ranging torpedoes has a great advantage. Thus, suppose a fleet possessing torpedoes capable of running 3000 yds. engages an equally powerful fleet whose torpedoes only ran 2000, the latter must either keep outside 3000 yds. or run the risk of being torpedoed without the possibility of reply.

So far as is known, no nation has at present any considerable number of torpedoes capable of running much more than 2000 yds. Moreover the speed of existing torpedoes at this range is low, some 20 knots at the most. Experiments are, however, being pushed on, and we hear that Whitehead at Fiume has constructed experimental 18-in. torpedoes which run 3300 yds., and it is presumed will load in existing tubes without overmuch alteration. The U.S. Admiralty have on order two 21-in. torpedoes to run at 26 knots for 4000 yds., and there are reports as to modifying the existing 18-in. torpedo so as to enable it to run 3000 yds. The modification will take the form of increasing the size of the air chamber, raising the pressure to which it can be charged, and possibly of superheating the air by an alcohol flame whilst on its way to the engines. This latter device has given good results in the United States, and seems likely to be generally adopted.

Up till comparatively recently, very little attention has been paid to the rate of fire from a torpedo tube. Not only were the loading arrangements clumsy and complicated, but there was generally no provision for placing another torpedo quickly in position ready for loading. During the last year the question of the rapid reloading of submerged tubes has received a great deal of attention, and as a result it is stated by the *Naval and Military Record* that in the Mediterranean as many as three torpedoes have been fired in two minutes from a single tube. In fact, the torpedo, like the gun, is becoming a long range quick-firing weapon. A natural concomitant of long range is more or less wild shooting, and there will be many more misses than heretofore. From this it follows that a larger supply of torpedoes will have to be carried.

The argument that as the torpedo increases in importance ships will decrease in size by no means follows. As long as ships fight on the surface the gun will always be important. Guns entail armour to protect them, and big guns and thick armour require large displacement. The large ship is more likely to survive the effect of a torpedo than is a small one. Moreover, two small ships having the same fighting-power as one large one present a larger target collectively to the torpedo. On the whole, therefore, the increasing importance of the torpedo is not likely to entail any immediate alteration in the present type of ship.

As to the relative advantages of submerged and above-water tubes, the former are greatly superior when the ship using them has to sustain a heavy fire before she reaches torpedo range. On the other hand, at night or in a fog above-water tubes will be well-nigh as efficient as those submerged. Both sets of tubes are therefore valuable, and for any vessel that acts as a scout and is likely to find herself on a dark night in close proximity to an enemy who is probably unaware of her presence, it is of the highest importance to be able to fire as many torpedoes as possible in a short time. The lightness and simplicity of the above-water tube enables small craft to carry several, and whilst submerged tubes will continue to be fitted to large ships, every small vessel will have as many above-water tubes as she can carry, whilst there will be many ships which will have both above-water and submerged tubes. All fleet scouts should certainly have both types of fitting.

Advertisements as History

HADFIELDS LTD., HECLA WORKS, SHEFFIELD, ENGLAND

The Largest Manufacturing Capacity in the World for

PROJECTILES

Contractors to the British Admiralty & War Office, Colonial Offices, The United States, Japanese, and other Foreign Governments.

PROJECTILES IN FORGED STEEL,
CAST STEEL AND CAST IRON
Of all Types and Calibres up to 16-inch.

ARMOUR-PIERCING, COMMON, SHRAPNEL, & H.E. SHELL
Either Empty, or Filled and Fused, together with Gun Charges.

Managing Directors :
Sir ROBERT HADFIELD, M.Inst.C.E.
ALEXANDER M. JACK, M.Inst.C.E.

Supt. Ordnance Dept. :
Major A. B. H. CLERKE, late R.A. (Director).

Telegrams :
"HECLA, SHEFFIELD."
Telephone No. : 1050 Sheffield.

London Office :
Norfolk House, Laurence Pountney Hill, E.C.

P. B. BROWN, M.Inst.C.E. (Director).

Telegrams :
"REQUISITION, LONDON."
Telephone No. : 2 City, London.

HADFIELD'S "HECLON" CAPPED ARMOUR-PIERCING PROJECTILES, 13·5-INCH, 14-INCH AND 15-INCH CALIBRES, AFTER PERFORATING K.C. (KRUPP CEMENTED) ARMOUR-PLATES OF THE LATEST TYPE, FROM 12 INCHES TO 15 INCHES IN THICKNESS. ALL THE PROJECTILES WERE FITTED WITH THE JACK PATENT CAP.

—— *Sole Makers of* ——

HADFIELD'S PATENT "ERA" STEEL

which has been adopted by the British Admiralty and War Office, also by other Powers. The Supreme Material for Ammunition and Communication Tubes, Conning and Director Towers, and all Armoured Parts exposed to shell fire ; also for Gun Shields for Naval and Land Service.

"ERA" STEEL COMBINES HIGH RESISTANCE WITH GREAT TOUGHNESS.

STEEL CASTINGS AND FORGINGS of *Every Description.*

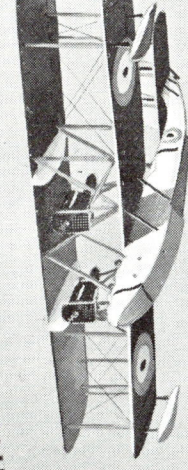

ACCURACY MEANS ECONOMY OF FORCE

Nothing is more costly in terms of naval strength than the projectile which misses its mark or the aircraft that fails to find its target.

SPERRY

Naval Fire Control Systems

Marine and Aeronautical Navigational Equipment

THE SPERRY GYROSCOPE CO. LTD., GREAT WEST ROAD, BRENTFORD, MIDDLESEX

PART II: 1905–1914
Anglo-German Rivalry and Continuing Technical Advance

Introduction	59
15. 1906	
Steam Engineering: The Turbine	60
G. R. DUNNELL	
16. 1907	
The Tactical Qualities of the Dreadnought Type of Battleship	62
LIEUTENANT-COMMANDER W. S. SIMS, USN	
17. 1909	
The Naval Expansion of Germany	67
J. LEYLAND	
18. 1909	
Speech on the Naval Estimates	72
R. McKENNA, FIRST LORD OF THE ADMIRALTY	
19. 1912	
Naval War Staffs	76
J. LEYLAND	
20. 1912	
Recent Changes in Warship Design	80
SIR WILLIAM WHITE	
21. 1913	
The Progress of Naval Aeronautics	84
ANON	
22. 1914	
Armour and Ordnance	89
C. N. ROBINSON	

Introduction

With the establishment of the *Entente* with France in 1904, British attention focused on the increasing naval challenge from Germany. Sir John Fisher, as First Sea Lord, dedicated to improving the Navy's readiness for war, concentrated the modern battlefleet in home waters and introduced the all big gun battleship, the *Dreadnought*, to meet the threat in the North Sea. He claimed that the *Dreadnought*, in both its battleship and battle-cruiser form, made all its predecessors obsolete. There were many critics of his policy, including A. T. Mahan himself. But all the great naval powers were caught up in the *Dreadnought* fever and the American government put up W. S. Sims, the future head of his country's wartime naval mission to London, to refute Mahan's arguments.

Even the most ardent advocate of the big battleship recognised the need for constant improvements in its protection against shells and mines and torpedoes. War was to show that Germany excelled in this.

In Britain there was great uneasiness among informed opinion that the Navy had concentrated so much on material that it had neglected systematic operational and administrative planning. This was the view of Winston Churchill, and after his appointment as First Lord of the Admiralty in 1911 he established the War Staff in a belated attempt to remedy the shortcoming.

Recommended Reading

Arthur J. Marder, *From the Dreadnought to Scapa Flow*, Vol. I (1961).

15. 1906

Steam Engineering—The Turbine
G. R. Dunnell

The attention of marine engineers is at the present time so largely devoted to the great change in propelling machinery now taking place that no excuse is needed for devoting the whole of the space available in the *Naval Annual* for engineering matters to the steam turbine, especially as there has been little change in other branches of late. The water-tube boiler—the other great revolution in ship propulsion—remains in much the same position as when we last wrote; for war vessels it is practically universal, but in the mercantile marine it has made comparatively small progress. The fitting of small-tube boilers into some of the largest ships in the Royal Navy is a notable feature, but these types of steam generator have already been fully dealt with in former issues of the *Naval Annual*.

It is understood that the Admiralty has decided—and doubtless the fact will be publicly announced before these lines are in print—that practically all vessels in progress for the Royal Navy, and not yet engined, are to be fitted with steam turbines as a means of propulsion. It is a step that constitutes one of the most important incidents in the records of naval construction, and may be coupled with that hardly less striking fact in the history of the mercantile marine—the placing of steam turbines in the two new Cunarders, each of about 70,000 H.P., now under construction on the Tyne and on the Clyde. As the warships that are thus to be propelled by turbine machinery include the new Dreadnought—of 18,000 tons and 23,000 H.P., the largest and most powerful battleship yet laid down—and the three first-class cruisers, Invincible, Inflexible, and Indomitable, and as the new Cunarders far surpass in size and power any vessels that have preceded them, it will be seen that the largest and by far the most important ships ever put in hand, either for war or commerce, are to have propelling engines of a type that a year or two ago was considered to be suitable only for small craft of special design, and a very few years earlier was not thought to be applicable for marine propulsion at all.

The advance of the steam turbine has been truly phenomenal, and we may fairly take pride in it being due to British ingenuity supported by British enterprise. With so little practical experience on a large scale, the decision to discard the time-honoured reciprocating engine may have seemed a bold one; and, indeed, it was a bold one, both for the Admiralty and the Cunard Company. Foreign naval authorities are proceeding more cautiously. The German Government are making the venture with two third-class cruisers—the Lübeck with Parsons turbines, and a sister vessel with Curtis turbines. The Americans are undertaking a similar experiment with three scouts. Happily, so far as experience has yet gone, it may be said that the boldness of the British Admiralty and of the Cunard Company has been justified. There are now a number of vessels, some of large size, fitted with Parsons turbines, and, generally, their performance has been successful. It is perhaps worth considering what corresponding degree of boldness would have been necessary had by chance the steam turbine been the original marine motor, and had it been proposed to substitute for its simple and continuous rotary motion the reciprocating movements of the heavy pistons, slide-valve, crossheads, and connections of the older steam-engine—masses of metal which have to be brought to rest and restarted twice in every revolution, thus involving serious alternations of stresses on crank-pin brasses and main bearings.

* * * * *

16. 1907

The Tactical Qualities of the Dreadnought Type of Battleship

Lieutenant-Commander W. S. Sims USN

* * * * *

Concerning the advisability of building "all-big-gun" ships—that is, discarding all smaller guns, except torpedo defence guns—and designing the ships to carry the maximum number of heavy turret guns (these alone to be used in battle), I think it can be clearly shown that Captain Mahan is in error in concluding that it would add more to our naval strength to spend the same amount of money that the big ships would cost for smaller and slower ships carrying the usual intermediate guns—6-in., &c.—and that it is only a question of speed. This error is probably due to the fact that much important information concerning the new methods of gun-fire was not considered by the author in preparing his article.

In order that this question of gun-fire may be clearly understood, it will be necessary to mention the fundamental principles of fire-control. In order to hit at long ranges you must have the sight-bar range within small limits—that is, one half the danger space. No existing range-finder will measure such long distances within these limits. With the ammunition carried by ships in service (that is, ammunition that is not new) this sight-bar range usually differs for each index of power, even for guns of the same calibre, and the sight-bar range for one calibre is rarely the same as that for any other calibre. Therefore there must be a separate fire-control for each calibre on board, which greatly increases complication and difficulty of control when there is more than one calibre. Since the range-finder can never be relied upon, it follows from the above that the sole means of bringing the sights on to an enemy, and keeping them there, is by observing the splash of projectiles from aloft by the vertical method. If a ship has guns all about one calibre, fire-control is comparatively easy, because each splash is necessarily made by that calibre, and the fire of the ship is controlled by one fire-control officer (directing one fire-control party). When the distance is over 5000 yards you cannot follow the projectiles throughout their flight and identify their splashes, unless the projectiles of the different calibres differ greatly in diameter—as 12-in. and 6-in.—and consequently make splashes that are markedly different in size and appearance.

Therefore we have to decide what the calibre for each class of ships should be, a decision which should present no special difficulty provided it be first determined how and where to defeat the enemy—whether by the destruction of their ships by sinking them, or disabling their guns, or by the destruction and demoralisation of their *personnel*. In this connection the following facts should first be clearly understood—namely, turrets are now for the first time being designed that are practically invulnerable to all except heavy projectiles. Instead of having sighting-hoods on the turret roof, where sight-pointers and officers are exposed to disablement (as frequently happened in Russian ships), there will be prismatic sights projecting laterally from the gun-trunnions, through small holes in the side walls of the turret; and the gun ports will be protected by 8/in. armour-plates, so arranged that no fragments of shells can enter the turrets. On the proposed "all-big-gun" ships the heavy armoured belt will be about 8-in. thick above the water-line, extending from end to end; the conning-tower, barbettes, &c., will be of heavy armour, there being no intermediate battery (which could not be protected by heavy armour on account of its extent). It follows that in battle all of the gunnery *personnel*, except the small single-fire control party aloft, will be behind heavy armour, and that therefore neither the ship nor her *personnel* can be materially injured by small calibre guns.

Considering, therefore, that our object in designing a battleship is that she may be able to meet those of our possible enemies on at least equal terms, it seems evident that it would be extremely unwise to equip our new ships with a large number of small guns that are incapable of inflicting material damage upon "all-big-gun" one-calibre ships of our enemies, or upon the *personnel* manning their ships.

* * * * *

Captain Mahan characterises the sudden inclination of all navies to increase the size of the new battleships from about 15,000 tons to about 20,000 tons as "wilful premature antiquating of good vessels," a "growing and wanton evil." It seems to me that the mere fact of there being a common demand for such vessels is conclusive evidence that there must be a common cause that is believed to justify the demand. This common cause is undoubtedly the common belief that the same amount of money expended for large war vessels will add more to a nation's naval power than the same amount expended on smaller vessels.

It should never be forgotten that the credit for the inception of the epoch-making principles of the new methods belongs exclusively to Rear-Admiral Percy Scott, Director of Target Practice in the British Navy, who has, I believe, done more to improve naval marksmanship than all the naval officers who have given their attention to this matter since the first introduction of rifled cannon on men-of-war.

From the point of view of naval efficiency, we should have nothing to fear from even a still further increase in the size of our battleships. For example, reverting to the supposed fleet of ten 20-knot ships, the "S" fleet above described, there can be no doubt that the same sum—100 million dols.—expended for a less number of still larger battleships would produce a superior fighting fleet. For the same sum we could doubtless build eight ships, each having a broadside fire of ten 12-in. guns instead of eight, with 1 knot more speed. Such a fleet would be 1·7 miles long instead of 2·1 miles, with the concentration of forty-eight heavy guns per mile to oppose to the twenty-one per mile of the long "L" fleet (3·9 miles) of small vessels, not to mention the increased superiority of its manœuvring qualities, and the superiority of its protection against gun-fire and torpedoes. One of the great advantages with a large vessel is that the under-water hull may be so designed that the ship cannot be materially damaged by one torpedo.

A brief explanation is now required as to why Prussian and American naval officers are practially a unit in advocating the "all-big-gun" one-calibre ship, while some other navies, discarding the small guns—6-in., &c.—still retain in new designs large turret-guns of two calibres, as 12-in. and 8-in. and 11-in. and 9·2-in., &c.; the larger turrets being in the bow and the stern, and the smaller ones on the sides. The reason is, I believe, that they do not yet understand the great difficulty of controlling the fire of guns that approach each other in calibre. From the facts and arguments presented I derive the following main conclusions, formed upon what I believe to be fundamental theoretical principles of gun-fire and tactics:—(1) That in consideration of the fact that the ultimate object of the fleet is that in the event of war we may be able to overcome our possible enemies on the sea, we should so design our battleships that they will at least equal those of our possible enemies in all of their fighting qualities—speed, gun-power, height of gun positions, protection, &c. (2) That, subject to the above requirements, it is always desirable to increase the speed a certain reasonable amount. Incidentally it may be remarked that this indicates the advisability of developing maximum speed with minimum coal consumption by placing all similar vessels in continuous competition in steaming, much in the same manner that we utilise the competitive principle in developing their maximum gun-power. (3) That it is always desirable to substitute heavy turret guns, such as 12-in., for the equivalent weight of the usual intermediate guns (6-in., &c.); in other words, that the "all-big-gun" one-calibre ship affords the greatest possible capacity of effective hitting. (4) That in order to simplify fire control, and attain its maximum efficiency, all of the main-battery guns of ships of whatever type should be of the same calibre. (5) That for similar reasons all of the torpedo defence guns should be of the same calibre. (6) That very important tactical advantages are obtained by the concentration of many heavy guns on each large vessel of high speed, and the consequent intense concentration of heavy-gun fire due to the tactics of the fleet. (7) That the tactical advantages of size, speed, and diminished numbers are of much greater importance than any advantages to be obtained from the increased number of smaller and slower vessels that may be built at the same total cost.

In addition to the superior individual and tactical advantage of large vessels, they also possess the following minor advantages:—(1) A fleet of ten 20,000-ton ships, such having a broadside fire of eight 12-in, guns, or eighty in all, would cost about 100,000,000 dols. (2) A fleet of twenty smaller vessels, each having a broadside fire of four 12-in. guns, or eighty in all, would cost about 120,000,000 dols. or 130,000,000 dols. (though I previously assumed the cost of these fleets to be equal, in order to accentuate the tactical value of large ships). (3) It requires less men to man the main-battery guns on an "all-big-gun" ship than a mixed battery ship; for example, it requires less men to serve the ten 12-in. guns of the Dreadnought than the four 12-in. and sixteen 6-in. guns of the Missouri. (4) It will require no more men for the Dreadnought's crew than it would for the Missouri's, if she had the full complement of men (as measured by European standards), which neither she nor any of our battleships have. (5) The complement of officers of the Dreadnought is not as great as should be that of the Missouri or Louisiana, because the former requires one fire-control party, while the latter ships require respectively two and three parties, as well as more officers to command the guns. (6) Then assuming 800 men and twenty combatant officers in each ship, it would require 8000 men and 200 officers for ten "all-big-gun" ships, and about 16,000 men and 400 officers for the fleet of small vessels having the same broadside fire. (7) It will cost nearly twice as much to dock as the ten large ones and the latter can be docked in one-half the time, which is a great advantage in time of war. (Captain Mahan notes that the absence of a big ship for docking, coaling, &c., reduces the strength of the fleet more than the absence of a small one, but he neglects to note that with twice as many ships in the fleet there will be twice as many absentees in a given time.) (8) From the above it is clear that the cost of maintaining a fleet of small vessels having the same broadside fire as a fleet of large ones (of double the individual broadside fire) will be nearly twice as much as that for a fleet of large vessels of about the same total man-power. (9) I understand that the cost of maintaining a battleship is near 1,000,000 dols. a year; therefore the yearly maintenance of a fleet of ten large vessels would cost about 10,000,000 dols. less than that of the twenty smaller ones. (10) The final conclusion is, that for the sum that it would cost to maintain the twenty small battleships we could maintain a fleet of ten large ones, that would be greatly superior in tactical qualities, in effective hitting capacity, speed, protection, and inherent ability to concentrate its gun-fire, and have a sufficient sum left over to build one 20,000-ton battleship each year, not to mention needing fewer officers and men to handle the more efficient fleet.

17. 1909

The Naval Expansion of Germany

J. Leyland

The growth of the German Navy is for the British Empire the most significant fact in the recent history of the world. Forty years ago Germany was no more than a geographical expression, but her powerful states awaited the supreme moment when they should be welded together by high statesmanship into one of the great Powers of Europe, distinguished above all others by its vast military organisation, and the spectacle that it presents of a nation trained to arms. Its military system has been extolled as the very type and exemplar for the organisation of our own. The regeneration of Prussia after the disasters of 1806, by the great soldiers who shaped the steps of progress, and the demonstration of organised efficiency in subsequent wars, had impressed Europe with the conviction that in the countries of Germany, and more especially in Prussia, the true secret of military efficiency was known. But those who look back even twenty years remember that the German people betrayed even then no instinct for the sea. All their trust was in the Army, which was a part of themselves. Some prescient persons, like Prince Adalbert of Prussia and Albrecht von Stosch, knew, nevertheless, that Germany had also her place at sea, but for the mass of the people, and for their representatives in the Reichstag, there was no consciousness of the necessities of the Navy that was beginning, and time after time the sums demanded for the building of vessels were struck out of the estimates by majorities distrustful of the purposes of the Government. Since that time we have witnessed a revolution both in German public opinion and in the views and, in some measure, of the objects of the Government, and a work is being accomplished which has had no parallel since the time of Colbert. No longer content to be a powerful military state, Germany is resolved to rank with the great maritime Powers of the world, and the German Government has told our own, as Mr. Asquith has explained in the House of Commons, that the German shipbuilding programme is to suit the needs of Germany, "and will not be influenced by anything that we may do".*

*House of Commons, Feb. 18th, 1909.

It used to be said that the Navy was a hobby of the German Emperor's, and it has often been asserted that Germany is impelled to her naval expansion by some active hostility towards ourselves. To assert either of these things is to misunderstand the conditions altogether. Nations are moved by economic necessities infinitely more than by national antipathies, which, indeed, as all history shows, are the result, and not the cause, of the conflict of interests and the clash of arms. The German Emperor, in proclaiming his conviction that the future of Germany lies on the water, and that a fleet is her bitter need, expressed a fundamental truth, and the untiring energy with which he has laboured to bring home to his people the objects he has in view, which are their own objects more than his own, has been crowned with complete success, and has won the unstinted admiration of Englishmen. He has proved that he possesses the power of inspiring and leading a nation, and his mistakes are forgotten in the contemplation of what has been accomplished. The German Navy is becoming an immense potentiality. The object is stated in terms that admit of no misunderstanding in the preamble to the Navy Law of June, 1900. There is only one means, we are told, of protecting Germany's sea trade and colonies, viz., that she must possess a fleet of such strength that, "even for the mightiest Naval Power", a war with Germany would involve such risks as to jeopardise that Power's own supremacy. For this purpose, it was explained, the German Fleet should not necessarily be as strong as that of the great Sea Power implied, because, in general circumstances, such a Power would not be in a position to concentrate all her forces against Germany, and even if she should oppose a superior force to the German Fleet, the consequence to herself would be such a considerable weakening of force that even if she proved victorious her supremacy would not for some time be effective. This plan has been assumed to express some active hostility towards this country, but the truth, as Germans are never tired of repeating, is that the object is to provide security and defence for German commerce and enterprise. They recognise at the same time, as Mr. Asquith said on the occasion referred to, that "it is natural for us to take what steps we think necessary to protect our own interests".

* * * * *

In no country save our own, and in no time before our own, has a great Power been able to maintain with success her position as a strong Naval Power and a strong Military Power at the same time. The military preoccupations of the Spaniards were very largely the cause of their naval unreadiness in 1588, and it was military pressure that ultimately led to the decline of the Netherlands as a Sea Power. A double burden is now borne by Japan, but the conditions prevailing in the Far East are not strictly comparable to those existing in Europe, and Japanese naval and military outlay per head of the population is much less than half that in Great Britain, Germany, or France. It is the wealth of Great Britain that has enabled her so far to bear a stupendous taxation for Imperial Defence, in which military expenditure counts for a sum approaching that devoted to the Navy. Whatever the financial resources of European countries may be, there lies inevitably before Germany the need of maintaining her great Army as well as the Navy she is developing. The wealth of the country has largely increased, and however serious may be the burden, nothing will be stinted on the Army or the Fleet. Upon her military expenditure in 1909–10 she does, indeed, hope to save something more than £1,000,000, but the margin is small, and the reduction is strongly opposed.

* * * * *

It is the highest interests of both countries that good relations should be maintained, and the German Emperor, when he visited the Guildhall in 1891, and on other occasions, has expressed his desire to preserve the peace of the world unbroken. All that he has said has been re-echoed and emphasised by King Edward in Berlin. But the Emperor did not imply, nor could he have implied, that the policy of his country had undergone any change. He knows, as do German statesmen, that Germany cannot speak aloud in the affairs of the world unless behind her diplomacy is the strong, long arm of naval power. Prince Bülow, in his capacity as Minister-President of the Prussian Diet, speaking on January 19th last, showed that no change has passed over the views of the Government on the subject of naval defence. "For the foundations of our welfare and our greatness, of our might and our security, for the Army and the Navy," he said, "it is the best that is just good enough. We cannot, and dare not, save money at the expense of our readiness for battle and the peace of the country; our geographical position is too unfavourable for that."

This significant statement was made subsequently to the public declaration of policy of the British Prime Minister on behalf of the Government that he accepted the two-Power standard of naval strength as meaning a "preponderance of ten per cent. over the combined strengths in capital ships of the two next strongest Powers." A great impression was caused in Germany by this unequivocal statement, and an inspired writer, "v. R.," in the official *Marine-Rundschau*, said, in discussing the question, that Germany was witnessing the opening of a new stage in British naval policy. The Liberal Government had assumed as its own the attitude of its predecessor, and was supported by public opinion. The object of the writer was to question or dispute the basis of that standard. He declared once again that the German Navy was solely for the defence of German interests, while we were prepared to make England superior to any two Powers in all circumstances and at all times. He averred that a serious doubt was raised as to the standard by introducing the phrase "capital ships," and the argument was apparently directed to induce us to establish the standard upon the basis of the power of ships and not upon their number. Again, he urged that we should consider our superiority as consisting in part in the single direction to be given to our Fleet, as contrasted with the dual control of an allied fleet that would be opposed to us. The strength of our individual ships and the inherent weakness of a coalition were the arguments adduced to suggest to us a smaller basis for the translation of our formula into terms of capital ships, and he did not hide from his readers his contention that, though we were building a two-Power Navy, we were, in fact under the ineradicable belief that German naval preparations were directed solely against ourselves. To this it might be answered that it is for the Board of Admiralty to advise the Government both as to the number and character of the ships required to establish and maintain the standard adopted, and that whether the British and German Fleets are actually built to encounter one another or not, it is certain, if they should ever come into collision like the brazen and earthen pots floating down the stream, that the weaker would sink on impact with the stronger.

With this thought in their minds, no doubt the Germans are increasing their fleet at a surprising rate, and building "capital ships" intended to rival, in every respect, their British prototypes. There is no finality in the present situation. German policy is disclosed in successive steps of naval development. The present programme is declared by many German writers to be inadequate, and the German Navy League is urging a further increase. It is maintained that the limit of age of capital ships, placed at twenty years, is too great, and the consequence of the adoption of such a view would be a fresh acceleration of the pace of construction, and an increase in the number of ships without altering the programme, in the sense that a reserve of ships would be formed, which, if no longer new, would still possess considerable fighting value.

What is best worth noticing in this progressive expansion is the thoroughness, consistency and confidence with which it has been pursued. When the time came for adding cruisers to the programme, they were added, and when the greater power of capital ships became evident, the financial provisions of the scheme were augmented. The evolution is not in the material of the Fleet only. The provision of larger shipbuilding resources, the improvement of the ports and harbours, the construction of slips and docks, the increase in the working-power of the naval establishments—all these have kept pace with the larger requirements of the growing fleet. The Kaiser Wilhelm Canal is to be widened and deepened to admit of the passage of the largest vessels, and the charges will not be laid on the Naval Department. The expansion is universal, and touches every side of German naval life and activity. It does not end with the enormous development of the ports, in communication by internal waterways with the great industrial centres of the country. *Pari passu* with the growth of the Government establishments, we find an enormous advance made by private shipbuilding yards and factories, which, by sane and judicious measures, are all provided with an abundance of work that strengthens and develops subsidiary industrial establishments throughout the country.

* * * * *

The first part of this chapter was devoted to an exposition of the fundamental causes which lie at the root of German Naval policy, and to the inevitable nature of the growth of the German Fleet. It has concluded with a sketch of the policy translated into action, and of the material resources by which that action is made possible. The subject might have been pursued further in an account of the influence of the German Navy League and of the various patriotic associations in the creation and direction of public opinion. The material side of the question might have been extended to an investigation of the numerous subsidiary establishments, which, by their efficiency, contribute to the rapidity and excellence of German warship building. But enough has been said to show the steady, thorough, and increasingly rapid development of the German Navy. Close, meditative, laborious, indefatigable was the process of inception, expansive the spring of inevitable development, and remarkable in the highest degree is the success with which the work has been put into execution. The policy of Germany was declared long ago by Frederick William I.: "Wenn mann in der Welt etwas will decidiren, will es die Feder nicht machen, wenn sie nicht von der force des Schwertes soutenirt wird." This pregnant phrase was repeated by the present German Emperor at the Zeughaus in Berlin on New Year's Day, 1900. The force behind the diplomatic pen of Germany, deciding her place in the world, is the strong Fleet which she is bending all her efforts to create and maintain.

18. 1909

Speech on the Naval Estimates
R. McKenna, First Lord of the Admiralty

* * * * *

The Safety of the Empire

But there are occasions when even the most determined economist is willing to make a sacrifice. The safety of the Empire stands above all other considerations. No matter what the cost, the safety of the country must be assured. As the House will have already seen in the statement which has been furnished with the Estimates, the particular item of increase in 1909–10 is the Vote for new construction. Financial provision is made for laying down two large battleships in July and two more in November. This of itself, without regard to the further contingent order, of which I have spoken, is already a great advance upon the programme which was accepted last year by the House of Commons. What has happened in the interval to lead to such an increase of the scheme of shipbuilding that was accepted a year ago by Parliament as adequate, and proposed with general acceptance by this Government? I will answer this question in a moment. But before I do so let me make one general observation on which I do not think there can be any disagreement. It will be regarded as axiomatic that our island position, the extent and dispersion of our Empire, and the magnitude of our trade, oblige us, so long as we are equal to the task, to maintain a Navy adequate in strength to insure our shores from invasion, our Empire from hostile attempts, and our trade from destruction in war. It follows from this that we cannot determine in advance any definite limits to our Navy. These limits for us must be fixed by the progress of foreign Powers. We cannot take stock of our Navy and measure our requirements except in relation to the strength of foreign Navies. I am, therefore, obliged to refer to foreign countries in making estimates of our naval requirements. Several of the Powers are rapidly developing their naval strength at this moment; but none at a pace comparable with that of Germany. If in what I have to say now I select that Power as the standard by which to measure our own requirements, the House will understand that I do so only for what may be called arithmetical purposes, and without presuming upon the expression of feeling or opinion of my own—except it be one of respectful admiration for administrative and professional efficiency.

* * * * *

Accelerated German Construction

Last year we were not in a position to make any possible forecast of the probable construction of foreign countries. The difficulty in which the Government find themselves placed at this moment is that we do not know—as we thought we did—the rate at which German construction is taking place. We know that the Germans have a law which, when all the ships under it have been completed, will give them a Navy more powerful than any at present in existence. We know that, but we do not know the rate at which the provisions of this Act are to be carried into execution. We now expect that the four German ships of the 1908-9 programme will be completed, not in February, 1911, but in the autumn of 1910. I am informed, moreover, that the collection of materials and the manufacture of armaments, guns, and gun-mountings have already begun for four more ships which, according to the Navy Law, belong to the programme of 1909-10. Therefore we have to take stock of the new situation, in which we reckon not nine but thirteen German ships may be completed in 1911, and in 1912 such further ships, if any, as may be begun in the course of the next financial year, or laid down in April, 1910. We may stop here and pay a tribute to the extraordinary growth of the power of constructing ships of the largest size in Germany. Two years ago, I believe, there were in Germany, with the possible exception of one or two slips in private yards, no slip capable of carrying a Dreadnought. To-day they have actually no less than fourteen such slips and three more under construction. And what is true of the hull of the ships is true also of the guns, armour, and mountings. Two years ago any one familiar with the capacity of Krupp's and other great German firms would have ridiculed the possibility of their undertaking the supply of all the components parts of eight battleships in a single year. To-day this productive power is a realised fact, and it will tax the resources of our own great firms if we are to retain the supremacy in rapidity and volume of construction.

* * * * *

The Life of a Battleship

We cannot afford to run risks. If we are to be sure of retaining superiority in this by far the most powerful type of battleships, the Board of Admiralty must be in a position, if the necessity arises, to give orders for guns, gun-mountings, armour, and other materials at such a time and to such an amount as will enable them to obtain delivery of four more large armoured ships by March, 1912. We should be prepared to meet the contingency of Germany having seventeen of these ships in the spring of 1912 by our having twenty, but we can only meet that contingency if the Government are empowered by Parliament to give the necessary orders in the course of the present year. I can well imagine that this method of calculating in Dreadnoughts and Invincbles alone may seem unsatisfactory, and even unfair to many persons. They may say:—"What has become of the Lord Nelsons, the King Edwards, the Duncans, and the Formidables, and the earlier battleships on which our naval superiority had been so constantly reckoned? Is no account to be taken of our powerful fleet of armoured cruisers, numbering no less than thirty-five?" Yes; the Board of Admiralty have not forgotten these ships. They still constitute a mighty fleet. The Dreadnought has not rendered them obsolete, and many of them would give a good account of themselves in the line of battle for many years to come. But, though they have not been rendered obsolete by the Dreadnoughts and the Invincibles, yet their life has been shortened. Let me explain what that means. To determine the value of a battleship in relation to the value of ships of a newer and better type is a problem of the same kind as that which confronts the manufacturer whose plant is getting out of date, and who has to determine the precise moment when it would pay him best to scrap his old machinery and to lay down new. Every new improvement, every new invention, every improvement in the method of construction shortens the life of a manufacturer's plant. If he is to compete successfully with his rivals, he must keep his machinery up to date. A battleship must be regarded as a machine of which the output is fighting capacity. All improvements in the designs of ships which increase the fighting capacity necessarily shorten the life of earlier battleships just as in the case of any other machine. The greater the value of the improvements, the sooner the earlier ships become obsolete. Though the upkeep of a Dreadnought costs little if anything more than the upkeep of earlier types of battleship, its fighting capacity is greatly superior, and it follows that the advent of this new improved machine has materially curtailed the profitable life of our previously existing Fleet.

Our Battle Strength in 1912

There is, however, a further consideration to be borne in mind. As the years go by the scrapping of older ships is inevitable for another reason. I have seen many forecasts recently of what our Battle Fleet strength would be in 1912. The framers of these forecasts have assumed that we may have sixteen Dreadnoughts and Invincibles in commission in that year, or twelve more than we have at the present moment. To these sixteen they have added the whole of our existing Fleet of battleships, and have produced a startling total, whether reckoned in numbers or in tonnage. Those who, quite naturally and properly, regard this vista of incalculable increase with alarm may be reassured by the reminder that if twelve more Dreadnoughts and Invincibles are put in commission in 1912, twelve other large ships must have passed out of commission. The only condition on which they can all be retained in the Fleet at the same time is that we should greatly increase our *personnel* and our dockyards, at an expense which would be truly staggering, and with a resultant fighting capacity which would not be worth the cost. We have, then, in making our comparison with 1912, to reckon only such ships as will then be on the active list. The House will not expect me to go through our ships in detail, nor could I attempt to give the fighting value of each. Suffice it to say that on the present scale of our Navy, our numerical strength in battleships which could be placed in the fighting line, not including Invincibles, is roughly about fifty, consisting of fully-commissioned and nucleus crew ships, ships in the Special Service list with no more than seventy men on board, and ships in dockyard hands. With this limit to our total numbers it is obviously essential that we should not fall behind in the most powerful type of battleships. There will come a day when by an almost automatic process all ships of an earlier type than the Dreadnought will be relegated to the scrap-heap. The maintenance of our superiority will then depend upon our superiority in Dreadnoughts alone. I have given reasons for believing that the German power of construction of this type of ship is at this time almost, if not fully, equal to our own, owing to their rapid development in the last eighteen months, and we cannot be assured of retaining our superiority at sea if ever we allow ourselves to fall behind in this, the newest and best class of ship.

* * * * *

19. 1912

Naval War Staffs

J. Leyland

There is reason to expect that the organisation and character of the Naval War Staff, as outlined in the First Lord's Memorandum, dated January 1st, 1912, and further explained to the Fleet in a circular dated March 11th, will commend themselves generally to the good judgment of the Naval Service. That the scheme has been accepted with reluctance by some officers is no doubt true. Certain safeguards were and are necessary, such as that of protecting the Navy from the danger of the rise of a distinct and privileged class of officers for whom commands and appointments would be reserved. Such a result could have no other effect than to weaken the spirit of comradeship in the Fleet and to discourage a large class of deserving and meritorious officers. But if this consequence be averted, the wisdom of what has been done will be generally recognised. The best feature of the scheme is that it is the outcome of organic growth within the Admiralty departments, which have developed and changed to meet the new conditions and complexities of the Naval Service. The existence of the Board of Admiralty is a potent and highly beneficial factor in the creation and sphere of action of the new Staff. It may be contended with reason that the War Staff implies nothing that is really new. Certainly all its functions have been executed—and executed with unexampled success—in the past, and when its organisation and duties are examined it will be seen that nothing more than a new and better form and an enlarged system are given to things which existed already. It is also a paramount merit of the scheme that it is based neither upon military nor upon foreign parallels.

The command, leading, and conduct of troops in the field, if they do not differ in all respects in kind from the command and handling of ships and fleets, differ profoundly from them in degree. The differences, indeed, may well be so great as in their consequences to be fundamental. The campaign of Hawke which ended at Quiberon Bay, the long blockades of Brest and the Atlantic ports, and Nelson's blockade of Toulon, showed that the business of supplying and maintaining a fleet demands both experience and knowledge, but, as the First Lord's Memorandum stated, war on land varies in every country according to numberless local conditions, involving the thinking out of a whole series of intricate arrangements and elaborate processes. In other words, the sea service has nothing to do with problems arising in the transport and supply of various military units, as affected by muddy roads or no roads at all, flooded rivers, broken bridges, and a hundred other circumstances of land warfare. "The sea, on the other hand, is all one, and, though ever changing, always the same; every ship is self-contained and self-propelled."

Still more important is it to observe the manner in which the system of the British Naval War Staff, or Admiralty War Staff, as it is styled in the Navy List, differs from the systems prevailing abroad, where complications arise from the want of any organisation answering precisely to the Admiralty Board in this country. These differences are mainly the subject of the present chapter, but it is first necessary to show the gradual development of the British Naval War Staff from earlier organisations in order to explain some dangers that have been avoided. It may be argued with reason, that after Sir James Graham had abolished the Navy and Victualling Boards, and absorbed the Civil Departments in the Admiralty, the Sea Lords became inevitably more and more engrossed in the complexities of a vast material business, and consequently had fewer opportunities of studying problems of war and war training than had their predecessors, until, at last, the constitution of a Naval War Staff became imperative. This great transfer of business took place in 1831, Sir Thomas Masterman Hardy being at that time First Sea Lord, but it was not until 1883 that the Foreign Intelligence Branch came into existence. It had a modest beginning and was not regarded with much favour, and in 1886, when some reduction was projected, Lord Charles Beresford, on that and other grounds, resigned his seat on the Admiralty Board, and proposed the institution of a Naval Intelligence Department.

There is no intention of recording the history of the Intelligence Department here. It continued to do useful work, conducted in a few dusty and inconvenient rooms in Whitehall, endeavouring to master the significance of every fresh development of naval science, and formulating plans for use in the event of the outbreak of war. It was the agency always available to the Admiralty for duties of this kind, and those who were acquainted with its work knew that, as time passed on, the title of Intelligence Department became a misnomer, intelligence, as such, forming the least important part of the duties of the Department. The conspicuous success of the German Great General Staff of the Army, under the guidance of Moltke, had impressed itself upon the minds of thinkers in every country. The Hartington Commission of

1889 toyed with the subject, and it was first brought prominently to public notice by Mr. (now Professor) Spenser Wilkinson, in a little book entitled "The Brain of the Navy," 1895. It may be questioned, however, whether that writer has greatly influenced the changes which have subsequently taken place. He regarded the Board of Admiralty as a "legal fiction"; it recorded nothing and was altogether subservient to the First Lord. What he desired was a Moltke for the Navy—the best naval strategist in the Service—and no one was to stand between him and the Cabinet, as represented by the First Lord of the Admiralty. "If you have a first-rate strategist, with an office of picked and trained officers as assistants, to work at the arrangements for a possible war, it would evidently be absurd to put another man as a buffer or telephone between him and the Cabinet which needs his advice." Evidently, then, in this conception of the case, the high strategist could be no other than an invigorated and responsible First Sea Lord.

The developments which have taken place have not led to this result, and the First Sea Lord stands between the Chief of the Staff and the First Lord. The Admiralty Board had a clear view of one vital necessity. There must be a direct line in the naval hierarchy from the Staff up through the Board and the First Sea Lord to the Cabinet Minister. The latter as a civilian is, by the very nature of things, incompetent to decide between two distinct lines of policy advocated by responsible naval authorities. It was of the utmost importance to guard against this manifest danger. Any other arrangement would be contrary to the highest traditions of the Service, and fraught with insecurity and the promise of disaccord. As to the view expressed in some quarters that the Chief of the Staff should present an annual report to Parliament, thus superseding the Board of Admiralty and overriding his senior officers, there could, of course, be no parley with a contention so palpably absurd.

The sub-committee of the Committee of Imperial Defence, which was assembled to investigate the grave charges of naval unpreparedness made by Lord Charles Beresford, in a letter to the Prime Minister, dated April 2nd, 1909, finding that there were differences of opinion amongst officers of high rank regarding important principles of naval strategy and tactics, stated in their report, dated August 12th, 1909, that they looked forward "with much confidence to the further development of a Naval War Staff," from which the naval members of the Board might be expected to derive common benefit. Two months later a change was made "in further development of the policy which has actuated the Board of Admiralty for some time past of organising a Navy War Council." The Naval Mobilisation Department was brought into being under the direction of a flag officer (Rear-Admiral H. G. King-Hall), and took over that part of the business of the Naval

Intelligence Department and the Naval War College which related to war plans and mobilisation. Under the presidency of the First Sea Lord the officers directing the Naval Intelligence and Mobilisation Departments and the Assistant Secretary of the Admiralty were to form a standing War Council, with which the Rear-Admiral commanding the Naval War College might be associated when the business was such as to require his presence.

Surprise was expressed in some quarters at the leisurely manner in which those distinguished officers, Lord Fisher and Sir Arthur Wilson, proceeded in this matter of organising a Naval War Council or Staff. Perhaps the explanation of the circumstance is to be found in a wise remark made by Moltke in the course of a comment on German Generals and the Army Staff. "There are generals," he said, "who need no counsel, who deliberate and resolve in their own minds, those about them having only to carry out their intentions." "But such generals," he added, "are stars of the first magnitude, who scarcely appear once in a century." This judgment of the great German soldier suggests a further reflection touching the British Naval War Staff. These "stars of the first magnitude," themselves finding a staff a luxury or superfluous, may have foreseen the rise of lesser luminaries at some future time to whom a staff would prove a necessity.

Mr. Churchill's Memorandum on the constitution of the Naval War Staff is printed elsewhere in this volume, and the organisation and duties of the Staff will not be described here, but the diagram given below will illustrate the relations and lines of responsibility and authority in the several departments. It will be seen that the only relations which can properly exist—and the point is of great importance—between the Chief of the Staff and the First Lord must be through the channel of the First Sea Lord. The provision that "the First Lord and the First Sea Lord will, whenever convenient, consult the Directors of the various Divisions, or other officers if necessary," seems, however, to present some risks against which precautions should be taken.

First Lord of the Admiralty.
"Delegate of the Crown in exercising supreme executive power."

First Sea Lord.	Other members of the Admiralty Board, directing Departments.
For certain purposes holding the "position of Commander-in-Chief of the Navy."	

Chief of the Staff.
"Primarily responsible to the First Sea Lord."

Intelligence Division.	Operations Division.	Mobilisation Division.
"War Information."	"War plans."	"War arrangements."

20. 1912

Recent Changes in Warship Design
Sir William White

* * * * *

Essential Differences between pre-Dreadnoughts and Dreadnoughts

The essential differences in the designs of the Dreadnoughts as compared with their predecessors may be summarised as follows: first, higher speed; second, a principal armament of ten 12-in. guns for battleships instead of four 12-in guns, and of eight 12-in. guns instead of four 9·2-in. guns for armoured cruisers; third, the absence of any secondary armament (7·5-in. or 6-in. guns); fourth, an important change in the distribution of the side-armour. In consequence of these changes, it became inevitable that the dimensions, displacements and costs of the new types should be greater than those of their predecessors.

Most fortunately for naval architects, the genius and perserverance of Sir Charles Parsons placed at their disposal the marine steam turbine at the time when the increase of speed was decided upon. Higher speeds, of course, necessitated the development of greater engine power. Steam turbines provided a means of obtaining a greater development of engine power in proportion to the weight of propelling apparatus—because they proved to be more economical than reciprocating engines in their consumption of steam and coal at or near maximum powers. Consequently, for a given horse-power the use of turbines secured economies of weight and space in boiler rooms; and, although turbines required somewhat greater floor-space than reciprocating engines, the total floor-space needed for turbines and boilers was not much larger than that required for reciprocating engines and boilers giving the same power. Turbines could be placed lower in the ships, and occupied less height, leaving above them considerable clear space, which would have been occupied by the cylinders of reciprocating engines. Their lower situation in the ship also gave better protection in action. The adoption of the steam turbine, therefore, in the

Dreadnought and Invincibles greatly facilitated the attainment of higher speeds on smaller displacements and dimensions than would have been possible had reciprocating engines been employed, as they necessarily were in earlier battleships and armoured cruisers. The principle hereby illustrated is of general application, and has received endless illustrations in ship-design both for war and commerce. Ships of later date always benefit by the march of improvement in science and manufacture; and the fact must not be overlooked when they are compared with vessels built at earlier periods. Not only in propelling apparatus but in materials of construction and naval ordnance the Dreadnought and Invincibles necessarily gained upon their predecessors, and are at some disadvantage as compared with later ships—the so-called super-Dreadnoughts.

Radical changes in the character of the principal armament of the Dreadnought and Invincibles, although named as the second cause of increased dimensions and displacements, had really the most potent influence on the designs. The use of a much greater number of 12-in. guns, of course, involved considerable increase in weight of armament; five armoured stations had to be provided for the ten guns, as against two such stations in earlier ships; in order to secure large arcs of horizontal command for more numerous heavy guns, some of them were placed at greater heights than heretofore, and this fact necessitated increase in the weight of barbettes and protecting armour. On the other hand, there was a saving in weight by an abandonment of the secondary armament and of the battery or turret armour used to protect it; but, after allowing for this fact the adoption of single-calibre big-gun armaments was necessarily accompanied by a large proportionate increase in weight. In all ship-designs the principles are recognised that increase in the load to be carried at a given speed must involve an increase which is many times greater in the displacement, and that as the maximum speed to be attained becomes higher, the proportion of the increase in displacement to the increase in load will become greater. In the Dreadnoughts, therefore, the cumulative effect of higher speed and greater load of armament and protective armour was serious and had to be provided for by the naval architect.

* * * * *

In regard to the value of the higher speeds with which the Dreadnoughts and Invincibles were endowed authorities differ widely. The official view was expressed as follows: "The greater the mobility the greater the chance of obtaining a strategic advantage. This mobility is represented by speed and fuel endurance. Superior speed also gives the power of choosing the range. To gain this advantage the speed designed for the Dreadnought is 21 knots." The speed trials were made at normal draught, and the speed attained was about 2 knots higher than had been reached by preceding battleships. It has since been demonstrated conclusively that such a difference in speed does not and cannot exercise any important effect in determining the range at which a fleet action will be fought. As to the strategic advantages of superior speed much may be said, but such a discussion lies outside the scope of this paper.

For the Invincibles the maximum trial speed was fixed at 25 knots; the speed attained at normal draught was about 26 knots, showing an excess of about 2 knots above the trial speeds of preceding armoured cruisers. No British armoured cruiser of earlier date had been armed with guns exceeding 9·2 in. in calibre; but the Invincible class was designed to carry eight 12-in. guns in four armoured positions. A few foreign cruisers had been armed with four 10-in. or 12-in. guns, in addition to a good secondary armament. The step taken in the Invincible class was therefore most notable; it involved the creation of vessels which were originally classed as armoured cruisers but were obviously intended for the line-of-battle, and are now officially designated battle-cruisers. The installation of a heavier armament, concurrently with the provision of propelling machinery of 43,000 H.P.—an increase of 40 per cent. above the engine-power of the swiftest armoured cruisers of earlier date—necessarily involved a large increase in length and displacement for the Invincibles. Their armour protection was weak relatively to that of contemporary battleships, especially in that section of the defence which was devoted to the heavy gun stations. Opinions differed, and still differ, in regard to the policy of building such large and costly cruisers, and of endowing them with very high speed, if they are primarily intended to take part in fleet actions. There is, however, no reason for supposing that smaller vessels could have been produced which would have fulfilled the governing conditions of speed, armament, defence and fuel-supply laid down by the Admiralty for the guidance of the Director of Naval Construction and his staff.

* * * * *

Dreadnoughts and Post-Dreadnoughts

The development of British armoured ships since 1905 has taken place along lines, starting, respectively, from the Dreadnought, classed as a battleship, and the Invincible, originally classed as an armoured cruiser, but now officially designated a battle-cruiser. The latter class are superior in speed to battleships, but inferior in armour defence and in the number of their heavy guns. These battle-cruisers, it is said, are

ANGLO-GERMAN RIVALRY AND CONTINUING TECHNICAL ADVANCE

intended to act as the swift divisions of fleets; but many high authorities on naval strategy and tactics take exception to the fundamental ideas on which the designs have been based. The value of exceptionally high speed is especially doubted although its attainment has involved great additions to dimensions and cost.

* * * * *

The first fact to be noted respecting "post-Dreadnought" battleships is that their maximum speeds on contract trials have been maintained at 21 knots, the estimated speed of the Dreadnought. The number of heavy guns has remained the same as in the Dreadnought—namely ten—and these guns have been mounted in pairs. The disposition of the heavy-gun stations adopted in the Dreadnought was repeated in six of her successors, laid down in the period 1907–8; three later ships (Neptune class) have their heavy guns disposed on a different system (*see* Plate 2); and in subsequent battleships (Orion class) laid down in 1909–10, still another disposition is adopted (*see* Plate 1). The 12-in. guns mounted in the Dreadnought and her three immediate succesors were 45 calibres long; the next six post-Dreadnoughts (up to and including the Neptunes) carry 12-in. guns, 50-calibres in length, and of greater weight and power. In the Orion class 13·5-in. guns, 45 calibres in length, were introduced. This type of heavy gun is understood to be still favoured, improvements having been made in the designs of later weapons. Rumours are afloat to the effect that still larger calibres will be introduced. Opinions differ as to the desirability of abandoning the 12-in. calibre, which was adopted about fifteen years ago after full consideration, and in the light of actual experience with 13·5-in. and 16·25-in. guns. During the long period while the 12-in. calibre was in use the designs for successive types of 12-in. guns had been greatly improved, and they had been adopted as the principal weapons mounted in all battleships, except those of the German Navy, where 11-in. guns had been preferred. It is a significant fact that about the time when Germany was moving on to the 12-in. calibre the Admiralty should have adopted 13·5-in. guns. In this paper it is not proposed to deal with the arguments for or against increase of calibre. The responsible authorities have decided to make that change, and our present task is to show how great has been the effect produced thereby upon the sizes of ships and their cost.

* * * * *

21. 1913

The Progress of Naval Aeronautics
Anon

The first steps towards equipping their aerial squadrons have now been taken by the principal Naval Powers. Germany in particular has been quick to see how admirably suited to her purpose an air-craft of an offensive type, and with great steadiness of purpose and determination, in the face of repeated disaster to her early vessels, has continued strenuously to evolve her rigid airships. This activity on the part of Germany has led to the development in Great Britain of aircraft of an offensive-defensive type—hydro-aeroplanes, and a series of stations for these craft are being established round the coasts of the British Isles. Intermediate between the two types of air-craft above-mentioned there is the non-rigid airship, which has been considerably developed on the Continent, and with which some trial has been made in Great Britain.

For the purposes of this article, these three types may be classified according to their respective functions on similar lines to sea-going ships—

(1) Torpedo air-craft (aeroplanes, hydro-aeroplanes, flying boats).
(2) Scouting and mine-laying airships (non-rigid airships).
(3) Battle airships (rigid airships).

Though the construction and functions of these types are generally well known, it may not be out of place to give a short description of each, with the armament they carry, and the duties they may be expected to perform in war.

Both monoplanes or biplanes are at present in use as torpedo aircraft, and as far as the wings and the engine are concerned do not differ from the ordinary aeroplanes in use ashore. For naval purposes, however, the under-carriage is modified considerably. It is possible to remove the wheels of an ordinary machine and secure floats in their place, but this does not make either a neat or light job, and the usual practice is to build the floats on to the structure. Several satisfactory types of floats are now in use, one of the best having two long narrow floats set well apart, a system which gives good lateral stability on the water. In another system there is a single large float placed centrally, with two smaller supplementary floats under the wing tips, to prevent them dipping under water should the machine over-balance. Some hydro-aeroplane builders have replaced this central float with a good-sized boat, and in some cases the passenger, pilot, and engine are placed in this boat with, on the whole, fairly satisfactory results.

The functions of all the heavier-than-air types of machines are analogous to those of sea-going torpedo craft. It may be expected that the best machines will be capable of flying at a rate of 60 miles an hour with a 100-H.P. engine, and of lifting a load of some 900 lb. The figures may, of course, be varied. Greater horse-power or less may be used with a corresponding increase or decrease in the fuel required, and so on. But these figures have been taken as being typical of machines now being built for various Powers. If, then, the consumption of fuel is taken to be 0·7 lb. per H.P. hour, a good idea may be obtained of what air-craft of this type are capable of doing. Two men will weigh about 250 lb. Wireless equipment for 100 miles range requires 100 lb. Thus 550 lb. remains to be used as desired.

If all this can be utilized as fuel, the vessel can stay, approximately, eight hours in the air, which is equivalent to a distance of 480 miles. If floats are fitted for resting on or rising from the water, they will weigh from 200 lb. to 400 lb., and the amount of fuel, which means radius of action, or other weights, must be reduced accordingly. If the average weight of the floats is taken as 300 lb., this will allow of about three and a half hours' flight with two men and wireless equipment, or a radius of action of about 100 miles under these conditions. If, however, floats designed to let the machine descend on the water only, and not capable of rising from it, are fitted, the weight can be much reduced. Such floats will weigh about 100 lb., therefore allowing 450 lb. of fuel to be carried in addition to wireless equipment and two men. The vessel will be able to remain about six and a half hours in the air and have a radius of action of some 200 miles. Instead of a portion of this weight of fuel it may be desired to carry a gun of explosives, and, if so, for every 70 lb. of weight of this material one hour's fuel must be sacrificed.

Torpedo air-craft should carry the armament suited to the work on which they are employed. Thus for purely scouting duties no armament need be carried. For action against other aerial torpedo craft, and possibly against scouting and mine-laying airships, a gun armament would be necessary. Whilst for action against battle-airships, submarine boats, and for attacks on Naval bases, etc., a torpedo armament will probably be found most suitable. This armament consists of small mines to be dropped, or discharged from a tube, or of an explosive charge which may be towed on the lines of the old Harvey torpedo. The last-named plan is expected to give the best results.

Torpedo air-craft can be used to assist the port defences and for patrolling the coast between their different bases. They may also be carried in ships, just as second-class torpedo-boats were carried in the Anson class of battleship, or in special mother ships in a similar way to that in which the torpedo-boats were once carried by the Vulcan.

One of the principal functions of torpedo air-craft will, it may be presumed, be the denial of certain areas to the enemy's air-craft, and whilst it is doubtful if they can do this unless they are in great numbers, and the enemy attacks in the daytime, still, they can assure that if he comes at all he must come in force, and at considerable risk to himself. For this purpose a mixed gun and torpedo armament would have to be used, for whilst the gun is a most effective weapon for use against other torpedo air-craft, and against scouting and mine-laying airships, it is, as will be shown later, practically useless against the battle airship.

Possibly one of the most useful employments for torpedo air-craft will be that of locating and attacking submarines, for which purpose they should be particularly suited. It is easier for air-craft to locate submerged objects than it is for surface craft to do so, and the submarines may then be attacked by dropping small mines on them. Probably 20 lb. of explosive, if dropped within a certain limit, would be enough, but as weights up to 300 lb. have been dropped from aeroplanes there will be no difficulty about increasing the charge if necessary. Somewhat similar methods must be used for attacking the battle airship, but here the air torpedo craft is in a similar position to surface torpedo craft attacking a battleship using a very short range torpedo, and her chances of success are similar.

* * * * *

Further Progress

In regard to future development, it may be expected that a great advance will be made, especially in this country, as the coming of the air-craft may more profoundly affect Great Britain than any other nation. Hitherto the inhabitant of these islands has had the comfortable feeling that before he could be personally inconvenienced in any war in which his country might be engaged, it was necessary for the Navy to be defeated. The main protection must still be the Fleet and the Army, but the development of the air-craft for fighting and raiding purposes has made possible a danger of a different kind. The public have not yet realised what this may mean to the non-combatant, because they have never had an opportunity of seeing a battle airship and very little of any other type.

There is a general impression that the aeroplane is the antidote to the airship, and sufficient for the needs of our island, but this has not been proved. There is no doubt that for Army work, where scouting up to a distance of some 200 miles as a maximum is required, the aeroplane is most suitable, because it is easy of transport, and for patrolling narrow straits and stretches of coast-line the hydro-aeroplane will be very effective. There are many experts who doubt the power of aeroplanes, in whatever number, to stop half a dozen battle airships with their attendant mine-layers, just as it is doubted if the torpedo craft of the sea can annihilate the Dreadnought. It is the business of airships to remain outside the radius of action of aeroplanes by day and to attack by night or in thick weather. Owing to the clearness with which objects are visible from airships at night, it should not be difficult for them to locate their quarry, but whether their bombs will have much effect has yet to be determined by experiment. The Gamma passed over Westminster Bridge and within one hundred yards of the Houses of Parliament in a fog without being seen, though she was only about 200 ft. up at the time.

The power of air-craft may be easily exaggerated. From the way in which some writers expound the subject it might be imagined that air-craft in the future must be omnipotent. There is no doubt that if we could make our water fleets fly we should be delighted to do so, because their great power would not then be confined to the sea and a narrow stretch of coast-line, but we should be able to use it to strike straight at the enemy's heart. Decisive battles are seldom or ever fought at sea. It is the successful sea action that makes the decisive land battle possible. What air-craft can do is to avoid water ships and do damage on land. But to subdue a country it is necessary to occupy it, or be able to occupy it, and it is not feasible that air-craft should be able to carry sufficient men for the purpose.

Air-craft can be of great assistance to a fleet in locating the hostile battle squadrons. They can carry out raids on dockyards and magazines. They may even threaten the destruction of a town or its principal buildings. To meet this air menace means the building of air-craft of all types and their consequent development. This development will probably mean increase in size for all air-craft for overseas purposes, except those that may be carried in ships. At present the weight of the floats in hydro-aeroplanes is a serious handicap, but the larger the machine the smaller need the weight of the floats be in proportion, as with increase in size the displacement increases faster than the weight. Large floats are also more seaworthy.

The same principle applies to airships of all classes. The larger the ship the smaller proportion the weight of the hull bears to the total weight lifted. For technical reasons, which it is not necessary to go into here, the rigid airship gains more than the non-rigid when large ships are considered. It may be anticipated that before long experiments will be carried out to see if it is not possible to anchor air vessels like ordinary ships; and there seems no reason why they should not, if sufficiently heavy anchor gear can be carried. In order to carry this extra weight the size of the ship must be increased or the fuel capacity diminished. Then it is highly desirable that an airship should have high speed and large radius of action, not only for war purposes, but to enable her to ride out a gale in the air when it is too rough for her to descend, or to run away out of the storm's path. It is calculated that if the diameter of the latest Zeppelin were increased from 49 ft. to 60 ft. (the Schütte-Lanz is already of 60 ft. diameter), her length increased from 510 ft. to 625 ft., and she were engined in proportion, she would have a maximum speed of some 60 miles an hour, or 52·8 knots, and could maintain this for about forty hours; or if she proceeded at 30 knots she could steam for about 190 hours, provided all the weight she can carry was put into fuel. It is worth noting that the shed at the Zeppelin works at Friedrichshafen has been lengthened to 630 ft., so larger ships may be flying before the year is out.

22. 1914

Armour and Ordnance
C. N. Robinson

Students of the art of attack and defence by sea must have noticed that the past year has been one of unusual and significant activity and change, both at home and abroad. There has been manifest a widespread inclination for experiment and essay with new means and new methods—a desire for something different, if not always for something better, than that which is old and well-tried. Partly, this movement may be accounted for by the continued achievement and advance which has been made with the seaplane and the submarine. The feats performed both by the air-craft and by the under-water vessels during the past year have plainly demonstrated that they may have a very real and increasing value for war purposes. It would not be surprising, therefore, if these new munitions of war should have a distinct influence on warship design. In some quarters, at least, it is now felt that the new weapons, together with the increased range and efficiency of the torpedo, constitute an insistent menace, in the near future, to the supremacy of the gun and its carrier—the battleship.

It has, indeed, been said that the struggle for supremacy between the torpedo and the gun, considered relatively as elements of offence and defence, now equals, if it does not transcend, in importance, that between the gun and the armour, which for so many years provided the most perplexing problem for the solution of naval artillerists and architects. It is now well-nigh admitted by all authorities that even the best and thickest armour afloat does not, at decisive ranges, and measured by peace tests, provide absolute immunity against the gun attack. Even at the distances made necessary by the development of the torpedo, it is likely that the larger calibre guns are sufficiently powerful to put their projectiles through any of the armoured positions of all battleships now afloat. Possibly Mr. Churchill had this in mind when he compared an action between battleships to a couple of eggshells striking each other with hammers. That all nations continue to increase the area of the ship which is armoured, and the thickness of the armour, seems to be because they are inspired with a hope that at long ranges the angle at which the shell strikes, and the movement of the target through the water, will make direct impact difficult and prevent penetration. On the other hand, against the improved torpedo, in its various forms—mines, bombs, etc.—and their carriers, which include not only destroyers and submarines but the latest development of air-craft, much of the armoured protection of the big ship can be of no avail. It is useless to plate barbettes or casemates, or to belt the side, against an attack which will be delivered lower down on the hull. It will be possible, doubtless, to lessen the destructive effect of bombs flung from aloft by the method advocated by Sir Trevor Dawson—viz., by increasing the thickness and giving a larger curvature or whaleback formation to the armoured deck. To minimise the explosion of a mine recourse must be had to further internal sub-division and protective bulkheads. Nevertheless, these precautionary measures will not keep away the danger, and the bursting of high explosives, either against the curved deck or the hull below water, will almost certainly result in a breach. Only the gun, then, can fend off the torpedo-carrier, by destroying it before it can arrive at a decisive range. At present the gun for this purpose is carried in the battleship, but it seems possible that before long what is now called the secondary or auxiliary armament may be placed in separate vessels. The recent decision to discard torpedo-net defence, trusting for safety to high speed, with quick and frequent changes of course, may be regarded as an indication that naval opinion is tending in this direction. There are other signs of a similar character.

If in the future the big ship should only carry big guns and be armoured sufficiently to keep out big shell, then it may be that her displacement can be reduced. But as matters are, the guns in both the primary and secondary batteries are increasing in size and power, and in some cases in numbers also, while the demand for thicker armour more widely distributed, not to mention higher speed, inevitably means heavier displacement. This tendency, moreover, is traceable over the whole field of warship construction. Even in the destroyer, which by some is regarded as a moribund type, larger vessels are the rule, and the same is the case with the submarine. When we turn to the actual practice of war, there are no later lessons than those of Tsushima or the other conflicts of the Russo-Japanese War, yet since that time the advance in almost every direction has been tremendous and startling.

* * * * *

There appears to be a growing diversity in the practice of the naval Powers in regard to the main armaments of their big ships. It was remarked in this article a year ago that the consensus of opinion among naval artillerists favoured an increase of calibre rather than an increase of numbers. Possibly this may still be true, for the example of Great Britain and Germany, who have both adopted a main battery of eight 15-in. guns for their latest battleships, appears to be regarded as one to be followed by many of the other Powers, but difficulties connected with the manufacture of these larger weapons have prevented their adoption in other cases. Concurrently with the mounting of larger calibres of guns, however, in new battleship types the movement for adding to the number of weapons carried appears to have gained in favour during the year. This movement finds its most advanced expression among the French constructors who have designed a vessel to carry sixteen 13·4-in. guns in four quadruple turrets. This policy, the advocates of which advance the claim that in a given interval of time a number of 13·4-in. guns can deliver as many tons of projectiles as a like number of 15-in. guns, and that it is better to capture a ship than to sink her, is certainly against the trend of opinion of the Dreadnought era. There are no signs at present that it will be generally or even largely adopted on military or tactical grounds.

* * * * *

The Man of War in Transition

HMS IMPERIEUSE

H.M.S. "ROYAL SOVEREIGN."

H.M.S. "DREADNOUGHT."

H.M.S. "DARING."

PART III: 1915–1919
The Great War

Introduction	95
23. 1915	
The World War. Narrative of Naval Events and Incidents	96
C. N. ROBINSON	
24. 1915	
First Lord's Statement in the House of Commons, November 27, 1914	100
WINSTON CHURCHILL	
25. 1915	
German Declaration—War Area, February 2, 1915	102
ADMIRALTY STAFF MEMORANDUM	
26. 1919	
The Enemy Navies	104
J. LEYLAND	
27. 1919	
The Submarine War on Merchant Shipping	107
COMMANDER L. H. HORDEN RN	

Introduction

As mentioned earlier (p. viii), Brassey's was published in a limited and restricted form during the war and therefore its treatment of the great occasions such as Gallipoli and Jutland was not particularly illuminating. However, the following selection of articles does show how the Navy reacted to the novel elements in the war. The sinking of the *Aboukir, Cressy* and *Hogue*, three old cruisers, in September 1914 dramatically demonstrated the prowess of the submarine. Henceforward the deployment of all surface ships, including the mighty Grand Fleet itself, was to be severely inhibited by even the reported presence of U-boats, as well as by extensive German minelaying.

Germany's first declaration of unrestricted submarine warfare against merchant shipping in February 1915 marked the beginning of what, unexpectedly to both sides, turned out to be the German navy's most dangerous threat to Britain. Equally unexpected and long resisted by naval and commercial orthodoxy was the imposition of mercantile convoy in 1917 and its belated recognition as the only effective anti-submarine method.

Recommended Reading

Arthur J. Marder, *From the Dreadnought to Scapa Flow*, Vols. II–V (1962–1970).

23. 1915

The World War. Narrative of Naval Events and Incidents

C. N. Robinson

* * * * *

It is hoped that a concise review such as follows may prove of value to all naval students, and furnish to the future historian material which, as it shows how daily events presented themselves to bystanders, should be of more than ephemeral interest. It has been thought well not to set down every event in its chronological sequence of time, but to group together certain occurrences having a similar relation to phases of sea warfare and the principal theatres of action. Nor has it been considered necessary to set forth the sea strengths of the great Powers which are engaged in the struggle. These will be found tabulated on later pages of the *Naval Annual* as they stood at the outbreak of war. Although it is a hundred years since this country has been engaged in a naval war on such a huge scale, it is a fact to be remembered that our Navy was relatively stronger and better prepared for action than it ever was at the beginning of any of our wars in past times. This gave us an initial advantage which has not been lost, while every month since the war began has seen additions made to our Fleet, increasing its dominating influence. Thus it has been that our supremacy at sea has affected the issues of the conflict, and this even whilst there has been no decisive battle between the main fleets. Sea power has its victories by silent or static pressure as well as its successes by the exercise of dynamic force. It has been the former operating cause which has had, and is still having, a throttling effect upon the economic condition of Germany.

It is well that the prodigious influence on the War exerted by the Fleet should be fully realised, both by our own people and by our Allies. Owing to the presence and latent potentiality of that Fleet, "lost to view amid the northern mists," we have been able to send to France and Flanders an army of larger dimensions than any heretofore employed by this country. We have been able, also, to raise and train a still larger army for use when the propitious moment arrives. And it is also because that Fleet keeps the seas that the country has been secured from invasion, the danger of famine and financial ruin has been averted, and the social and industrial life of our people has proceeded without dislocation.

We have been able, moreover, to increase our output of the munitions of war and materially to assist our Allies in the same direction. Further afield, the patriotic aspirations of the Dominions could be given full play, reinforcements from Australia, Canada, and other colonies crossing the seas unmolested, while as the story unfolds itself it will be seen how contingents of Australians and New Zealanders wrenched from Germany her oversea possessions. Throughout the world our ports are free, our commerce still covers the oceans, while the Fleet of the enemy has been forced to withdraw into his fortified harbours, and his merchantmen, numbering nearly 5000 ships, quite one-half of them steamers, have been captured or driven to take refuge in neutral ports. The external commercial activity of Germany has entirely ceased. The completeness of the results of the eight months of sea warfare has been made possible by the protection afforded by the Grand Fleet, in the ships of which our seamen are still eagerly awaiting the opportunity for a battle in which they may emulate the glorious deeds and achievements of their predecessors. Never before has there been such a striking manifestation of the relation of Sea Power to Empire.

* * * * *

Within three weeks of the loss of the *Pathfinder*,* Germany's submarines made their biggest coup of the first nine months of the War, whether judged by the number of lives lost or the tonnage of the vessels destroyed. This was the sinking of the three cruisers of the Cressy class, on the morning of September 22nd, off the Dutch coast, by U 9, Lieutenant-Commander Otto Weddigen, an officer who, until he was destroyed with his crew in U 29 in March, was the most successful submarine commander on the German side. On the afternoon of September 22nd the Admiralty announced that the Aboukir, Captain John E. Drummond; the Hogue, Captain Wilmot S. Nicholson; and the Cressy, Captain Robert W. Johnson, had been sunk by a submarine in the North Sea. The Aboukir was torpedoed, and whilst the Hogue and the Cressy had closed, and were standing by to save the crew, they were also torpedoed. Three days later, on September 25th, the Admiralty published the reports of the commanders of the Cressy and Hogue, these being prefaced by the following Memorandum:

*Light cruiser sunk 5 September 1914 (*Ed.*).

The facts of this affair cannot be better conveyed to the public than by the attached reports of the senior officers who have survived and landed in England. The sinking of the Aboukir was, of course, an ordinary hazard of patrol duty. The Hogue and Cresy, however, were sunk because they proceeded to the assistance of their consort, and remained with engines stopped endeavouring to save life, thus presenting an easy and certain target to further submarine attacks.

The natural promptings of humanity have in this case led to heavy losses, which would have been avoided by a strick adherence to military considerations. Modern naval war is presenting us with so many new and strange situations that an error of judgment of this character is pardonable. But it has been necessary to point out, for the future guidance of his Majesty's ships, that the conditions which prevail when one vessel of a squadron is injured in a minefield, or is exposed to submarine attack, are analogous to those which occur in an action, and that the rule of leaving disabled ships to their own resources is applicable, so far at any rate as large vessels are concerned. No act of humanity, whether to friend or foe, should lead to a neglect of the proper precautions and dispositions of war, and no measures can be taken to save life which prejudice the military situation. Small craft of all kinds should, however, be directed by wireless to close the damaged ship with all speed.

The loss of nearly 60 officers and 1400 men would not have been grudged if it had been brought about by gunfire in an open action, but it is peculiarly distressing under the conditions which prevailed. The absence of any of the ardour and excitement of an engagement did not, however, prevent the display of discipline, cheerful courage, and ready self-sacrifice among all ranks and ratings exposed to the ordeal. The duty on which these vessels were engaged was an essential part of the arrangements by which the control of the seas and the safety of the country are maintained, and the lives lost are as usefully, as necessarily, and as gloriously devoted to the requirements of his Majesty's service as if the loss had been incurred in a general action. In view of the certainty of a proportion of misfortunes of this character occurring from time to time, it is important that this point of view should be thoroughly appreciated. The loss of these three cruisers, apart from the loss of life, is of small naval significance. Although they were large and powerful ships, they belonged to a class of cruisers whose speeds have been surpassed by many of the enemy's battleships. Before the war it had been decided that no more money should be spent in repairing any of this class, and that they should make their way to the sale list as soon as serious defects became manifest.

* * * * *

A German account of the affair appeared in the *New York World* of October 11th, when the story of the commander of U 9 was allowed to be published by the permission of the German Navy Office. He said:

It was 10 minutes after 6 in the morning when I caught sight of the cruisers, I was then 18 miles north-westerly off the Hook of Holland. I had travelled more than 200 miles from my base. I had been going ahead, partly submerged, with about five feet of my periscope showing. Immediately I caught sight of the cruisers I submerged completely and laid my course so as to bring up in the centre of the trio. I got another flash through my periscope before I began action. Then I loosed one of my torpedoes at the middle ship, which I later learned was the Aboukir. There was a fountain of water, a burst of smoke, a flash of fire, and part of the cruiser rose in the air. I submerged at once. The Cressy and the Hogue turned and steamed to their sister ship. As soon as I reached my torpedo depth I sent a second charge at the Hogue. I had scarcely to move out of my position, which was a great aid since it helped to keep me from detection. The attack went true, the Hogue half turned over and then sank. The third cruiser stood her ground as if more anxious to help the many sailors, who were in the water than to save herself. When I got within suitable range I sent away my third attack. This time I sent a second torpedo after the first to make a hit doubly certain. My luck was with me again, for the enemy at once began sinking by the head. All the while her men stayed at their guns looking for their invisible foe. They were brave, true to their country's sea traditions. Then she turned turtle.

* * * * *

24. 1915

First Lord's Statement in the House of Commons, November 27th, 1914
Winston Churchill
(*Extract.*)

I am going in a few words, if the House will permit me, to draw the attention of the House, and through the House the attention of the country, to some of the larger aspects of the naval situation at the present time.

The British Navy was confronted with four main perils. There was first the peril of being surprised at the outbreak of war before we were ready and on our war stations. That was the greatest peril of all. Once the Fleet was mobilised and on its war stations the greatest danger by which it could be assailed had been surmounted.

Then there was the danger, which we had apprehended, from the escape on to the High Seas of very large numbers of fast liners of the enemy, equipped with guns for the purpose of commerce destruction. During the last two years the sittings of the Committee of Imperial Defence have been almost unbroken, and we have been concerned almost exclusively with the study of the problems of a great European War, and I have always, on behalf of the Admiralty, pointed out the great danger which we should run if, at the outset of the war, before our cruisers were on their stations, before our means of dealing with such a menace had been fully developed, we had been confronted with a great excursion on to our trade routes of large numbers of armed liners for the purpose of commerce destruction. That danger has, for the present, been successfully surmounted. Our estimate before the war of losses in the first two or three months was at least 5 per cent of our Mercantile Marine. I am glad to say that the percentage is only 1·9, and the risks have been fully covered under a system of insurance which was brought into force, the premiums on which it has been found possible steadily and regularly to reduce.

The third great danger was due to mines. Our enemy have allowed themselves to pursue methods in regard to the scattering of mines on the highways of peaceful commerce that, until the outbreak of this war, we should not have thought would be practised by any civilised Power. And the risks and difficulties which we have had to face from that cause cannot be underrated. But I am glad to tell the House that, although we have suffered losses, and may, no doubt, suffer more losses, yet I think the danger from mining, even the unscrupulous and indiscriminate mining of the open seas, is one the limits of which can now be discerned, and which can be and is being further restricted and controlled by the measures, the very extensive measures, which had been taken, and are being taken.

Fourthly, there was the danger from submarines. The submarine introduces entirely novel conditions into naval warfare. The old freedom of movement which belongs to the stronger Power is affected and restricted in narrow waters by the development of this new and formidable arm. There is a difference between military and naval anxiety, which the House will appreciate. A division of soldiers cannot be annihilated by a cavalry patrol. But at any moment a great ship, equal in war power, and a war unit, to a division of an army, may be destroyed without a single opportunity of its fighting strength being realised, or a man on board having a chance to strike a blow in self-defence. Yet it is necessary for the safety of this country, it is necessary for the supply of its vital materials, that our ships should move with freedom and with hardihood through the seas on their duties, and no one can pretend that anxiety must not always be present to the minds of those who have the responsibility for their direction. It is satisfactory, however, to reflect that our power in submarines is much greater than that of our enemies, and that the only reason why we are not able to produce results on a large scale in regard to them, is that we so seldom are afforded any target to attack.

Those are the four dangers. I do not include among them what some people would perhaps wish to include as a fifth—the danger of oversea invasion, although that is an enterprise full of danger for those who might attempt it. The economic pressure upon Germany continues to develop in a healthy and satisfactory manner.

* * * * *

25. 1915

German Declaration—War Area, February 2, 1915

Admiralty Staff Memorandum

The German so-called "blockade" arising from the proclamation of a war area round the British Isles, which culminated in the destruction of the Lusitania on May 7th, had its origin in December last, when it had become clear to the German naval chiefs that the attack upon commerce by cruisers had failed. On December 22nd, Grand Admiral von Tirpitz, in an interview which appeared in the New York "Evening Sun," threatened a submarine war against England. During January and February attacks were made upon several vessels, including the hospital ship Asturias, and were received by the German papers as fulfilling the Grand Admiral's threat. On February 2nd an announcement was made in the official "Reichsanzeiger," signed by Admiral von Pohl, then Chief of the Admiralty Staff, in which peaceful shipping was urgently warned against approaching the coasts of Great Britain owing to the serious danger it would incur. Two days later the announcement was issued. The following was the operative part of the Memorandum:

"Germany hereby declares all the waters surrounding Great Britain and Ireland, including the entire English Channel, an area of war, and will therein act against the shipping of the enemy. For this purpose, beginning February 18, 1915, she will endeavour to destroy every enemy merchant ship that is found in this area of war, even if it be not always possible to avert the peril which threatens persons and cargoes. Neutrals are, therefore, warned against further entrusting crews and passengers and wares to such ships. Their attention is also called to the fact that it is advisable for their ships to avoid entering this area, for though the German Naval forces have instructions to avoid violence to neutral ships, in so far as they are recognisable, in view of the misuse of neutral flags ordered by the British Government, and the contingencies of naval warfare, their becoming victims of attack directed against enemy ships cannot always be avoided. At the same time it is especially noted that shipping north of the Shetland Islands, in the eastern area of the North Sea, and in a strip of at least 30 sea miles in width along the Netherlands coasts, is not in peril.

[To this instruction a long statement was prefixed arguing that since the beginning of the War Great Britain had carried on a mercantile war against Germany "in a way that defies all the principles of international law." "Since the shutting off of food supplies has come to a point when Germany no longer has sufficient food to feed her people, it has become necessary to bring England to terms by the exercise of force. She is in a position where her life depends on her putting into effect the only means she has of saving herself." It was said that we had renounced the Declaration of London in its most important particulars, although British delegates had recognised its conclusions as valid in international law. We had wrongfully put certain articles on the contraband list, and were also accused of abolishing the distinction between absolute and relative contraband.]

"All these measures have the obvious purpose through the illegal paralysation of legitimate neutral methods, not only to strike at German military strength, but also at the economic life of Germany; and finally, through starvation, drive the entire population of Germany to destruction. The neutral Powers have generally acquiesced in the steps taken by the British Government. Especially they have not succeeded in inducing the British Government to restore the German individuals and property seized in violation of international law. In certain directions they have also aided British measures which are irreconcilable with the freedom of the sea, in that they have, obviously under the pressure of England, hindered, by export and transit embargoes, the transit of wares for peaceful purposes in Germany. The German Government has in vain called the attention of the neutral Powers to the fact that it must face the question of whether it can any longer persevere in its hitherto strict observance of the rules of the London Declaration if Great Britain should continue in the same course, and the neutral Powers continue to acquiesce in these violations of neutrality to the detriment of Germany."

26. 1919

The Enemy Navies
J. Leyland

* * * * *

There is no parallel in naval history for the surrender, of the German High Seas Fleet to impotent internment and seizure at Scapa Flow, and to inevitable destruction or decay, nor to the giving up of the whole of the submarine flotilla with its salvage vessels. Grenville would not die without his fight in the Revenge. Drake, Hawke, Howe and Nelson never acted, and we think never could have acted, as the German officers acted when they brought their ships to the Forth. The surrender seemed strange and almost incredible. In some respects the German ships were better than our own. That had certainly been the case in 1916. Lord Jellicoe says that it was a rare thing for British capital ships mined or torpedoed, to survive, whereas the German ships, after being hit by more than one torpedo, were able to get back to port. The Seydlitz made a wonderful recovery after her battering by twenty-eight shells and one torpedo in the Jutland battle. The König, Grosser Kurfürst, Markgraf, and other ships were very badly injured, but all reached port and were quickly repaired, as well as the Ostfriesland, which went home with a large hole torn in her side by a mine.

The German officers in the High Seas Fleet were men of high scientific and professional attainments, and did not lack courage. They had acted with what seemed curious reticence and discretion throughout the war, but they were so circumstanced that they would run few risks. The dangers were too great. Lord Jellicoe has shown very clearly what they were. The Dogger Bank and Jutland battles were not of German seeking. In some respects German officers did certainly seem wanting in enterprise, but their passive attitude was almost certainly dictated by the policy of the German High Command. All the evidence that came to light after the Armistice was signed showed that the officers had been ready to strike a blow, that they were willing and even eager to fight a battle, but that the spirit of discouragement and mutiny had sunk so deeply into the minds of the men that they were powerless to overcome it. Let us not forget that in 1797 mutiny left Duncan with only two ships, during some months, to watch the Dutch at the Texel.

There was a root of division in the German Fleet. The aristocratic officers who had crowded to the quarter-deck looked with ill-veiled contempt upon the officers of the Engineer branch, and the resulting situation was dangerous to the Fleet. The artificer class brought in the spirit of the big towns and the discontents of the workshops. Volunteering for the submarine service had come to an end, and the men were drafted to this duty as to any other. Great numbers of them never returned, and there was bitter discontent in the barracks and depôts, while parents deprived of their sons, and widows and orphans, made loud lament in the ports. The men did not habitually live in the ships, which were built not as places of habitation but as engines of war. They were housed in barracks or in accommodation hulks, and went on board merely for their duties. The ships' companies were raised by conscription with all the weaknesses of that system for the manning of a fleet. The consequences were disastrous. Discipline weakened, discontent grew, and open way was made for the evil influences and social propaganda of the shore. A feeling of impotent discouragement also prevailed, arising from other causes. Captain Persius, whose animus against the German Naval Authorities sometimes, it must be admitted, overbalanced his judgment, declared that for a year before the Armistice was signed practically no German High Seas Fleet existed. Be that as it may, it is a known fact that the Battle of Jutland had taken the heart out of the men. Lauded as a German victory, they saw each day more clearly thereafter that every movement made by their Fleet was fraught with peril and pointed straight to disaster. The Fleet could not stir without the British Fleet knowing that it was on the move. The Germans were watched up to the very edge and in the intricacies of their own minefields. The sea sense, deepened in long weeks of cruising in storm, fog, peril and watchfulness, the steady discipline and unshaken nerve of the British Fleet, were against them.

The Grand Fleet and the flotillas and patrols based on Dover, Harwich and other ports hemmed them in. One German officer confessed that the officers of the High Seas Fleet had endeavoured to trick the men into going to sea on the eve of the Armistice. How different was the spirit from that which existed when the splendid volunteers mustered and trained for the attack upon Zeebrugge! Captain Persius alleged in the *Berliner Tageblatt* on November 18th that the German seamen of the lower deck "had been brought by bad handling at last into a state of despair!"

The surrender of the German Fleet was, therefore, a recognition of the inevitable. To fight a battle might have been heroic. There would have been great destruction and slaughter, the whole German Fleet would have disappeared, and the action would not have availed. It was a sad hour for the better class of German officers when the High Seas

Fleet, the source and the expression of so much national pride, was surrendered, and no officer of any disciplined force could derive any satisfaction from the ignoble spectacle of the day.

* * * * *

27. 1919

The Submarine War on Merchant Shipping
Commander L H Horden RN

* * * * *

The submarines we had to contend with were of several types. The smallest was the U C—a minelayer of about 160 ft. long, which could remain out from seven to fourteen days, and usually operated in the North Sea and English Channel. This class had a maximum speed of 12 knots on the surface and 7 knots submerged. The U B Class was rather larger, about 180 ft., and could remain out fourteen to twenty-four days, which enabled them to operate further from their base. These submarines carried about ten torpedoes, and were armed with a 4-in. gun. Their maximum speed was 13 knots on the surface or 8 knots submerged. The U-boat was a large class, mostly fitted with torpedoes, but some for laying mines. They were about 220 ft. long, with a maximum speed of 16 knots on the surface and 9 submerged. They could remain out twenty to thirty days, and were armed with a 4-in. gun. The converted mercantile submarine cruisers were armed with two 6-in. guns, were over 300 ft. long, and could remain at sea for three or four months, but their speed was no greater than that of the U-boats. Considerably larger U-boats and cruisers were coming along at the end of the campaign. The maximum submerged speed given above could not be kept up for more than an hour, and could only be used in practice for a much shorter time, as if the batteries were run down the submarine would have to come to the surface. The usual under-water speed seldom exceeded 3 or 4 knots.

Contrary to the popular idea, a submarine spends most of its time on the surface and only submerges when there is danger or she wishes to make an attack. This operation could be carried out in less than a minute, and under favourable conditions it would be possible to remain under water up to forty-eight hours, if necessary.

Passages were necessarily made on the surface, and the usual speed was about 9 knots. When on the hunting grounds a submarine would usually lie on the surface with engines stopped or running slowly, for by keeping a good look-out they could be sure of seeing before being seen. The smoke of a steamer can be seen in clear weather up to 15 miles, and of a convoy somewhat further, so that after sighting it was comparatively easy to get ahead of the target on the surface without being seen. When the desired position had been reached, and the steamer was 4 or 5 miles off, so that there was a risk of the submarine being seen, she would dive, and after that would only show about 2 ft. of her periscope for 5 or 10 seconds at a time. As this part of the periscope is only about 2 in. in diameter it will be realised that the chances of seeing it before the submarine is within torpedo range is slight.

It should also be remembered that a submarine is not out to fight. If there is appreciable risk she will wait for another opportunity. Of course, some commanding officers would take more risks than others, and would succeed where others failed, but knowingly taking risks was confined to a few. Towards the end of the time there were signs that the *moral* of the crews was not what it had been, and, indeed, this was only to be expected, for our methods of dealing with them were continually improving, and their losses were increasing.

Having now dealt with the changing circumstances which had to be met at different periods of the war, and with the craft we had to deal with, it is advisable to review the various counter measures which we employed to combat the menace to trade and shipping. These may be divided into major operations and local measures, or the tactical employment of various apparatus.

The four major operations which had effect over large areas were:

1. The arming of merchant ships.
2. The introduction of the convoy system.
3. The mine or net barriers in the Straits of Dover and Otranto, and in the North Sea.
4. The blocking of the Belgian ports.

The effect of 3 and 4 falls within the scope of this chapter; the actual operations do not.

The arming of merchant ships was commenced in 1915, but at first few guns, and those only of small calibre, were available. By the end of 1916, 1337 vessels had been armed; a year later this had increased to 3193, and in September, 1918, it had risen to 4139. During the early part of the time, in order to make the guns go as far as possible, it was the practice to remove them from outward-bound vessels at certain ports well outside the submarine area and mount them in homeward-bound vessels. Later on, as larger guns became available, they were placed in the larger ships which had previously been fitted with small guns, and the latter were then mounted in smaller vessels. By the date submarine warfare ceased, some 60 per cent. of the guns mounted in merchant ships were of 4 in. and above. If a comparison be made between British vessels, which were mostly armed, and foreign vessels, which were mostly unarmed, it will be seen that armed vessels had a considerable advantage. Taking the period from February, 1917, to October, 1918, the figures are:

	British Per cent	Foreign Per cent
Average percentage of attacks by gun to total attacks	13	27
Average percentage sunk of attacks made by torpedo	59	73
Average percentage sunk of attacks made by gun	25	66
Average percentage of total sunk to total attacked	54	72
Average percentage of escape without loss or damage	34	20

It would seem from this that armament increased the ships chances by something like 20 per cent.

Before the convoy system was introduced steamers trading oversea were specially routed, but the time required to inform them of a change of route and to divert them was necessarily considerable. In effect, when a submarine had discovered the new route she had ample time as a rule to do a considerable amount of damage before it could be changed again. The first sailings in convoy under protection were vessels engaged in the French coal trade, and during the war nearly 38,000 vessels have sailed under these circumstances with about 30,000,000 tons of coal for France, and with a loss of only 0·14 per cent. A second route which was developed on slightly different lines was the Scandinavian trade. Between 10,000 and 11,000 vessels were convoyed during the war on this route with a loss of 1·2 per cent. mainly caused by two attacks by surface vessels.

In May, 1917, the first homeward Atlantic convoy was brought in, and the system was gradually extended until eventually over 90 per cent of the oversea trade was convoyed. During the first six months of "ruthless" warfare the losses in British and foreign vessels, trading oversea with the United Kingdom, averaged sixty-seven per month in the Atlantic and approaches to the western coast of the British Islands. Six months later they had declined to eleven, a saving of 83 per cent. on those routes. There can be little doubt that this system entirely upset the enemy's plan to starve us out. The amount of organisation required and the difficulties of putting the system into practice were very considerable, and can hardly be realised except by those who actually carried it through. There were at first some fears that the interference with trade would do more harm than good, by delaying sailings and by preventing the faster ships making use of their speed. Moreover, as the masters had no experience of working in squadrons the danger of ships steaming in company without lights was thought to be considerable, especially during simultaneous changes of course while zig-zagging, or if attacked. The manner in which Mercantile Marine officers surmounted these difficulties deserves high praise, especially when it is considered there were nearly always foreign vessels in the convoys. Instruction courses were organised at the principal ports through which a large number passed, and this was of great advantage.

The effect of the convoy system on the submarine was (a) to make it difficult for her to find a target, and (b) to add considerably to her danger when making an attack. As a corollary to (a) the operations of the larger submarines were greatly interfered with by compelling them to work nearer the coast. Owing to the ease with which ships could be routed, it was henceforth of little use for a submarine to wait where she had seen a ship pass in the expectation that within a few hours others would follow. Each convoy was surrounded by a destroyer screen for a considerable distance out from the coast, and this screen was supplemented nearer in by patrol vessels and, when weather permitted, by aircraft. An enemy attacking a convoy was therefore certain to be counter-attacked in a very few minutes after showing himself, and his difficulties were correspondingly increased in making an attack. Large submarines dislike the risk of diving in shallow water, and prefer to keep well away from the land. Before this they had been able to proceed outwards along the trade routes and attack ships 300 and 400 miles from the coast. Convoys, by ignoring trade routes and taking different courses each voyage, made it impossible for submarines to find them so far out except by chance, and the latter had therefore to close in where the routes must converge, where they were more liable to be counter-attacked, and their larger size was a disadvantage. That they would be driven closer in was, of course, foreseen, and the distance it was necessary to provide a destroyer escort for convoys was arranged accordingly.

The actual formation of the convoys, the disposition of escorts and the methods of zig-zagging, were modified by experience, and the whole system was brought to a most efficient state in a remarkably short time; but there were always some ships which could not be brought into the system, and a considerable number who, for part of their voyage, were unconvoyed owing to their Home ports being scattered.

The grand total of vessels convoyed during the war under British organisation reached nearly 90,000, with a loss of 436 vessels, or less than 0·5 per cent. a sufficient proof of the efficiency of the system.

* * * * *

So much has been published in the papers describing attacks on submarines by one or other of these methods that nothing new could be added. Such actions were red letter days in the otherwise dreary round of watching and waiting, which continued throughout the war with occasional alarms, true or false, the protecting of merchant ships or saving their crews, sweeping up mines and endeavouring to get to close quarters with the foe. Through all this and supporting it was the feeling which every fisherman knows, the expectant hope that the next cast will hook the big fish.

From the above it will be seen that there was no lack of energy in combating the submarine menace, and that, notwithstanding the novelty of this form of warfare, we were able in the end of defeat the forces the enemy could bring against us. It is, perhaps, too much to say, that such methods as the Germans used will never be employed again, but it need not be feared that any civilised nation would adopt them. If any other should attempt to do so our accumulated experience gained during the war will enable us to meet them on terms which would probably compel their abandonment.

The Great War—Victors and Vanquished

Elswick 21-in. Side-Loading Torpedo Tube.

H.M.S. "CRESSY."

GERMAN ARMOURED CRUISER "SCHARNHORST."

GERMAN BATTLE-CRUISER "VON DER TANN."

H.M.S. "QUEEN MARY."

PART IV: 1920–1935
Disarmament and the Air Power Controversy

Introduction	**115**
28. 1920–1921	
Foreign Navies: United States and Japan	**117**
J. LEYLAND	
29. 1921–1922	
The Capital Ship	**121**
REAR-ADMIRAL SIR REGINALD BACON	
30. 1921–1922	
The Possibilities of the Torpedo	**126**
REAR-ADMIRAL S. S. HALL	
31. 1921–1922	
Air Power and Sea Power	**132**
MAJOR-GENERAL SIR S. W. BRANCKER	
32. 1934	
Foreign Navies	**139**
CAPTAIN E. ALTHAM RN	
33. 1935	
Japan and Her Navy	**144**
COMMANDER S. TAKAGI IJN	
34. 1935	
The Fleet Air Arm	**153**
"VOLAGE"	

Introduction

Naval thought and policy in the post-war years was full of controversy. The international scene was dominated by demand for disarmament to prevent a recurrence of the arms race which was widely, although erroneously, held to have been the cause of the Great War. This demand, especially in Britain and the United States, was reinforced by a conviction that drastic reduction of arms expenditure was a prerequisite of economic recovery. That navies rather than armies and airforces were at the centre of the disarmament drive was due partially to spreading doubts on their utility in the new conditions. With the drastic reduction of the German navy after Versailles there remained no major challenge to the overwhelming might of the United States and Royal Navies, and it was universally accepted that they would never fight each other. Their relations, however, were complicated by the sometimes strident proclamations of the United States that she would never accept anything but parity with Britain and would outbuild her if necessary. British statesmen and responsible naval opinion had to accept this hard reality, although inevitably there were regrets that the long era of global maritime supremacy was over, and more cogent arguments that Britain, as the centre of a world-wide empire and dependent on international trade, needed a bigger navy than the United States. Another complication was America's apprehension of Japan's growing naval power and her insistence that an essential part of substantial naval disarmament depended on the ending of Japan's alliance with Britain.

The Washington Conference of 1921–2 took very substantial steps towards naval disarmament. Its central provision was the imposition of stringent limitation on the tonnage and numbers of future battleship construction, combined with the agreement that the ratio of battleship strength between Britain, the United States and Japan should be 5:5:3. The 1927 Geneva Conference sought to set similar limits on cruisers, but disagreement between Britain and the United States on acceptable gun calibres and individual ship tonnages produced a deadlock. Attempts to resolve this were resumed at the London Conference of 1929–30, but failed to achieve complete success due to

Franco-Italian friction over relative naval power in the Mediterranean. The Admiralty felt strongly that their political masters, led by Ramsay MacDonald, had offered too many concessions, especially in undertaking to reduce from 70 to 50 the number of cruisers they judged necessary for trade protection. Although Britain, Japan and the United States reached some agreement, Japan was bitterly resentful of the continuing assumption that she must always remain inferior to the other two. On the wider international front, Japan's attack on Manchuria in 1931, Hitler's accession to power in 1933 and Mussolini's increasing belligerence, suggested that the climate would soon become unfavourable for further disarmament.

There were also significant controversies in more narrowly naval matters, especially on the future of the capital ship. Increasing importance was given to its vulnerability to submarine and air attack, upon which naval opinion itself was divided. As the growing importance of maritime air power was accepted in Britain, violent disputes arose over who should control it—the Navy or the Royal Air Force. A series of compromise solutions were to prove highly unsatisfactory.

All the above issues were ventilated in Brassey's but there was one surprising omission. From Washington onward, Britain, mindful of the near success of German U-boats in 1917, argued unsuccessfully for the total abolition of the submarine. This was resisted by lesser naval powers, especially France and Britain had to be content with the acceptance by virtually all the naval powers of an agreement severely limiting the methods of submarine attacks on merchant shipping, limitations which were quickly abandoned in 1939. All this is given virtually no attention.

Recommended Reading

S. W. Roskill, *Naval Policy Between the Wars*, Vol. I (1968).

28. 1920–1921

Foreign Navies: United States and Japan

J. Leyland

* * * * *

United States

The United States Navy issued from the war more powerful than any Navy in the world with the exception of our own. It is in progress of rapid expansion through the execution of the shipbuilding programme of 1916, which has been described as "the first far-reaching constructive programme in the history of the Republic," though it suffers from serious difficulties in the raising and maintaining of its personnel. How far, under new political direction, the impetus to material expansion will continue it is as yet impossible to determine. Under the programme of 1916 sixteen capital ships are under construction, ten of them battleships and six battle-cruisers, and these, with nineteen existing Dreadnoughts, and at least twelve pre-Dreadnoughts, will give a total of nearly fifty capital ships in 1925. It is important to notice that, of the Dreadnought fleet of thirty-five units, sixteen ships will mount 16-in. guns and eleven others 14-in. guns. The sixteen ships mounting the heaviest guns will be enough when they are completed to ensure to the United States Navy the first position among the Fleets of the world, unless there should be a new British programme carried out expeditiously. At present Great Britain has the ten Royal Sovereigns and Queen Elizabeths and the Hood to compare with them; the two Renowns will be unequal to the later, heavier, and better-armed Lexingtons.

When Mr. Josephus Daniels, Secretary of the Navy, made his report on the year 1919–20 he spoke in glowing terms of the future. The American Navy, he said, had become incomparably stronger and more powerful than before, and was far in advance of any other Navy, in ships, in men, and in every element of strength. The organisation of the Fleet in two great divisions had given ample defence both in the Pacific and the Atlantic. With battleships in service equal or superior to any then in commission in any Navy, and twelve battleships (including the California and Tennessee, the last-named since commissioned, of the 1915 programme) and six battle-cruisers under construction—some of them larger than any yet in commission, and to be armed with 16-in. 50-calibre guns, more powerful than any afloat—the Navy, he said, was pressing forward to greater things, justifying in peace and war the country's firm confidence in its "first line of defence." He added that the great fleets in the Pacific and Atlantic were well officered and manned, and gave guarantee of protection and of readiness "to serve our country and the world." At Charleston, West Virginia, on August 31, on the occasion of the third anniversary of the commencement of large armour and projectile works, he stated that the delay in the ratification of the Versailles Treaty had prevented any cessation in naval expansion. There had therefore, he added, been no change in the plans for American naval development. "We are not only completing this great plant, but are building enormous docks and other needed shore facilities elsewhere, and are constructing 18 Dreadnoughts, with a dozen other powerful ships, which, in effective fighting power, will give our navy world primacy."

The General Board of the Navy prepared a programme for a further expansion of the Fleet, to be initiated in 1921, "with able argument," said the Secretary, in pursuance of his views and of its own. This plan is still in abeyance. The Board recalled the fact that since 1915 it had maintained that "the Navy of the United States should ultimately be equal to that of the most powerful maintained by any other nation of the work," and that "it should gradually be increased to this point by such a rate of development, year by year, as may be permitted by the facilities of the country, but the limit above defined should be attained not later than 1925." The plan was presented by Rear-Admiral Badger, who said that the scheme was conceived with the idea of giving the United States "the largest Navy in the world." The Board proposed the immediate laying down of two battleships, one battle cruiser, ten scout cruisers, five flotilla leaders, and some small craft, while upon aircraft construction, including experimental work, a sum of 27,000,000 dollars should be expended. Admiral Badger argued that Dreadnoughts and battle-cruisers of that class would remain the backbone of the Navy, with a tendency towards increased size. It was stated that the intended Dreadnoughts would outclass the British Hood and similar ships planned by the Japanese. The battleships, said Admiral Badger, would displace 44,000 tons, and the battle-cruiser 32,000 tons, both classes carrying 16-in. guns, while the scout cruisers should be of 10,000 tons with 8-in. guns. * * * * *

Japanese Naval Policy

With regard to the policy which underlies the Japanese shipbuilding programme, the Minister of Marine, describing its features to the Budget Committee of the House of Representatives in July, said it was not directed against any potential enemy, but was the measure of the necessities arising from the country's insular position, although the possibility of the despatch of foreign strength to the Orient had not been disregarded. At an earlier date the *Jiji Shimpo*, discussing the supplementary programme, of which the general character had become known, referred to its relation to the disarmament clause of the Covenant of the League of Nations. The news that Japan contemplated a new programme of shipbuilding for the Navy would not, it was apprehended, be received favourably by foreign Powers. The fact was, however, that the question of the programme of sixteen capital ships, which Japanese naval authorities regarded as the minimum of strength necessary for national defence, had been pending ever since the time of the Russo-Japanese War, and the whole nation desired its fulfilment. Progress had been interrupted by political and financial considerations, but the question was not new. The programme, therefore, said that *Jiji Shimpo*, should not make Japan an object of suspicion to foreign Powers. "Even if all the Powers of the world carry out the limitation or reduction of national armaments, to reduce the naval strength of Japan below the 8-8 standard would be inconsistent with her national safety and her geographical situation and circumstances." But when Admiral Kato, Japanese Minister of Marine, was in England in September, 1919, he had said Japan would like a general agreement for disarmament, but was waiting to see what was being done in that way by the other great Powers.

The programme of the Japanese Navy, hampered as it is by financial depression, will be followed with great interest. It might have been supposed that the plan of 1916-17 was conceived at the time when nearly every one in Japan believed Germany would never be beaten, and that therefore the Alliance could be of little protective value to Japan, but the new programme, brought forward at the time when the Alliance was about to be renewed, forbids such a conclusion. Undoubtedly, therefore, if the programme receive effect, it can only be because of national necessities which compel the Japanese Government to establish a situation which will give the country, not only a Fleet of sixteen very modern ships, but a considerable force of powerful ships a little less modern. Every effort is being made to maintain the efficiency of the Fleet. The Naval Manœuvres in October, 1919, were the most extensive in the history of the Japanese Navy, and included several actions and aeroplane attacks on the coast towns. The operations were shadowed by an explosion which occurred in the battleship Hyuga in Tokio Bay, whereby fourteen men were killed and thirty injured. A gun-turret was blown overboard.

29. 1921–1922

The Capital Ship
Rear-Admiral Sir Reginald Bacon

* * * * *

In discussing the place which large ships will occupy in naval warfare of the future, we must recognise that the points involved are so many and varied that confusion of ideas is bound to arise unless the various considerations are kept clearly in mind and each dealt with on its own merits.

Two main questions arise either of which, if answered in the affirmative, would mean the abolition of the large ship. Firstly, Do the widely-spread geographical positions of the present first class maritime Powers preclude the useful employment of large ships in naval warfare? and secondly, Has modern war experience proved the big ship to be incapable of keeping the sea owing to the development of mines, submarines, and aircraft? The first question is largely strategical; the second is a tactical proposition, which can only be answered in the light of our experience in the late war.

To deal with the first question, we must appreciate that continental Europe, from the naval point of view, is exhausted. The Triple Alliance has vanished; France and Italy cannot, for financial reasons, rebuild their navies; Russia is no longer a Power. The theatre of naval operations has therefore shifted from the narrow waters of the Channel, the Mediterranean, and the North Sea to the Atlantic and Pacific Oceans. How will this affect naval construction?

* * * * *

The Future of the Capital Ship

In order to answer the question whether the capital ship is able to keep the sea in the face of submarines, mines, and aircraft, we have primarily to depend on the experience of the past. Were this all, the task would not be difficult, and opinions among naval officers would differ only in the exact reading of the lessons past events have provided. There are, however, two factors which have, in addition, to be considered which materially add to our difficulties. First, the question has to be answered whether the late war, which after all is the greatest of all our sources of information, can be taken as typical of future wars, and if in such wars similar dispositions at sea will obtain, and whether the conditions at sea during the Great War gave full scope to modern weapons and tried them to the uttermost? The second problem is to forecast the development of naval armaments, both offensive and defensive, then to foresee probable improvements in each, and to balance these against each other in offence and defence— guns against armour, submarine against large ships, aircraft against floating vessels, the torpedo against modern ship construction. It is in this attempt to forecast the future that our main difficulty lies, but the difficulty of the problem is no excuse for not making the attempt.

The most marked lesson of the late war, so far as fighting ships were concerned, was the revelation that thought and foresight had provided means to defeat the newer weapons that had appeared among naval armaments. It is well to examine this more closely. The history of torpedo warfare is a classic example of how weapons, untried in war, may, in peace time, appear to be invincible and unanswerable, and yet, under fighting conditions, be found not to preponderate in the way expectation had led us to anticipate. In fact, the history of the torpedo is a standing warning against assessing too highly, in peace time, the power of weapons of which no experience had been gained under the conditions of war. In the late 'eighties of the last century, the torpedo carried by small torpedo-boats working from shore bases became a threat to ships at anchor. The breakwaters at Portland, Plymouth, Dover, Malta and Gibraltar completely annulled the threat. Meanwhile the destroyer had been designed and built to chase torpedo-boats; these were given torpedoes and, being seaworthy enough to accompany the Fleet, were able to remain at sea and were no longer dependent upon harbour refuge. The range of the torpedo grew from hundreds to thousands of yards, and therefore the torpedo developed into a weapon for use by destroyers in a day fleet action or at night on the open sea if the enemy's fleet had been marked down. The reply to his threat was an increased secondary armament of the ships, and the tactics of the fleet in case of attack. The gun, in the meantime, had developed in range no less than the torpedo, and, since the efficiency of the gun is the main factor in deciding the range of a fleet action, the increased range at which the opposing battle fleets would undoubtedly choose to commence the fight, give time for a fleet to turn "away" or "towards" the attacking vessels and so to defeat the attack. Now before the war, the writer, like many others, was one of the most ardent upholders of the torpedo. It seemed inconceivable that if destroyers were near an enemy's fleet, either in daylight or dark, a tithe of the ships should not be reaped; but events proved the opposite. The counter stroke by other destroyers, the visibility of the wake of the torpedo, the watchfulness of the Admiral, and at night covering darkness and uncertainty if ships met with were friend or foe, robbed, in stern war, the torpedo of its anticipated success, so that the destroyer and torpedo, now that they have passed through the ordeal of a great war, must merely take their place in the category of weapons which largely affect naval dispositions and exercise a restraining control on the tactics of fleets, but they cannot be looked on as determining factors in large ship actions, since that control, when exercised, provides an antidote which robs them largely of actual material achievement.

Submarine and Aircraft

As with the destroyer, so with the submarine. Before the war, prophets loudly proclaimed that fleets would have to remain in harbour or be sunk. This was proved not to be so. Our Grand Fleet steamed in the North Sea a distance equivalent to a triple circumnavigation of the globe without the loss of a single ship from attack by a submarine. Again, defensive precautions robbed the new weapon of its terrors. It was found merely to exercise certain minor limitations on the free action of the Fleet.

With aircraft, the case was somewhat different, in that the war may be considered as having given birth to aircraft. Before the war they were purely experimental; during the war they grew rapidly and waxed strong, but in offence against ships were proved to be futile. On the Belgian coast, up to the end of 1917, scores of attacks were made on our vessels and hundreds of bombs dropped, but only one bomb struck a ship under way. The submarine mine was no more successful, only one capital ship of the Grand Fleet, and a very small percentage of other men-of-war other than the vessels employed in sweeping operations, were lost by this new weapon. So that, summing up this brief review, we are forced, almost with surprise, to the conclusion that in the late war all the new weapons of offence failed to realise the expectations formed of them. We say almost with surprise, since the verdict arrived at is greatly against our previous anticipation.

Now had these weapons a fair chance in the late war? Were there special conditions unlikely to recur in the future which hampered them and reduced their offensive value? The ardent advocate will always find excuse for failure; zeal is apt to gloss over the difficulties which militate against realisation. But let us briefly review the past conditions and see where the failure of each lay and whether future improvement will banish these and raise the weapon superior to the causes of defeat.

* * * * *

Development of the New Weapons

The new weapons, therefore, have their counterparts in attack and defence with ordinary surface warfare. They have no claim, at all events in their present state of development, to exceptional powers. The question still remains: Can these powers be augmented so as to make them really deadly?

The limitations of the submarine, which make it inferior to the destroyer, are matters of speed in attack and inability to work in co-operation with its fellows. In order greatly to increase submerged speed, some new form of propelling energy is required; until this is invented further criticism is unnecessary. So also with concerted action. To a human being in a vessel travelling at high speed nothing can approach the efficiency of untrammelled vision; no instruments however ingenious or efficient can make up for partial blindness. The two fundamental limitations of the submarine, which make it inferior to the destroyer, seem likely to remain undiminished for many decades to come, whereas its one advantage—viz. invisibility during approach—can never be improved: it never can be more invisible than it is to-day. For these reasons, one need not expect the submarine to be more deadly against large ships in future wars than it is to-day.

With aircraft the same line of reasoning applies. Defensive measures with protecting squadrons of fighters will advance with offence. No reason exists why the war in the air should not progress on the same lines as that on the water. Keep the analogy of the surface attack by destroyer and the aerial attack by the bomber clearly in mind, and see how the defensive measures against the one on the surface of the water indicates the defensive measures to be taken against the other in the air. But one disability in attack must be noted. The bomber must be less agile than the fighter and, therefore, more at its mercy, unless supported by fighters also. To translate this so as to obtain a strict analogy, we must imagine heavy, slow, torpedoing destroyers protected by light fighting destroyers, a condition distinctly unfavourable to successful destroyer attack.

So far, therefore, we may confidently expect the large ships to continue reasonably safe from destruction by the newer vessels of war. Over and above all the preceding considerations, we have the factor of improved naval construction. Large ships can now be built so as to be practically immune from the explosion of a torpedo, and in fact from successive explosions of several torpedoes. This, of itself, goes far to abolish real danger in torpedo attack. Such vessels are very large and exceedingly expensive. This increased cost necessitates building fewer numbers, but does not argue building none at all. It may be claimed that the explosive charge of the torpedo can be increased so as largely to annul constructive defence. In this we are up against the old gun and armour problem—defence and offence progressing and each in turn obtaining the mastery. The subject is too technical to deal with here, but increase in explosive charge of the torpedo to an amount necessary to defeat modern construction will mean a very considerable increase in weight of torpedoes, and consequent increase in size and complication of the vessels carrying them. The increase in size of the torpedo may well prove more difficult and expensive than the *pro rata* increase in ship protection. With the exact methods of constructional defence we are not concerned—whether the bulge should merely exist at the waterline or be carried up the height of the ship is a matter to be decided by conditions of stability, weight, and expense. Suffice it for our purpose that adequate protection against reasonable torpedo attack can now be provided. We are therefore forced to the conclusion that the newer weapons provided by progress in science and construction, have not in any way pronounced the doom of the capital ship. Nor has modern redistribution of sea power reduced the necessity for our building large ships to protect our world-wide interests.

30. 1921–1922

The Possibilities of the Torpedo

Rear-Admiral S. S. Hall

* * * * *

The Moral Influence of the Torpedo

The torpedo has never been, and never will be, a weapon of precision, whether fired at short range at a single ship or in numbers against a fleet of ships. The always uncertain factor of "speed of enemy" and the small difference in speed between that of target and torpedo will always cause this to be so.

In the first case, the attack and accurate judgment of enemy course and speed are inherently difficult; in that of long-range massed fire "into the brown" the slow speed of the torpedo is the cause, both of these and the visibility of the tracks of existing torpedoes rendering avoidance possible by alteration of course, were the principal reasons in the last war.

The universal adoption of a high sea-speed, destroyer screens, and the practice of always steering a zig-zag course have tended further to decrease the chances of a direct hit, even allowing for the fact that the number of torpedoes that can be fired from one submarine in one "salvo" has increased from two to four and in many cases to six. Though it is possible that this balance may be redressed in the near future, as will be described later in forecasting probable improvements in material, the dominating influence of the torpedo is undoubtedly a moral and an indirect one. It is not always realised what a tremendous influence this is and what a phenomenal effect it has had upon all naval operations, due entirely to the advent of the torpedo-carrying submarine. The moral effect of a heavy gun is great, but the gun must be in sight; when it is not the influence disappears, and it is non-existent for all ships possessing speed greater than that of heavy-gun-carrying ships.

In the case of submarines, however, it is only necessary for one belligerent to possess quite a small number of such vessels, which need not even be at sea, to cut down automatically the endurance of the whole of the enemy's surface fleet to that obtainable at high speed on a zig-zag course. Their potential value is such that their mere existence is sufficient to reduce the sea-activity of all surface ships by varying amounts, usually more than a half of what it would be if there were no submarines opposed to them. Capital ships and valuable cargoes are further restricted by having to be protected by large numbers of destroyers, and a universally restrictive effect is thereby produced on all the naval operations of surface war vessels which is at times so great as to render them impossible. The Japanese attack on Port Arthur, and the Allied attack on the Dardanelles are two cases in point; neither of these operations would have been undertaken in the face of modern submarine opposition.

In the last war this restrictive effect of the torpedo on our Grand Fleet was not sufficient to prevent it carrying out a distant blockade of the German surface fleet, owing to the small distance separating the two fleets; but in the future if our enemy, as appears certain, is much more favourably placed geographically, the effect, coupled with that of the torpedo-carrying aeroplane and mining operations on the scale that war has shown to be possible, will be so great that, in the opinion of many, it will have the result of rendering the present capital ship useless.

* * * * *

The Ultimate Naval Problem of the Future

Though it appears, at first sight, a hopeless task to forecast the influence of the torpedo in a future war that is so vaguely defined, it is not difficult to find certain factors which are definite and sufficiently non-controversial as to indicate the future with some certainty.

It is not necessary to define with whom we have to contend for our existence, for that of any of our possessions, or in the defence of any treaty to which we may be a party. Whatever the cause and whoever the enemy, it may be taken as certain that we shall not go to war again on the grand scale without the ultimate stake eventually becoming our continued existence as an empire. Before the last war, it was insisted by some, and correctly so, that if Germany wished to impose her will upon us, there was one and only one road open to her, and that was by cutting and keeping severed our sea-communications.

Invasion, it was argued, had become more and more difficult. Her surface fleet was inadequate to enable it to compete with ours, and so there remained only one way of bringing us to defeat and that was by a submarine war on our trade. These arguments have as much force now, and apply with equal certainty in the future. The already immense difficulties in the way of invasion have been increased manyfold by the advent of tanks, aircraft, and mechanical transport, to mention only three of the complications which the last war produced. There is only one other way in which a future enemy can reduce us to defeat to the extent of surrendering even a part of our empire, and that is by the means recently attempted by Germany though not necessarily by precisely the same methods.

A local gain of command of the surface in the Far East would not suffice, supposing, for example, the possession of Australia was at stake. This whole continent might in fact be occupied, but the matter would not end there. Ultimately it could only be held if the local command was extended to these islands in sufficient strength to cut off our supplies of food and other necessaries.

The question of the persistence, or otherwise, of capital ships, and fleet battles for a partial command of the sea, is still unsettled, but it is not a dispute that whether capital ships are built or not, whether they do or do not fight, and whether ours win or lose the fight, attacks on our sea trade can proceed as they did in the last war; and this is for us "the last ditch."

The vulnerability of merchant ships in the face of new weapons is beyond question the vital matter for us concerning a future war, and their protection should be our first thought. The day has passed when our object could be achieved by "seeking out and destroying the enemy's fleet," or even by an effective blockade of his surface ships.

We may or may not be able to achieve either of these objectives—perhaps aircraft will provide the means. If we cannot do either, we shall have to compete with attacks on our trade from surface ships as well, in which the influence of the torpedo will not be felt, but whether we do so or not, the real menace is the submarine one. Nothing that we can do can remove that; it will have to be faced whatever else may happen. It must always be more acute than the depredations of surface raiders, which can be much restricted if we keep a strong submarine and cruiser fleet.

As in the last war, therefore, the submarine is certain to be the cause of all our troubles in the future, because it is by far the most efficient weapon for striking us in our tenderest spot, and since the submarine is essentially a torpedo-carrier, the torpedo, or rather the opportunities for using it, will predominate.

* * * * *

Future Development of Torpedoes

Though the future influence of the torpedo depends mainly upon whether the opportunities for using it will be even more frequent than in the last war, rather than in any great improvement in the weapon itself, there are certain lines of development indicated which should be mentioned, as they will, when perfected, remove many of the disabilities under which torpedoes have hitherto been used. There is no reason to expect any great advance in speed or weight of explosive carried, but, if the lessons of the war are correctly read, a throwing open of torpedo manufacture and design to a wider field should result in greater reliability and improved depth taking and depth keeping.

It may be anticipated that the wake of the torpedo will be rendered far more difficult to detect if not quite invisible, and since the milky track left by German torpedoes was very marked, even in a broken sea, this will have its result on the numbers of ships which were saved by prompt use of helm when the track of a torpedo was first seen. It will also add greatly to the difficulty of using depth charges effectively. It is an extraordinary state of affairs now that the torpedo is a daylight weapon and particularly from the submarine point of view, that a highly complicated vessel specially constructed to deliver an attack depending for success and for its own safety upon invisibility, should be handicapped just as the final act is reached and at the most dangerous moment, by a large upheaval of air on the discharge of the torpedo from the tube. This remains for some time to give away the submarine's position, rendering the submarine liable to be depth-charged by any nimble escort, whilst the track of the torpedo assists also to this end, in addition to giving the enemy a fair chance of avoiding the torpedo.

It is such an obvious matter that no doubt it has already received attention. It is urgently required for submarines, since it will greatly increase the confidence in attack, to know that the torpedo will not be seen nor the direction from which it has come be visible, that it will be worth even a considerable reduction in "speed of torpedo."

The means of detonating the torpedo, both German and our own, were inefficient and remained so during the whole of the last war. Considering the trouble and difficulty of getting such a delicate mechanism as a torpedo to hit a target probably at high speed, it is wonderful that those responsible should be content with a pistol which, even theoretically, will not fire the charge for 30° out of a possible 180° of striking angle.

A ship's side is not a flat plate and as the torpedo may strike the swell of the bow, the counter, the curve of the ship's bottom, or one of the many irregularities on it, it is impossible to guarantee the present pistol operating whatever may be the angle between the torpedo's attack and path of target, an angle always tending to become acuter by reason of the target attempting to dodge the torpedo. If, as often happens, the target happens to be going faster than the torpedo, it is an even chance if the torpedo explodes even against a flat side. It is presumed that this defect will be remedied and that our future enemy will not be able to take home the warheads of our torpedoes to be made into flower pots.

The Germans employed in the last war a magnetic pistol and it must be assumed that this will be in general use before long. This will not have great effect in attacks on merchant ships, though it should slightly raise the percentage of hits to misses by eliminating all misses "under." In torpedo attacks on big ships, it will to a great extent discount the value of the blister and necessitate more protection to the double bottoms of these ships.

The number of bow tubes fitted in submarines has now risen to six, and therefore salvoes of at least four torpedoes must be expected against single valuable ships; under certain conditions it may be desirable to work submarines in company to obtain larger salvoes against a squadron of ships. There is no difficulty in controlling, by under-water signals, two, or even three, submarines, submerged at different depths to avoid all chance of collision, when salvoes of eighteen torpedoes can be fired. This is a serious menace and will be more so when the torpedo wake has been eradicated.

The Torpedo-Carrying Aeroplane

So far this chapter has dealt mainly with the influence of the torpedo when carried in submarines, because the latter are, at present, the most efficient carriers and are likely to remain so for some time on the high seas, but a rival is appearing in the aeroplane which bids fair to become, in its own sphere, even more formidable.

When very fast and handy torpedo-carrying aeroplanes are produced—the type is already evolved—machines that can dive and flatten out on to their point of discharge, just as a fighting plane does now, a further heavy restriction will be placed on the operations of all surface ships within their radius of action. They require a very fast torpedo, and, since the efficiency of the attack depends upon speed and not invisibility, a wakeless torpedo is of no moment.

There can be no doubt that these aircraft accompanied by a proportion of swift planes carrying machine-guns and some equipped with smoke-screen apparatus, will render obsolete all forms of fixed permanent defences. No squadrons of ships could withstand successive attacks by them for long, even though able to bring strong destroyer escort, for the upper decks of these vessels would be shot to ribbons, leaving a clear road for the torpedoes against the bigger ships.

They are ideal in conjunction with submarines, or even unassisted, for the defence of outlying bases, or for the attack of ships or fleets in harbour wherever they can be reached. In the last war, for example, they would have stopped the reduction of Kaiou Chow and prevented the attack on the Dardanelles; the safety of ships in Italian ports, in Dover, Portland, Plymouth, all East-Coast harbours and roadsteads, including even Scapa Flow, would have been imperilled; either by day or bright moonlight these anchorages would have become unsafe.

The future influence of these torpedo-carriers will be such as further to restrict the opportunities for using guns and so will add to the dominance of the torpedo already outlined.

The balance of advantage in the probable future development of weapon technique, lies with under-water weapons as compared with any means yet in sight for countering them or with guns.

Quite apart from, and independent of, this advantage, the restrictive effect on the operations of all surface warships caused by under-water weapons of all kinds—which became such a marked feature in the concluding stages of the last war—will be accentuated by the development of aircraft.

The greater distances that for many years appear certain to separate opposing fleets of capital ships, which must always rely upon the gun as their principal weapon, will preclude distant blockade, the sole remaining function left to these fleets in the last war, and by so doing will cause the opportunities for obtaining decisive results by means of gun-fire to diminish still further if not to disappear.

This will confine naval warfare to attack and defence of sea communications, and since submarines and aircraft are the most efficient for this, the result will be that the torpedo will be the dominating naval weapon of the future.

31. 1921-22

Air Power and Sea Power
Major General Sir S. W. Brackner*

It has frequently been claimed that the advent of aviation will revolutionise completely the conduct of all operations of war. During the last great struggle, the effect of the new army was felt most strongly in land operations; in spite of its influence, however, the armies continued to fight on the principles and with the organisation laid down during peace.

In co-operation with the Army, although aircraft proved to be the only efficient means of reconnaissance and artillery observation in trench warfare, and although it gave an additional power to strike with high explosives vital points behind the enemy's lines and beyond the range of artillery, it did not seriously affect the strategy, the tactics, or the equipment of the men on the ground. The Army of to-day goes on without much thought of aviation, beyond its usurpation of the principal functions of cavalry, reconnaissance, and shock tactics against formed troops; and its ability to reach with explosives important points within the zone of operations, but outside artillery range. The Air Force of the future will, however, considerably reduce the measure of importance of the Army in any scheme of National Defence, though it may not alter the principles of its leadership, or change seriously its organisation or equipment.

On the other hand, far less experience of the working of aircraft in co-operation with the Navy and their effect on naval operations is available. On the outbreak of the Great War, the development of seaplanes was considerably behind that of aeroplanes; this, coupled with the fact that the whole of our Army was thrown into the front line almost immediately, led to the great bulk of our aerial resources being applied to the service of land operations in the early stages. The experience thus gained begat further pressing and vital demands, with the consequence that it was not until well on in the war that naval

*Later Air Marshal (*Ed.*).

requirements and the possibilities of aircraft working with the Fleet were fully considered. Thus we must rely more on conjecture and argument in considering the effect of aircraft on naval warfare than in the case of land operations; but it seems probable that the effects on the organisation and equipment of the Navy will be more far-reaching than on those of the Army.

Aircraft in relation to naval warfare may be divided into two categories: firstly, those working in close co-operation with the Fleet under the direct orders of a senior Naval officer, and secondly, those working independently on missions of destruction. The operations carried out in the first category are wholly the concern of the Admiralty, whilst those in the second will be controlled by the Air Ministry, and will often be carried out by aircraft, which, at other times, will be working against military or civil objectives. It is difficult to lay down an exact demarcation between the functions of these two categories. Some division is necessary, however, and this is not the place to discuss the benefits and evils of an independent Air Ministry and the exact scope of its responsibilities.

No fleet of the future will put to sea without a large number of aircraft—some flying with the fleet, some carried in large, speedy ships designed for the purpose, and some carried on the capital ships themselves. The services of these aircraft to the Fleet may be classified as:

(*a*) Reconnaissance.
(*b*) Spotting.
(*c*) Defence against aerial attack.
(*d*) Anti-submarine operations.
(*e*) Aerial attack.

* * * * *

Aircraft-carrying Ships

From the foregoing, it is obvious that a modern fleet must be provided with several aircraft-carrying ships; these must be really big, fast craft, at least capable of manœuvring with the fastest capital ships. They should have a clear deck space of at least 600 by 100 feet, and be so designed that none of the funnels, superstructure, etc., seriously disturbs the air through which the aircraft must fly when alighting. Braking devices may reduce the length necessary for alighting, but as the ships must be large enough to carry at least twenty aircraft, must have great speed and fuel capacity, and must be able to defend themselves against destroyers and light cruisers, they must always be of considerable size.

A fleet must be provided with at least three aircraft-carrying ships; one for fighters, one for reconnaissance and anti-submarine aircraft, and one for torpedo carriers. Each has its urgent duties the moment two rival squadrons approach one another, and any attempt to make them share the same deck must lead to confusion and loss of efficiency. In addition, each capital ship will carry a small complement of gun-spotting aircraft, which must be capable of flying off the superstructure, and all of which, except the unit ready for action, can be kept unerected between decks.

Operations of Aircraft in Warfare

Whilst the fleet lies at anchor, long-range reconnaissance will be carried out by airships and large aeroplanes and flying boats working from shore bases. The fleet's reconnaissance aircraft will be patrolling for submarines, and small surface craft will be cruising out to sea to give timely warning of the approach of hostile aircraft; the fighters will be ready for instant action in case of hostile attack or reconnaissance. When the fleet puts to sea, the reconnaissance aircraft will increase their activities, and extend them ahead of the course ordered. As the hostile squadrons of capital ships are located and come within aerial range, the torpedo-carriers will take to the air and, in flight and squadron formations, will fly out and attack from as close a range as can be attained without reckless leading. These attacks will be continued relentlessly until either the action is over or all the torpedo-carriers are destroyed or incapacitated. Simultaneously, the fighters will commence patrolling above the ships and in the direction of the enemy, attacking all hostile aricraft which venture within a certain distance. As the fleets close, the gun-spotting craft will be thrown into action, and fire will be opened at the extreme range of the capital ships' armament.

During the action, and in the ensuing pursuit or withdrawal, all available aircraft will be hotly engaged in the neighbourhood of the opposing squadrons. The fighters, bombers, and reconnaissance aircraft will return and alight on the decks of their carriers as their petrol runs out or their missions are completed. These aircraft-carriers will be a most vitally important and vulnerable unit in the fleet, and it seems likely that they will cruise at some distance from the battle line so as to be as far as possible out of harm's way. The gun-spotting machines, as they conclude their tour of duties, must either alight on one of the aircraft-carriers or drop into the water and chance being picked up. Casualties amongst them will be replaced and reliefs provided by aircraft hastily erected on the superstructure of their parent capital ships. Meanwhile, the long-range reconnaissance craft,

DISARMAMENT AND THE AIR POWER CONTROVERSY 135

working from their bases, will be covering all lines of approach to the locality of the battle, and will give warning of any naval movements outside the immediate battle zone; they will frequently have to fight with hostile craft in order to maintain their positions and carry out their work.

* * * * *

Effect of Aircraft on Naval Operations

The foregoing covers the activities of aircraft working in close co-operation with a fighting fleet. It is necessary now to consider what effect these new conditions will have on the conduct of naval operations, before proceeding to consider the influence of independent air operations. I write absolutely as an amateur in naval affairs, but venture to state that the following appear to be logical conclusions if the efficient operations of aircraft, as indicated above, are admitted to be possible.

(i) A considerable number of the surface craft at present considered necessary for reconnaissance can be dispensed with.

(ii) Information regarding the enemy's dispositions and movements will be more rapidly obtained and will be considerably fuller than has been possible heretofore. This new factor should give the opposing leaders a greater incentive to manœuvre, and enable the weaker force to avoid battle if it so desires.

(iii) If both Admirals wish to fight, the battle will begin earlier, and at much longer ranges, than previously, and for this reason a greater number of ships will be sunk or put out of action. The new factors of attack from the air, and improved accuracy in the observation of long-range gun-fire will increase the measure of destruction to an incalculable extent. The power and effect of attack from the air is the most uncertain factor in the whole problem, and its possibilities threaten the very existence of the capital ship in the future.

(iv) A measure of aerial superiority will give to the fleet possessing it an advantage which may completely eliminate defects arising from inferiority in the power and numbers of its surface craft. A powerful and efficient fleet, blinded by the destruction of its reconnaisance and spotting aircraft and attacked by numerous torpedo-carriers of sufficient power to brush aside the defence of its fighters, might well be defeated by squadrons considerably inferior in every respect except in the efficiency of their air units. It is this possibility which again calls into question the exact potential value of the capital ship.

(v) Finally, to sum up very briefly, aircraft will not affect the generally accepted principles of organisation and operations in naval warfare, except as regards the one great question—the power of aerial attack against capital ships. An endeavour will be made later to estimate the results of this new factor.

Independent Aerial Operations

It is now necessary to consider the effect of independent aerial operations on naval warfare. In accordance with our present policy, such operations would be carried out under the orders of Air Officers, and directed by an Air Ministry. The ultimate objective of an Independent Air Force will be the destruction of some vitally important organisation on sea or land, but air squadrons will certainly have to fight in order to carry their operations to a successful conclusion. Practically the only available experience of such operations was gained in the attacks by Zeppelins, and in a few cases by Gothas, on England, and our attacks on the Rhine Valley from the neighbourhood of Nancy. The destructive effect of the Zeppelins was small owing to their vulnerability and consequent very cautious handling, but the few ventures of the Gothas proved far more destructive. Our attacks on the Rhine Valley had hardly begun when the war ended, but considering the small number of aircraft employed, they proved remarkably effective. In all cases, the moral effect, and consequent dislocation of transport and manufacture, was very great indeed. Our experience points definitely to certain facts—the operations of a well-led and properly trained formation of high-speed powerful bombers are not seriously affected by anti-aircraft gun-fire. Such a unit can reach its objective in spite of any efforts that small and lightly armed aircraft can make to stop it; and it can operate efficiently by night. The most obvious defect of air attacks during the war was lack of accuracy in the actual dropping of the bombs.

In discussing the work of aircraft in close co-operation with the Navy, I have confined myself to the conditions of the present and of the very immediate future, but in considering the development and capabilities of an Independent Air Force, it will be legitimate to look somewhat further ahead. There will be many calls on the activites of the Air Force as an offensive weapon. Tempting objectives will be plentiful, and will include the enemy's seat of government, his great railway centres, his most important factories and industrial districts, the billets of his main army reserves, his principal aerodomes, his naval bases and submarine docks, and last, but not least, his fleet. The selection of the most important objective will be a matter of national policy, and must be made by the Government. It is only the last two objectives that affect our subject.

* * * * *

Aviation and the Work of the Navy: Twenty Years Hence

It is now necessary to try to visualise the effects of aviation on naval warefare as we know it. The Navy to-day has four main duties: to protect our shores from invasion; to protect our seaborne commerce; to transport our armies in safety to points on or near hostile territory; to destroy the enemy's seaborne commerce. Our Navy succeeded in carrying out all these duties during the Great War, although, at one time, the German submarine very nearly starved us by destroying our merchant ships. It is claimed that if the Germans had put more energy and a greater portion of their resources into the submarine campaign at the beginning of the war, they would have brought us to our knees in spite of our naval supremacy. The German Navy accomplished none of the duties enumerated above, and eventually failed to starve us; consequently Germany was beaten.

Let us imagine a situation similar to that of August, 1914, with the exception that both nations are supported by air power on a scale which could be arrived at twenty years hence, and that Germany has a definite and effective superiority in the air. Could our Navy do anything to prevent devastating attacks from the air on London or any of our big industrial centres? No. Could our command of the sea protect our seaborne commerce? Only when it was out of aerial range of hostile territory. Could our transports cross the Channel in almost perfect safety and with unfailing regularity? No; they would suffer most serious, and perhaps crippling, losses in spite of anything our Navy might do. Could the Navy destroy the enemy's seaborne commerce? This would be possible on account of Germany's geographical position, which is peculiarly susceptible to blockade by Great Britain. So even with command of the sea, our Navy could only perform one of its duties completely and one in part; but it is doubtful whether the command of the sea could long be maintained in such conditions. Our fleets could certainly no longer lie at anchor at Harwich, Rosyth, or Scapa Flow.

Let us be more optimistic and assume that we have a distinct superiority in the air. How far will this alter the position of the Navy? No invasion by Germany would be possible in face of our aerial supremacy. The Air Force would relieve the Navy of a considerable portion of its duties in protecting our seaborne commerce; in any case aircraft are certainly the most dangerous enemies of the submarine. The Air Force would also be a predominating factor in protecting the transport of the Army across the sea. It will be the most efficient means of destroying the enemy's seaborne commerce either from our own shores or from aircraft-carrying commerce destroyers.

This is merely an arbitrary example. Similar problems relating to various other possible adversaries can be considered. In all cases, it will be found that air supremacy is the dominating factor, and that aviation has usurped the functions of the Navy and Army to an extent which varies with the geographical position of the country considered. Thus, it seems inevitable that the offensive power, the supreme importance and the wide scope of the Navy as it exists to-day must be considerably curtailed by the development of aviation. The value of the capital ship becomes problematical. The maintenance of big naval bases on our eastern coasts is a dangerous policy if European war is still to be anticipated. The value of the submarine must be enhanced; it is the only seaborne craft than can hope to defy aerial attack. In fact, aviation has become, par excellence, the weapon of offence. No war can be won except by offensive action; and consequently the Navy and the Army must give way to the new arm in national importance, and in their claims on national expenditure.

Co-operation between the Services

Finally, the amphibious nature of aviation will demand a greater measure of close co-operation between the Navy, the Army, and the Air Service than ever before. This can only be obtained by expert executive, control in war. Such an organisation cannot be extemporised with efficiency, and so should exist in peace. This leads to the fact that the institution of a Minister for War and a War Staff with control over all three services is a logical development for the future. By such means, the three services can be developed to meet the demands of a really co-ordinated scheme of national defence, and the funds available for armaments can be apportioned according to the importance of the activities of each arm demanded by this scheme. The Power that neglects the Air in order to maintain the strength dictated by preconceived ideas for land and sea, will assuredly lay herself open to attack and defeat.

32. 1934

Foreign Navies
Captain E. Altham RN

It is now strikingly evident that the sole results of the Washington and London Treaties, and the prolonged Disarmament Conference, have been to cripple the British Navy and to impel every other Power with any pretensions to a fleet, except Germany, to build as they have never built before. Even Germany is straining at her bonds, and threatening to increase her Navy beyond the restrictions of the Peace Treaty. The years of deliberation at Geneva on disarmament have produced no agreement which would tend to check the growing threat to our sea security, nor does the impoverished state of their national finances prevent foreign nations from continuing to increase their fleets. Indeed, the United States, in the throes of a financial cataclysm, actually adopted a huge naval construction programme as one of their measures for promoting national recovery.

There is every indication that, on the expiration of the London Treaty in 1936, Japan will claim "parity in principle" in regard to naval strength as compared to the British Empire and the United States. It has been stated categorically that she will do so by a "spokesman" for her Foreign Office, while M. Naotake Sato, her delegate to the Disarmament Conference, opposed the inclusion in the proposed draft armament treaty of any extension of the Washington and London agreements, in order that their provisions should not be carried on after the expiration of the latter. A claim by Japan for naval parity will certainly meet with very strong opposition in America, while it is not likely to be viewed favourably by Britain. This is not the least of the many thorny questions which will arise at the international naval conference due to be held in 1935, unless by then the Disarmament Conference has arrived at any agreement which would make a naval meeting unnecessary—a highly unlikely contingency in view of the situation at the close of 1933. The United States show no signs of agreeing to reduce battleship tonnage from the 35,000 of the Washington Agreement to the 22,000 proposed by Britain; nor do they seem disposed to reduce the individual tonnage of their cruisers. Japan desires the abolition of aircraft carriers; but, again, the United States disagrees. Britain, the United States, and Italy support the total abolition of submarines; Japan, France, and a host of smaller naval Powers regard them as indispensable.

Should the termination of the London Treaty find such a diversity of views and claims as to make any further agreement between the Powers regarding limitation of various classes of men-of-war impossible—a condition of affairs which may well arise—it may be that Britain will be the ultimate gainer thereby. A revival of our shipbuilding industries will go far to solve unemployment. Once free to build the ships—especially cruisers—we sorely need, and to reestablish our naval prestige, vis-à-vis any and every foreign Power, the lessons of history all go to show that there will be an automatic increase in our Mercantile Marine, with all the prosperity that this would bring in its train.

* * * * *

United States

A comprehensive statement on naval policy was issued by the Secretary of the Navy in the summer of 1933. The following are the principal points of interest:

General Naval Policy.—To create, maintain, and operate a navy second to none and in conformity with treaty provisions. To develop the navy to a maximum in battle strength and ability to control the sea in defence of the nation and its interests. To organise the Navy for operations in either or both Oceans so that expansion will only be necessary in the event of war. To support American interests, especially the development of American foreign commerce and the merchant marine. To encourage civil industries and activities useful in war.

Fleet Building and Maintenance.—To build and maintain a fleet of all classes of fighting ships of the maximum war efficiency as permitted by treaty provisions; to replace over-age ships under continuing programmes. To prepare and maintain designs for new ship construction of all types; and to make superiority in their class the end in view in the design of all fighting ships. To provide great radius of action in all classes of fighting ship.

Capital Ships.—To replace existing capital ships when treaty provisions permit.

Naval Air Service.—To develop naval aviation primarily for operations with the fleet.

Heavier-than-air Craft.—To build and maintain aeroplanes to the full complements authorised for aircraft carriers and tenders, battleships, cruisers, and marine expeditionary forces.

Lighter-than-air Craft.—To maintain as necessary the rigid airships now built and building, to determine their usefulness for naval and other Government purposes, and their commercial value. To build non-rigid airships only for training purposes.

Conversion.—To maintain detailed plans for rapid acquisition and conversion of merchant vessels to naval use in time of emergency.

Fleet Operating.—To organise forces afloat so as to obtain maximum administrative efficiency, tactical and strategical flexibility and mobility, decentralisation and unity of command. To assemble the United States fleet for a period of not less than two months at least once a year. To keep in commission, fully manned and in active training, a maximum number of fighting ships. To exercise economy in expenditure compatible with efficiency.

Shore Establishments.—A system of outlying naval and commercial bases suitably distributed, developed, and defended is one of the most important elements of the national strength. To retain only those shore stations that would be of use in war. To further the development of outlying bases in the Hawaiian Islands and the Canal Zone. To further the development of two main home bases on each coast.

Personnel.—To maintain the personnel at the highest standard and in sufficient numbers to meet the requirements of naval policies. To assign officers to duty in foreign countries to broaden and perfect their professional education. To retain a reasonable excess of petty officers over those required for peace-time operation of the Navy. To avoid frequent shifting of personnel.

Information.—To acquire through naval and other agencies accurate information concerning the political, military, naval, economic, and industrial policies and activities of all countries. To provide protection against espionage. To keep the public informed of the activities of the Navy compatible with military secrecy.

As a statement of standards and ideals for which it is intended to work, the above cannot but command admiration, and such a robust exposition of policy may well merit imitation by our own authorities.

* * * * *

Japan

In spite of arguments and appeals by the League of Nations, Japan is pursuing a virile and independent course, and establishing herself as the predominant Asiatic Power. Bent on making her expanding interests no less secure on the sea than on the mainland of the continent, she is jealously watching the growth of a great fleet on the opposite side of the Pacific, and straining her resources to match it within the limitations at present imposed on her. Under revised regulations for the General Staff of the Navy, issued in September, 1933, the Minister of Marine remains responsible for administration, but the staff is given increased power in regard to policy and can convey the Imperial instructions to the commanders without requiring the sanction of the Minister.

The position as regards the original 1931–6 building programme towards the close of 1933 was:

Building.—2 8,500-ton cruisers: the Mogami and Mikuma; 6 first-class 1,400-ton destroyers: the Hatsuharu, Nenoi, Wakaba, Hatsushimo, Ariuke, and Yugure; 3 590-ton torpedo boats: the Manazuru, Chidori, and Tomazuru; 5 submarines; 3 minelayers; 2 mine-sweepers.

Projected.—2 8,500-ton cruisers; 1 minelayer; 6 first-class destroyers; 1 torpedo boat; 4 submarines; 4 minesweepers.

The small aircraft carrier Ryujo was commissioned at Yokosuka on May 9th, 1933.

Originally it was intended that a second naval programme should be imitated in 1934–5; but the Naval General Staff is urging that this should be started at an earlier date. It is represented that Japan's commitments in Manchukuo and the gravity of her international relations make it necessary to accelerate ship construction. According to a Press report from Tokio, the new programme would make provision for thirty-three vessels to be completed in three years at a cost of £31,000,000.

It includes: 2 8,500-ton cruisers; 2 10,000-ton aircraft carriers; 1 5,000-ton minelayer; 6 submarines; 14 1,400-ton destroyers; 4 torpedo boats; 4 submarine chasers.

This programme would be entirely in accord with the London Treaty, and it has been officially stated that no plans have yet been made for construction after the expiration of the Treaty. Officers and men contributed to a collection made on Navy Day—the twenty-eighth anniversary of the battle of the Japan Sea—the proceeds of which will be devoted to national defence.

Naval Bases

As part of her general naval policy, Japan is desirous of consolidating and developing her strategical position in the Far East. Port Arthur, which guards the Gulf of Chihli and communications with North China, has been formally constituted a minor Naval Station. Rear-Admiral Tsuda, who previously commanded the Second Foreign Service Squadron, was appointed to that base during the summer of 1933. The ships of his squadron are now included in the Port Arthur Command. One of the principal objects of the base is to afford repair facilities for vessels in Manchurian waters.

Japan having given notice to withdraw from the League of Nations, the question has arisen as to her position in the various Pacific islands mandated to her after the War. She maintains that her right to these islands, which were originally German property, is derived from secret agreements made during the War, effect to which was given in Art. XXII of the Versailles Treaty. This vested the mandate in the Emperor, and therefore she is not prepared to surrender any part of these territories to any other Power. In spite of the fact that both the League Covenant and the Washington Naval Treaty prohibit the fortification of the islands, or the establishment of naval or military bases there, and that it has been authoritatively announced that there is no intention of departing from these conditions, the Navy Department has made it known that they regard the islands as being of great value to Japanese naval strategy. Commercial bases, which are necessarily potential naval bases, are being developed in selected positions. The Vice-Minister of the Overseas Affairs Department has stated that, although the South Sea Islands are held "nominally" under League mandate, they have but "little connection with the Geneva organisation."

Japan welcomes the independence of the Philippines as a manifestation of American policy "to leave the peace of the Orient to orientals; a trust in which she regards herself as the leader."

At the beginning of 1933 Admiral Minco Osumi succeeded Admiral Keisuke Okada as Minister of the Navy, the latter having resigned on account of ill health.

Naval Air Service

The Minister of Marine has stated that Japan is now building aircraft equal in every respect, if not superior, to those constructed by foreign Powers, and that the Navy was doing its utmost to develop its air arm. He was strongly opposed to a separate Air Ministry, which, in his opinion, would "be more of an encumbrance than an advantage."

The proposed new naval programme includes an addition to the Naval Air Service of five flying-boat squadrons, at a cost of £9,000,000. Some three years ago Japan purchased the licence to build large Short K.F.1. flying boats. The last remaining airship in commission has been withdrawn from service, and it has been decided to abandon lighter-than-air craft for naval work.

* * * * *

33. 1935

Japan and Her Navy
Commander S. Tagaki IJN

The people of Japan have an inherent and ever-present fear of outside interference as the result of their experiences following such successive events as the negotiations they were obliged to enter into with other countries by the use of armed force at the time of the Restoration: the Sino-Japanese War, the interference of three countries on its conclusion; the Russo-Japanese War of 1904–5, and the difficult problems in the Pacific and Far East following the Great War. It is not to be wondered at, therefore, that the London Conference placed the Government of Japan at that time in an extremely awkward position.

During the ten years following the signing of the Washington Treaty the international situation underwent remarkable changes, and great strides were made in the science of warfare, particularly in regard to aircraft. It was the general opinion that the Washington Treaty, which was entered into when circumstances were anything but normal, was not suitable for the subsequent altered conditions. It was therefore strongly felt in Japan that the Treaty should be amended to meet these.

Nevertheless, the Japanese Government, being extremely anxious for a disarmament agreement to be concluded and to avoid a new race in naval construction, not only decided upon the continuance of the Washington Treaty, but went so far as to agree to the terms of the London Treaty, which was much more unfavourable to them in that it greatly restricted the construction of auxiliary naval craft, even if for a short period only. This created intense opposition in Japan, where the people's main thought was that the security of their country had thus been endangered for the sake of the financial policy following Japan's return to the gold standard.

In the national outburst of indignation which followed, several Japanese statesmen and financiers were murdered, whilst others were threatened with a similar fate, and for the first time in seventy years Japan experienced a state of internal dissatisfaction and anxiety. The following incidents in the history of Japan may be helpful in making more clear to those who take an interest in Japanese affairs the feelings of our people at that time.

* * * * *

Naval Plan following the London Treaty

From what has already been written, no difficulty should be experienced in understanding and appreciating the strong opposition raised in Japan to the terms of the London Treaty.

After the signing of the London Treaty the Japanese Government was in an extremely difficult position in its discussions with the Privy Council as well as in Parliament; whilst the Government party, though very strong in numbers, was forced to make concessions to the opposition in order to placate them and the people of Japan, and at the same time outline its policy for nullifying the deficiencies from which Japan suffered as a result of the Treaty. Moreover, the Minister of Marine, who had been one of the delegates to the London Conference, was forced to resign before the next meeting of Parliament.

In such an atmosphere was the first urgent naval plan of 1931 prepared. Even so, at that early stage it was condemned as being inadequate, and the new Minister of Marine had to announce that a new plan was necessary and would be prepared immediately.

The first programme was as follows:

	Type	No.	Tonnage
(a)	Restricted vessels:		
	B. Class Cruisers	4	33,800
	Destroyers	12	16,536
	Submarines	9	11,700
(b)	Unrestricted vessels:		
	Minelayers	4	6,150
	Torpedo Boats	4	2,140
	Submarine Tenders	1	10,000
	Other Craft	8	3,600
		42	83,926

Japan's Position in the Far East

The Manchuria incident which occurred in 1931 gave rise to numerous disputes which were felt throughout the whole world: most of them being fostered by abstract jurisprudence, or on account of countries interested each having its own particular axe to grind and being sadly misinformed as to the true state of affairs in the Far East, especially those directly concerning Japan; or being misled by demonstrations which were not based on the actual state of affairs. To judge without knowledge of the subject is at all times a dangerous undertaking. Rarely, if ever, is the decision reached in consonance with the matters in issue. Japan, though standing well out in the Western Pacific, is relatively but a short distance from the mainland of Asia, and in consequence must feel the repercussions of every movement happening in Manchuria, China, and Siberia, in the same way that England is influenced by the smallest political and industrial movement taking place in the countries bordering the English Channel and the North Sea.

* * * * *

Even after the 1917 revolution, the menace of the power and influence of Russia remains unaltered so far as the security of Japan is concerned. As a matter of fact, Outer Mongolia, with an area almost seventeen times larger than Great Britain, is under the complete domination of the U.S.S.R., and has reached such a state that even China herself no longer is permitted any voice in the government and military affairs of her own territory. Having suffered from bitter experience within living memory, it is only natural that Japan should look with some apprehension on the presence of the greatest forces in the world, all consumed with the fever of revolution, gathered along the Russo-Manchukuo frontiers. From her geographical position Japan is largely dependent upon foreign supplies and cannot allow the slightest interruption to her communications with the mainland. Consequently, to maintain uninterrupted contact and to protect her interests there she has to rely upon an efficient Navy.

According to reports from reliable sources, Russia has for years planned to have a large standing submarine squadron at Vladivostok, and even to-day it is known that there are in commission there several powerful 1st and 2nd class submarines, whilst amongst the numerous aircraft operating in the Far East are included a number of powerful long-range bombing planes ready for use.

Thirty-one per cent of the total Japanese imports in 1932 were from Asia, so it is only natural that Japan should observe very closely any movement likely to menace directly her communications with the mainland, which is regarded as of the gravest importance not only to her industries but to the very existence of the country. When Japan joined the League of Nations, created by the earnest efforts of President Wilson, she entertained no doubt that America would also join. As it happened, however, she did not do so, and Japan, left to bear the pressure from the two great countries outside the League—America and Russia—had to do the best she could, considering the feelings of even countries situated on the other side of the globe, to meet the troubles continually arising on the mainland.

Furthermore, Japan accepted the disparity ratio of 8:5 in naval strength laid down at the Washington Conference on the condition that the defences of the Pacific would be limited; but soon after the ratification of the Treaty, the fortifications of Hawaii and Singapore were strengthened under new schemes to such an extent as to render them almost impregnable bases of great strategic importance. Admittedly these two places are outside the areas covered by the Treaty, but to take advantage of that agreement to increase their fortifications round the area where peace has been promised, is certainly not the way to promote a friendly spirit among the Powers concerned.

* * * * *

The Second Naval Supplementary Programme

The explanation already given relates to only a part of the very complex situation in Eastern Asia, but it will easily be realised that the importation of foreign arms and ammunition by a nation whose militarists are continually in dispute, constitutes a grave danger to the peace of neighbouring Powers and is bound to create a feeling of serious insecurity and irritation. A matter which to a distant country may seem as a happening to another world, may, to a neighbouring country, be an occurrence seriously affecting her fate.

The export by European and American countries to the Far East of such warlike materials as machine guns, revolvers, ammunition, smokeless powder and the like, under such an innocuous label as "Kerosene," may not be contrary to any treaty. In the continent of China, with an area almost as large as Europe, there are many causes which keep the country in a continual state of unrest. Adding fuel to what may prove a disastrous conflagration really cannot be reconciled with the principle of the "Open Door" and is lamentable having regard to the peace of the world.

The situation in the Far East has changed considerably since the 1931 incident, and Japan, unfortunately, as a result of her views and those of other countries failing to be reconciled in what she considers to be for her a matter of life or death, has been left to endeavour alone to maintain peace in the Far East. Consequently, in the spring of 1934, the Diet was compelled to proceed with the prearranged naval programme, the so-called Second Naval Supplementary Programme, and the following recommendations were passed:

(a) Vessels under Limitation:

Type	No.	Tonnage
Aircraft-carrier	2	20,800
B. Class Cruisers	2	17,000
Destroyers	14	19,600
Submarines	4	7,480

(b) Vessels Exempt:

Type	No.	Tonnage
Oil Tankers	2	30,000
Seaplane Tenders	3	30,000
Repair Ships	1	15,000
Torpedo Boats	16	9,600
Total	44	148,680

Also miscellaneous small vessels.

Changes in Mode of Living in Japan

When Japan commenced her association with the outside world her population was 30,000,000, 80 per cent of which was the farming class. Sixty years later the population had doubled, yet the farmer class accounted for only 48 per cent. Owing to the mountainous nature of Japan, only 16 per cent of the land is fit for cultivation, almost the whole of which is now so employed. Because of the nature of the country and its population having reached saturation point, conditions of life had rapidly to be altered, and thus the people of Japan became industrialised. To meet these quickly changing conditions, resources from abroad had to be called upon, with the natural sequence that foreign markets for their products had to be found. Whereas previously Japan had been forced to associate with foreign nations, matters were now altered, and all over the world countries are closing their doors to the Japanese people and to Japanese products.

PERCENTAGE OF VARIOUS MATERIALS WHICH JAPAN HAS TO IMPORT

Material	%	Material	%
Cotton	100 per cent	Tin	78 per cent
Wool	100 ,,	Crude oil	74 ,,
Rubber	100 ,,	Soya bean	90 ,,
Nickel	100 ,,	Dye-stuffs and chemicals	68 ,,
Sugar	97 ,,	Pig-iron	57 ,,
Timber	96 ,,	Zinc	49 ,,
Heavy oil	93 ,,	Wheat	46 ,,
Lead	93 ,,	Leather	45 ,,

The principle of the "Open Door" can be maintained by the use of Dreadnoughts, but the doors of the world to markets and immigration, protected by customs barriers and immigration laws, cannot be opened by international law. Moreover, even in Manchuria, where Japan had treaty rights, the "Open Door" was almost to be denied her. But the relations between Manchuria and Japan were clearly understood by the Japanese people, and because of the great sacrifices made by Japan in Manchuria, and in view of her established position there, it is quite impossible for Japan to withdraw from Manchuria. It is not merely a question of dignity nor of prestige that Japan cannot withdraw from Manchuria; rather is it a matter of life and death for 90,000,000 people.

It is said that Japan has territorial ambitions as the result of her Imperialistic designs; but this is not so. It does not necessarily follow that greater territory will provide for the needs of her people: rather is it probable that such increase may result in intolerable burdens being placed upon the finances of the country. As a matter of fact, all that Japan really needs and earnestly desires are trade and security. Japan cannot therefore but be concerned about her Navy as a means of assuring security in the Far Eastern seas, and for the protection of the world-wide trade now necessary for her very existence.

Japan and Disarmament

Those who, whether deliberately or otherwise, failed to take account of the foregoing facts, pointed to Japan, after both the Washington and London Conferences, as the ringleader in the armament race. But when the relative nature of armaments is taken into consideration, it appears to be most illogical and the height of folly to charge a nation which has been shackled by treaty to a hopeless ratio, with the offence of being the chief instigator of the race in naval construction. To all projects looking towards the reduction or limitation of armaments Japan has always given unstinted co-operation; even after she was obliged by force of circumstances to give notice of withdrawal from the League of Nations, she pledged her continued support to all international undertakings calculated to enhance the cause of world peace. And that pledge she is now carrying out in the strictest of good faith, just as she has lived up to every obligation assumed under the existing naval treaties.

In an article contributed to a recent number of the American magazine "Foreign Affairs," Admiral Pratt stated: "Since we are not closely surrounded by real or imagined dangers, we look upon disarmament with a broad, generous outlook. ... We have made one serious error, however; and that is, having made arms agreements in the past, we have not lived up to them. ... Then when we find we are wrong we are forced to repair the damage. ..."

Admiral Pratt's observation is one which deserves our careful thought. Navies are not built, in practice, as a means of surveillance of smuggling ships or for other similar purposes. The inherent function of navies is to defend against dangers that threaten from overseas. It is, therefore, not the length of coastline or the number of seas which a country touches that determine the size of its navy. Rather is it the degree of danger by which a country might be threatened which must determine whether a navy shall be large or small; and it goes without saying that a country facing a difficult situation must always be equipped with a force adequate to cope with such a situation. Fortunate indeed is the country that can lay claim to a powerful navy simply to satisfy its desire to order the world, notwithstanding that it can be free from all external dangers.

The general public antipathy towards military expenditures should always be allayed in accordance with the forms and degrees of development of industries in the various countries. In a country so rapidly modernised as Japan, whose heavy industries are still in the initial stages of development, maintenance of the Navy plays an extremely important rôle in the wholesome development and prosperity of domestic industries. This fact becomes all the more obvious when it is considered that a large part of the military supplies in Japan is produced by small-scale factories distributed throughout the length and breadth of the Empire. How to save these manufacturing industries, virtually driven to the wall by the world-wide economic depression and the extreme retrenchment policy adopted by the Japanese Government round about 1930, became a question calling for urgent solution. The possibility of a suitable naval replacement programme to ease this industrial situation was naturally thought of; but this channel of relief was most effectively blocked by the restrictive provisions of the London Naval Treaty. It may thus be said that the London Treaty has been, in a sense, a serious blow to the industrial circles of Japan.

A Treaty Abreast of the Times

The argument advanced in some quarters seems to be that the authority of the existing naval treaties is so absolute that they should be maintained in perpetuity, wholly unaffected either by the passage of time or changes in circumstances. But that would be tantamount to warning independent Powers against the entanglements of international agreements. The existing naval agreements were doubtless based on the conditions prevailing at the time they were concluded, provision being made therein for steps to be taken for their abrogation when the treaties themselves should no longer be in keeping with changed conditions. It would, indeed, be impossible for any nation to enter into a treaty which imposes an unalterable and permanent obligation. In this connection, the utterance of Mr. Ramsay Macdonald, "Every treaty is holy but no treaty is eternal," is enlightening. This may be considered as the expression of an eternal truth.

There are some, also, who would point to Japan's desire to have the existing naval treaties abrogated or revised as presaging Japan's designs on sea areas outside the Orient. Nothing could be more groundless or absurd! The Japanese Navy has never entertained the idea of action in distant waters—its sole mission being the defence of home territory. It is difficult to understand why a Navy, upon whose shoulders rests the very existence of the nation, must be content with a strength distinctly inferior to that of navies whose chief concern is the defence of colonial possessions or trade routes.

The denial of the right of equality in national defence among civilised nations to-day is an injustice which can neither be explained away nor defended. Especially is this true at a time when even the provisions of the Versailles Treaty relating to military restrictions are being reconsidered with a view to their revision. In the past Japan acquiesced in the allocation of an inferior strength because of her lofty desire to facilitate the work of disarmament and to contribute thereby to world peace. But her sacrifices have all been in vain. Far from seeing their expectations materialise, the Japanese people have only suffered a further injury to their sense of security; and it is now clear beyond all doubt that the naval strength allowed by the past treaties is wholly inadequate for the maintenance of order and stability in the Far East.

Nowhere in history can we find an instance of the civilised nations enjoying peace for any great period as the result of one race being subjected to unequal treatment by another. In view of the inherent nature of naval armaments, the primordial need is to have all nations assured of their national security. All offensive armaments should be abolished in favour of defensive armaments; and a common limit should be fixed, within which all nations would be free to build to the point which they consider adequate in the light of their respective circumstances and needs. In other words, little contribution can be made to the cause of disarmament unless the nations can enjoy equality of security and autonomy in respect of national defence. A treaty which can have the result of enabling a country, for the purpose of acquiring markets abroad, to menace the very existence of another country located thousands of miles away cannot be considered as fair and just. Great Britain and the United States to-day possess 170 per cent of the naval strength of Japan; and yet there are some who contend that Japan's naval strength is excessive. Unless there is readiness on the part of nations with superior navies, or which face few difficulties or dangers, to make voluntary sacrifices, the lesser Powers will be left in a state of such constant restlessness and insecurity that whatever benefits might otherwise accrue from disarmament would be lost.

Conceptions of morality and justice vary with time, place, and peoples. If one race assumes the position of arbiter of world peace and morality, and attempts to impose its decisions on other races, if necessary even by the use of force, the world will never know respite from chaos and disorder. It is indeed a great thing for the world, as Senator Borah once said, that different peoples, different civilisations, and different political ideas can exist side by side.

34. 1935

The Fleet Air Arm
"Volage"

A decade has now passed since the creation of the Fleet Air Arm. The history of its development during these years shows that progress has been sustained in technique rather than in numbers. There has been no evidence of haste to reach a level of strength comparable with that of the other principal foreign naval Powers. But the omission has, *ipso facto*, permitted a greater deliberation in the choice of aircraft type and design, and in the development of ship equipment connected with the launching, flying off, and deck-landing of aeroplanes than would in all probability have been possible otherwise.

During the course of 1934, a general change in the policy of the Government with regard to our national defences has been made evident. Expectations of disarmament by foreign Powers, in harmony with that of Great Britain, having been disappointed, a claim upon public interest in the national safety has been made on behalf of our air defences. Although no specific mention was made of the Fleet Air Arm in the general announcement made by Mr. Baldwin in July last on the Government's new air defence policy,* a significant contribution towards the Navy's deficiency in air strength has been included in the 41 squadrons to be added to the Royal Air Force. It is intended that $3\frac{1}{2}$ of those squadrons are to go to the Fleet Air Arm.

Deficiencies

As in the case of all arms of the fighting services, there are two standards by which strength may be assessed. The first is by the comparative standard—the ship-for-ship, or plane-for-plane comparison between the British forces and those of a potential enemy. The other is by the absolute standard, i.e., one based on minimum requirements.

In respect of the number of ships required, the absolute standard is a difficult one to apply. In respect of ship-borne aircraft, however, the

*House of Commons, July 19, 1934.

problem is much simpler, given a fleet of already determined strength, because aircraft have now become a finite item of fleet equipment. The scale of air strength for a given fleet is a mathematical factor.

It is an unfortunate fact that during the last ten years, by either of these standards, British naval air strength has fallen to a very low point. By comparison with that of foreign naval Powers, it is overwhelmingly out-numbered, and the unfavourable ratio is growing every year. By an assessment of "absolute" ship-by-ship requirements it is also lamentably in default.

These deficiencies may be shown in detail under these categories:

(a) Shortage in the aircraft carriers.
(b) Lack of all or part of the equipment which would normally be carried in our existing capital ships and cruisers.
(c) Those further numbers for the operation of our present fleet, which cannot be embarked before the older cruisers are replaced by more modern types.

A careful analysis of our present Fleet would indicate that under these headings the Fleet Air Arm is shown to be at barely half its proper strength. Official information shows that there are at present 132 carrier-borne and 27 catapult-ship aircraft now in service. This year's estimates provide for a further 6 in each category. Ten capital ships still carry no aircraft at all, and the remainder but 1 apiece. Of the 49 cruisers now in commission, 2 only, the Exeter and the Achilles, are provided with 2 aircraft each, 18 have 1 and the remainder are not equipped. These numbers should be compared to the 2, 3, and 4 seaplanes regularly embarked in the battleships and cruisers of our Japanese and American contemporaries.

A detailed assessment of carrier deficiencies is more difficult to arrive at. No naval Power is willing at the present time to divulge the full working capacity of its carriers. But some inferences may here be drawn from the fact that 48 have on occasion been embarked by a single carrier in our Fleet, while the present complements of Courageous, Glorious, and Furious are shown as 42, 36, and 38 respectively. The Eagle, smaller in capacity than these three, has a complement of but 21, which may be assumed to be the largest number which she could satisfactorily operate. For the Hermes and Argus, no special provision of aircraft is at present authorised.

The decision to make good these deficiencies in the Fleet during the next five years, simultaneously with large developments in our home-based air defence forces, has led many people to believe that the Fleet Air Arm is an integral part of our air defence organisation rather than an integral part of the Fleet. The intended increase is none the less a very welcome decision and one which has long been waited for by all naval officers. Let us hope that this step will lead to the ultimate provision of an adequate and permanent quota of aircraft in the Fleet, the need for which has long been realised by those who have studied the technique of modern naval warfare and its bearing upon our vital food supplies.

* * * * *

In the four years immediately following the War, great developments were made in aircraft-carrier design, and the technique of practical deck-landing quickly followed. As a result, new types of aircraft suitable for deck-landing reconnaissance, and carrying torpedoes, came into being. Thereupon the situation which had led to uniformity in air effort automatically changed.

The duty of aircraft with the Fleet was no longer confined to flying from the deck of a ship to combat enemy attacks by air. It now involved the use of weapons with which to anticipate or reinforce attack by the gun or the ship-borne torpedo. Those directing these attacks must have a consummate knowledge of naval tactics and the use of the gun, in order that their effect should be properly concerted and not be wastefully haphazard. The pilots of these newly employed aircraft must also know their special tactical purpose, and besides having a sound knowledge of the functions of all fleet units, they must be intimately familiar with the appearance of all types of ships, and those subtle evidences of their activity which an officer of considerable naval experience alone can appraise. It need hardly be added that no less experience is needed by an observer who has to report all he sees, and also spot the effect of gunfire. To his repertoire very accurate navigation over the sea has also to be added.

During these years of development the supply, administration, and executive control in all matters of naval aviation, except ship design and the direction of operations, still rested with the Air Ministry. In 1924 the Admiralty, having strongly represented that this anomalous state of affairs was most adverse to the proper conduct of their assigned function, were granted a somewhat greater measure of control in naval air questions, and a maximum quota (70 per cent) of naval officers were permitted to hold commissions in that part of the R.A.F. known as the Fleet Air Arm. This is the point at which the F.A.A. administration, with a few minor adjustments, has been established during the last ten years. Material and operations have made even more progress since than before 1924. As will be described in the latter parts of this chapter, the aircraft themselves have now become specialised for naval work to a greater extent than ever, and the time has definitely been reached when land aircraft can no longer be operated at sea.

Ten years of progress in the tactical use of naval aircraft has, moreover, had a large influence on warship design, as well as on naval operations. The corollary is also true. The older methods of employing aircraft at sea are being replaced by newer ones evolved by the naval staffs who can, now that a very wide infusion of flying experience into the Navy has taken place, appreciate all aspects of naval air problems. In such problems the Navy is now fully competent.

* * * * *

The standard of efficiency of naval pilots is now very high, and improvements in the technique of flying on and off carriers have resulted in more practice than before being possible under adverse conditions of weather.

Half of the F.A.A. squadrons are now commanded by naval officers. These officers have all spent many years in the specialised flying duties required for the Fleet, and in this respect they may gain more experience than their contemporaries in the Air Force whose visits for service in the Fleet Air Arm are only intermittent.

No recent changes have been made in the organisation for providing and training observers. The general introduction of multi-purpose aircraft into the F.A.A., of which particulars are given later, has led to a far greater demand than hitherto for observers. The demand for pilots has also, of course, increased for the same reason. The crowded wardrooms of our aircraft-carriers already illustrate what a generous proportion of officers this service now carries.

In the Air Force it has for some time been realised that a more rational proportion of officers to men can be reached by training non-commissioned officers to fly. A few of these sergeant pilots are, in fact, at present serving in aircraft-carriers. The Air Ministry have now gone even farther, and introduced the rating observer, who will be trained in air navigation and in other branches of the land observers' technique. He will thus be able to relieve the Air Force officer of some of his duties in the air. It would seem likely that a similar scheme might be applied with advantage to the Fleet Air Arm, so that naval ratings as well as airmen could be employed as pilots and observers to reduce the now heavy requirements of officers. The high standard of intellect and professional knowledge which is attained by the naval ratings of to-day should amply justify their being given that responsibility.

* * * * *

Training

The appointment of Rear-Admiral to the command of the aircraft-carriers was a natural result of the growing influence of the Fleet Air Arm on fleet work as a whole. The responsibilities of this post are more

DISARMAMENT AND THE AIR POWER CONTROVERSY

connected with progress in training and development than with actual tactical command. With his flag flown in a Home Fleet carrier (the Courageous) he is able to watch and direct the applications of all new technique during its tests and trials. New types of aircraft, new catapults, accelerator and arresting gear for the carriers, new weapons and new methods of using them, all come under his immediate view while undergoing preliminary trial at sea. In close contact with the Commander-in-Chief, Home Fleet, he is able to make proposals for adapting them for their best use with the Fleet. He is likewise responsible for the sea training of the Fleet Air Arm and for its co-ordination with the Air Ministry arrangements, for which the Air Officer Commanding, Coastal Area, is responsible.

Much of the training in bombing and air fighting of F.A.A. units is carried out at R.A.F. practice camps, and for this purpose complete squadrons are landed from the carriers for periods of about a month at a time.

The work of training in other fleets is directed by their own Commanders-in-Chief through the commanding officers of the units concerned, but during each spring cruise period it is usual for the aircraft-carriers of both the principal fleets to carry out a comprehensive programme of combined training under the R.A. (Air). This period is a particularly important one, since it usually includes the several exercises in which the main fleets are in opposition to one another, and in which aircraft fulfil, of course, their war rôle on the largest possible scale. It is a period, in fact, which is looked upon as the climax of the year's training.

By means of the introduction of arrester and accelerator gear in carriers, flying has been made far more independent of adverse weather conditions than was formerly possible. Consequently, more flying can be done during any given period of time when varying conditions prevail, and a higher standard of practice can be reached.

While increasing efficiency in long-distance reconnaissance and improvements in method of bombing and torpedo attack occupy a prominent place in current training, night-flying has also made much progress during the past years. The problem is a great deal more difficult at sea than it is on land, because of the obvious objections to illuminating the deck of a carrier whose location it is hoped to conceal from the enemy. Similar objections in land aerodromes are not quite so vital, since the geographical position is nearly always accurately known. Navigation is also more difficult by night at sea, and deprived of many aids.

Exercises

Prominence has been given in the English Press to recent combined exercises in which the three Services have taken part. Conclusions on the relative value of sea and air power have even been drawn in these accounts, from the evidence of unofficial eyewitnesses. In making such deductions, however, consideration ought to be given to the real nature and circumstances of each exercise, and a tendency to judge without knowledge of this has been sometimes evident.

In actual fact, few, if any, of these exercises have yet been staged to represent the actual conditions of war. Co-operation between Air Force units and the other Services is still comparatively in its infancy and, as a rule, no attempt to exert any pressure at the same relative fighting level has been attempted. All three Services have been far more concerned with accustoming each other to the arrangements for co-operation which combined work demands—with teaching each other, by elementary steps, what to see and what to expect—than with pitting their forces by tactical skill in any kind of air *versus* gun contest.

The most important recent example of the kind was the exercise carried out near the Firth of Forth in September, 1933. In this the naval movements were artificially restricted, so as to ensure a contact with the air forces before the coast was reached. Conclusions on the efficiency of coastal reconnaissance, for example, drawn from this exercise are obviously of little value. Until many further exercises under progressively more realistic conditions have been carried out, the Services are wisely refraining from making any claims of relative predominance in coastal operations. The Navy knows that it has now the air factor to contend with in carrying out its age-old responsibilities, and that this factor may be either very great or almost non-existent owing to varying circumstances of place and weather. The Air Force knows its own limitations, and to what extent adverse conditions can be counteracted by improved training and technique. These exercises are not staged to dispel illusions, but to develop co-ordination and defence.

The Yorkshire coast was the scene of an important invasion operation in September, 1934, in which all three Services took part. Home Fleet ships, including the Courageous, Army units of importance, and five squadrons of the Royal Air Force took part, but details of the exercise, or the conclusions which may be drawn from it, are not yet available. It has been officially explained that the nature of the operation was mainly that of a staff exercise.

* * * * *

The Industrial Base

BORING MACHINE FOR HEAVY GUNS.

The Turbine Erecting Pits at the Naval Construction Works of Vickers-Armstrongs Limited, Barrow-in-Furness, with turbines of over 100,000 S.H.P. under construction.

Danger in the Far East

JAPANESE BATTLESHIP NAGATO.

(From a drawing by Arthur J. W. Burgess.)

H.M. SUBMARINE OBERON, BUILT AT CHATHAM DOCKYARD.

(From "Shipbuilding and Shipping Record.")

PART V: 1936–1939
The Approach to War

Introduction	161
35. 1936	
Relative Naval Strength	163
G. H. HURFORD	
36. 1936	
United States Naval Aviation	166
REAR-ADMIRAL E. J. KING USN	
37. 1937	
The Merchant Navy as a Factor in Imperial Defence	170
E. H. WATTS	
38. 1938	
British Naval Air Progress: The Fleet Air Arm Decision	174
H. G. THURSFIELD	
39. 1939	
The Interdependence of Home Defence and Commerce Protection	178
"SECURUS"	

Introduction

The Washington and London Treaties were so drafted that, unless new agreements were made before the end of 1936, all their restrictions would lapse and unfettered naval competition could begin. Accordingly a second London Conference was convened for 1936. It was marked by significantly closer agreement between Britain and the United States but doomed to overall failure by Japan's insisting on the acceptance in principle of her right to parity, which neither Britain nor the United States was prepared to accept. Japan withdrew from the conference and declared her intention of repudiating the Washington limitations. The remaining naval powers signed an agreement which included restrictions on the size and armament of major surface ships but also contained an escape clause which allowed signatories to compensate for increases by a non-acceding power.

In 1935, during the extended preparations for the London Conference, a separate treaty was agreed on by Britain and Germany by which Germany was permitted to build up to 35 per cent of Britain's surface warship tonnage and 45 per cent of her submarine tonnage with an escalatory clause allowing this latter to be increased to 100 per cent. In return, Germany agreed to accept the qualitative restrictions Britain would present to the forthcoming conference and whatever agreements on the control of submarine warfare were reached there. Britain's motivation in making this agreement was the conviction that Germany would rebuild her navy in any event and that it was preferable to take advantage of her present willingness to accept a definite limitation. The Admiralty in particular were very aware that the Royal Navy would not be strong enough to deal with the challenge it already faced from Italy and Japan if it had also to meet massive German strength in home waters.

The recognition of the importance of integrated air power for navies was now complete. The United States Navy had successfully maintained its own control of the new element and by 1938 the dispute between the Royal Navy and the Royal Air Force had largely been resolved in the former's favour. A further development as war approached was a timely recollection of the essentiality of a strong merchant navy—a factor always forgotten in peacetime.

Recommended Reading

S. W. Roskill, *Naval Policy Between the Wars*, Vol. II. (1976).

35. 1936

Relative Naval Strength

G. H. Hurford

Since the last issue of "Brassey's Naval Annual," the denunciation of the Washington Treaty by Japan, followed by the abrogation of the Treaty of Versailles by Germany, has brought questions connected with relative naval strength prominently before the peoples of the principal sea Powers. Over a period of fourteen years, conditions had been more or less stabilised by these treaties. There had naturally been many discussions as to the manner in which this or that Power was interpreting the terms of the pacts, how far each was short of its allotted treaty strength, and so on; but there always remained the maximum which might not be exceeded without due notice. The end of the treaties would mean the removal of a steadying influence on competitive building. No doubt the world, as the First Lord said in the House of Commons on March 14, 1935, has appreciated the inestimable benefits that were obtained from the Washington Treaty of 1922. It certainly brought some measure of calm to the face of the waters. Towards its continuance, no government made greater sacrifices than that of Great Britain, which accepted lower degrees of strength than had ever been known in modern times in order to facilitate agreement with other countries.

A salient feature of British official policy in regard to naval affairs during 1935 was the admission that no response had been made to the gestures of goodwill in the form of reduced British shipbuilding programmes. Rearmament had become essential. The form it would take was left to be determined by the Naval Conference called in December, 1935, a subject dealt with in another chapter. But it was made clear long before the Conference met that the ratio system inherent in the Washington and London Treaties would not be continued. The First Lord of the Admiralty, speaking in the House of Commons on July 22, 1935, announced that "the principle of ratio has had to be abandoned. We have had to give up any idea of ratio for the future because some countries felt it wounding to their national pride to have to accept a naval strength permanently inferior to that of some other country. We, therefore, have had to abandon the principle of ratio and, instead of asking naval Powers what the ultimate strength of their navy was going to be, we have to ask, what size navy do you propose to have in 1942? That in fact is the date we have taken. As I say, we have had to abandon all ideas of ratios in trying to perpetuate the state under which we have been living under the Treaty of Washington, and we have gone in for a system of programmes."

But while 1935 saw the disappearance of the ratio principle, so far as the five leading Powers are concerned it also witnessed its reappearance in a new direction. On June 18, 1935, Germany voluntarily entered into an agreement with Great Britain that the future strength of the German Navy in relation to the aggregate naval strength of the members of the British Commonwealth of Nations should be in the proportion of 35:100. In the course of a debate on the situation created by this agreement, in the House of Lords on June 26, 1935, Lord Londonderry, replying for the Government, said that, "taking France's present naval strength at about fifty per cent of our own naval strength, this agreement affords to France, at present levels, a permanent superiority of 43 per cent over the German Navy as compared with an inferiority of some 30 per cent before the War." In all the arguments about the agreement, none went to show that the naval position of either France or Italy would have been stronger had this agreement not been concluded. The Government's view was that in accepting the German offer they had acted for the best in a situation admittedly not free from difficulties. As Lord Beatty pointed out, supposing Germany had said 50 per cent, what should Great Britain have done? We could not have stopped her. But the gesture did away definitely and completely with all possibility of rivalry on the sea between the two countries. It does not limit the number of ships we may possess, nor prevent us from building what we like. "But it does ensure," said Lord Beatty, "that there will be no competitive building programme between us and at least one country in the world, which I maintain is something to be thankful for."

The revival of the German Fleet is, however, the outstanding feature of a review of relative naval strength at the present time. It removes one of the main arguments of the advocates of further retrenchment in the British Fleet. Three or four years ago, it was possible to contend that while the Royal Navy had been reduced considerably (from about 2,000,000 tons of the 1914 level to approximately 1,150,000 tons), yet on the other hand the German Fleet, its main rival in 1914, had been reduced from about 1,100,000 tons to 140,000 tons; that the Austro-Hungarian Navy, then allied to Germany, had ceased to exist; and that the Russian Navy had also virtually ceased to exist. The situation in both Germany and Russia is now radically changed, and the French and Italian navies, particularly the Italian, are far stronger relatively to the Royal Navy than they were immediately after the signing of the London Treaty of 1930.

The extent to which Germany implements her new agreement and the speed with which she builds the new units to which she is entitled will have an important bearing on the future naval position in Europe. Her effective tonnage in battleships on January 1, 1936, was five ships of 59,120 tons, or including the Admiral Graf Spee, commissioned on January 6, 1936, six ships of 69,120 tons. Three out of these six ships, of 39,120 tons, are thirty years of age and overdue for replacement. The effective tonnage is therefore only three "Deutschlands" of 30,000 tons. The British capital ship tonnage is 474,750, of which 35 per cent is 166,162 tons. This is sufficient for Germany to build immediately five more 26,000-ton battleships, including the two such vessels she laid down in 1934. As regards cruisers, 35 per cent of the British cruiser tonnage allowed by the London Treaty would give Germany 118,650 tons. She has now six ships of 35,400 tons. The difference is sufficient to allow her to build immediately eight 10,000-ton cruisers, including the two of this tonnage which she laid down in 1935. Moreover, any increase of British cruiser tonnage would increase the proportion available for Germany.

While Germany has only built about one-fourth of her allowance in cruiser tonnage under the agreement, in destroyers she has built only about one-sixth; 35 per cent of the British strength under the London Treaty would give her 52,500 tons, towards which she has already built only twelve vessels of 9,600 tons. The balance is sufficient to allow for the construction of 26 large destroyers of the 1,625-ton type, with five 5-in. guns, now in hand. But her greatest possible effort may be in submarines, of which she was formerly allowed none; 35 per cent of the British tonnage would allow her 18,445 tons, enough for 36 submarines of an average of 500 tons each. Actually in 1935, she put in hand 28 submarines of three different types, 250 tons, 500 tons and 750 tons, and 14 of the smallest class were completed before the end of the year. Only 31 British submarines have been laid down and completed in the 17 years since the War.

* * * * *

36. 1936

United States Naval Aviation
Rear-Admiral E. J. King USN

This article has been written in response to a request of the Editor for a review of the reasons that cause the United States Navy to have complete control of its own "air arm" and also for some account of the system in use to that end. It should be understood that only the writer himself is responsible for what is here written.

The basic principle of the national policy of the United States as to its armed forces is that they are maintained solely for defence. The first principle of naval policy is to maintain the Navy at sufficient strength to support the national policies, to safeguard the commerce, and to guard the continental and overseas possessions, of the United States.

* * * * *

Aviation has become a powerful factor in naval warfare. It is considered absolutely essential that the Navy provide, train and control its air component, and that the air component be an integral part of the Navy. There is too much at stake for the Navy to take for granted that an adequate air component is any other than one thoroughly trained in the particular tasks of Fleet air work. The United States Navy considers itself fortunate in having an air arm that has grown up with, lives with, speaks the language of, and is an integral and vital part of, the Fleet.

The general policy of the United States Navy as to aviation is, therefore, that naval aviation is an integral part of the Navy itself, and the functions of naval aviation are derived directly from functions of the Navy. All the capabilities and endeavours of naval aviation are directed towards the promotion of naval efficiency and are designed to enable the Navy better to perform its functions. In naval aviation the reliance of ships upon aircraft and of aircraft upon ships, the sameness of their objectives, the consequent co-operation needed between these two essential parts of the Navy, the resultant necessity for co-ordination of their employment through one high command—all require that naval aviation, as it is conceived to-day, and as it may be expanded tomorrow, be an integral part of the Navy. That this is so is due to the wisdom, foresight, and sound views of those who have controlled the development of the Navy since aviation became a factor in warfare— and to none of these more than to the late Rear-Admiral Moffett, who was Chief of Bureau of Aeronautics from its creation in 1921 until his death in the Akron in 1933.

Aircraft have not so far supplanted any other naval type, but they have added power and effectiveness to the existing types. They constitute a striking force that has gone far towards increasing the value of the United States naval forces. The Navy is the first line of defence, and naval aviation can now truthfully be regarded as the advance guard of this first line.

* * * * *

Experience has served to confirm the essential fact that airplanes in themselves are not capable of operating as independent units over the vast areas of the open sea for indefinite periods. Therefore, the naval force cannot avail itself of the advantages inherent in air operations unless it provides, in the Fleet itself, means for carrying, maintaining, and efficiently operating aircraft in whatever area it becomes necessary to exert sea power, and unless through constant training there are developed methods for harmonious co-operation between air, surface, and sub-surface activities. Through such constant training and indoctrination the methods of co-operation have now reached the point where aircraft in United States Fleet activities are accepted as a commonplace part of the day's work. Air and surface craft work in the closest harmony, each supporting and complementing the efforts of the other in resisting attack. The mission of the fleet aircraft is to assist the battle line to the maximum in defeating the enemy. With this aim always in mind various missions for aircraft have been so worked out and the various types of aircraft so designed as to carry out the assigned and delegated missions to best effect.

The heavier-than-air operations of naval aviation embrace three distinct fields, separate as to equipment and functions, but closely related in that each depends on the other, and the Navy as a whole, for its support and its effectiveness: first, the patrol type—the "flying-boat"; second, the seaplanes which are launched from the catapults in the battleships and the cruisers; third, the landplanes which are the complement of the airplane carriers.

* * * * *

The third, and one of the most important classes of naval aviation operation, is that of carrier aircraft. This is a relatively new weapon and takes its place in the Fleet as a means of carrying on an important offensive comparable to that of the guns and torpedoes of battleships, cruisers, destroyers, and submarines. The carriers themselves are but the bases from which the planes fly off and to which they return for re-servicing. The aircraft carrier is a good illustration of the intricate problems of adapting aviation to naval needs.

Carrier aircraft have a single function, the offensive; that is, to find and to strike the enemy, both on the surface and in the air. The types of planes employed are determined by this objective. First, the enemy must be located, so scouting planes are necessary to the complement. These planes, equipped with high-powered radio, search wide sea areas in the direction where the opposing force is expected to be found. When the enemy is so located, an attack group is sent out. This consists of bombing and torpedo planes whose purpose is to damage or destroy the enemy surface vessels exactly as do the guns of the battleships and the cruisers. Finally, there are the small high-performance fighting planes whose principal mission is to destroy enemy aircraft, but which may, as an additional function, be used to strafe the lighter ships and the upper works of heavy ships of the enemy with machine guns and light bombs.

* * * * *

Aviation Personnel

In order fully to understand United States naval aviation it is important to realise that there is no separate air corps in the Navy. The naval aeronautical organisation is composed of officers and men of the regular Navy, and service in the air component is voluntary. Naval aviators are chiefly line officers who have specialised in aviation. They are not only required to be expert aviators but they are also required to keep themselves in training by actual duty in surface vessels, to perform the general line duties of their rank so that at any time they may be detailed to general line duty aboard any ship. The Navy is, therefore, not faced with the problem of finding employment for aviation personnel when for any reason their service in aviation duties terminates.

* * * * *

Summary

The foregoing explanation of the subjects of organisation, personnel, and types of planes required and possessed by the naval aviation arm of the United States Navy is intended to outline the inherent relationship which exists between naval aviation and the United States Navy. To this policy is due, in no small measure, the high state of efficiency which has been attained. To loosen the coherence of this structure in any way would most certainly lessen the efficiency which has been so painstakingly built up from the days of the inception of naval aviation.

Not only in ordinary sea operations is integral naval aviation considered of vital importance, but when it comes to overseas operation aviation is particularly dependent on the surface navy to furnish logistic requirements necessary for continued and extended operations. It is difficult to visualise how aviation could function on its own as an independent force in an overseas campaign. Certainly if such a plan were followed it would court both a considerable wastage of effort and a probable failure in the attainment of desired objectives. All plans and operations of the air forces must be directed so as to fit into the various phases of the military campaign, step by step. The fundamental idea of cause and effect behind all aerial operations is the situation on the surface—land or sea.

The surface navy must be able to depend on its air force for the accomplishment of all the various tasks within the capabilities of aircraft. This dependence must be based on known capabilities of both the personnel and material of the air force and on the principle of unity of command. These tasks are so closely allied with those of the surface force and so vital to their success that they cannot be considered separately. Modern naval operations cannot be efficiently conducted without such services acting as integral parts of a single command. Moreover, the training of these air services that are to act with the Navy must be under the continuous direction and control of the command which is ultimately to use them in war. The surface navy must have within its organisation an air component adequate in types, numbers, and capabilities to meet all the likely situations that will arise.

One vital point which must be kept constantly in mind, in consideration of United States naval aviation, is that it is an integral part of the United States Navy. All efforts in the field of naval aviation development are designed to further naval capabilities and to fit the Navy better to perform its essential functions in national defence. Naval aviation in the United States Navy has reached the position to-day where it is very highly developed and efficient, and it has become so by reason of its being developed wholly by the Navy, of the Navy, and for the Navy. Those of us who are charged with the present and future development and expansion are convinced that its continued successful operation demands that it should remain as a component and inseparable part of the Navy.

37. 1937

The Merchant Navy as a Factor in Imperial Defence
E. H. Watts

The Merchant Navy is a service as well as an industry, and its importance in war time is so great that no commercial policy should be adopted without the fullest consideration of the needs of Imperial and national defence. Most great Powers have long recognised this, and have increased the strength of their Merchant Navies by protection, reservation of valuable trades and large building and running subsidies, none of which measures can be justified economically. Yet in spite of our Imperial and national needs being so very great, and the fact that we support a population which cannot even be fed without overseas transport, the official attitude—it can hardly be termed policy—towards our Mercantile Marine has been strictly commercial; indeed, it cannot even be justified commercially.

Those who for many years have criticised the Government on this point were gratified by the passage in King Edward VIII's speech at the opening of Parliament in which it was stated that the Government "was deciding what measures are required to secure the maintenance of a Mercantile Marine adequate for the needs of the country." Nevertheless, such an assurance would have been more timely four or five years ago, as unfortunately it is now no longer a matter merely of maintaining the Merchant Navy, but of building it up and of providing an adequate supply of cargo-carrying ships. Does this mean that the British Mercantile Marine requires large subsidies to enable it to regain its lost ground? In the initial stages, the answer is undoubtedly "Yes"; and more and more public money must be earmarked for this purpose until such time as the Merchant Navy is once more in a position to earn sufficient income to offset the losses and short-earned depreciation since 1929.

The present sharp rise in freights has made shipping a profitable venture once more, but the profits are not confined to British shipping. The British shipowner must set aside most of this profit to make good

THE APPROACH TO WAR   171

past losses and the depreciation for which no provision could be made during recent years; but many of his competitors are able to set them aside in preparation for the fight which will once more be renewed when the inevitable fall in freights takes place. The history of the last thirty years will once again be repeated. Unprotected British shipping will emerge from that slump far weaker than when it entered it, and its competitors will have absorbed more British ships and more British trade.

This chapter is confined solely to the British cargo-carrying fleet, in which term are included tramps, cargo liners and passenger liners with appreciable cargo space. It is by cargo ships alone that the population of this country can be fed, and they must bring in the enormous quantity of raw materials essential to the successful prosecution of war. To-day the cargo fleet is utterly inadequate for the responsibilities it would be called upon to shoulder in war time, when the necessity for adequate communications and transport ranks above any other.

It might be thought that, as we now have over 1,000 less cargo ships than in 1914, this fact was self evident; but that is apparently not the view of the Government. Two years ago, the tramp section of the Merchant Navy asked for a subsidy to enable it to carry on. Two million pounds were granted, and simultaneously the President of the Board of Trade produced his "Scrap-and-Build Scheme." Under this, to qualify for Government financial help in building cargo ships, two tons must be scrapped for every new ton constructed. In other words, it was evidently thought advisable to reduce the number of our cargo vessels drastically, and in spite of experience in the War it must have been supposed that the remaining "pocket" fleet would somehow or other prove sufficient. It therefore seems advisable to attempt an estimate of the size of our Merchant Navy necessary to make it self-sufficient and capable of fulfilling its duties as a vital factor in national and Imperial defence.

It is estimated that, allowing an average of three "round trips" a year for the various classes of cargo ships under the British flag, cargo space equivalent to 52 million deadweight tons' capacity is available to carry our annual requirements of 50 million tons of food and raw materials. Superficially, this may seem sufficient, but allowances have to be made for a number of factors:—

First of all, bunker space must be reserved, and an average of 10 per cent is a reasonable minimum. This, translated into deadweight tons, gives an immediate shortage of 3 million tons, or 150 ships.

Secondly, the provision of adequate bunker supplies throughout the world must be organised. The experience of the last War showed that some extra ships have to be detailed for this work, and allowing a minimum of 60 vessels, the total shortage becomes 210 ships.

Then nearly half of the Merchant Navy, including one-third of the cargo ships, burns oil fuel, and it is almost impossible to estimate what provision will be made for their supplies. The British tanker fleet can carry sufficient for the Royal Navy, but there will be little left over for the other fighting services, quite apart from the needs of road transport and power stations. Presumably, neutral tonnage is to be relied upon to fulfil these requirements, and provision will have to be made to supply our cargo vessels. The position would be desperate indeed if our oil-burning merchant ships proved useless owing to lack of fuel.

The next factor to consider is the measure of success likely to be achieved by enemy commerce raiders at the beginning of a war, during the period when the Royal Navy is gaining control of the situation and perfecting its war organisation. This may take about three months, and it would be surprising if less than 50 ships were lost before they could assemble for protection by the Royal Navy. This brings the shortage up to 260 ships.

Here I should like to cite a concrete example of the equipment of a first-class Power to carry out commerce raiding, should the need arise, and one which will serve to show how far we are behind others in our defence preparations. The Japanese are constructing a series of 20-knot, 10,000-ton deadweight, oil tankers. From a commercial point of view, these ships cannot be justified. The highest economic speed for a modern tanker is 12 knots, but the extra 8 knots would, obviously, be invaluable in war time. It would be possible to organise a striking force of warships, incorporating the 20-knot tankers, which could steam with the fleet and make it independent of bunker ports. On the other hand, these tankers might themselves be utilised as commerce raiders, and by burning the fuel carried in their 10,000-ton capacity they could steam round the world, independent of any port of call, and, indeed, untraceable by ordinary naval standards.

Further allowance must be made for submarine activty. Although anti-submarine devices have greatly improved, some measure of success is likely to be achieved, and surely it would be no overstatement of the case to allow 50 ships being lost from this quarter. The shortage is now 310 ships.

It is pure conjecture to estimate what the casualties may be from enemy aircraft, but the bulk of our grain and cargo discharging facilities are situated on the East Coast, and to reach them merchant ships would be extremely vulnerable to attack from the Continent. The losses might be as high as 150 vessels in the first year. This brings the shortage up to 460 ships.

What is perhaps the greatest danger to our food supplies is damage to ports by air attack. If the grain-discharging facilities of the Port of London were put out of action, incalculable delay would be involved in landing cargoes at other ports in the country. Such a dislocation would create an immediate need for at least 200 more ships.

The last point I have to make is that, as it stands to-day, the British Merchant Navy is not a balanced fleet. It has been driven by foreign competition from the smaller type of ship into the larger one. As an example, British ships are, for the most part, too large to enter into the grain trade from the Danube. During a war, we could not supplement our grain supplies from this area without utilising some foreign ships for their transport.

In view of all these circumstances, the final estimate of the shortage in our cargo-carrying fleet is approximately 700 ships. This is unquestionably optimistic, and some shipowners would allow 1,000, even 1,500 as being nearer the figure.

It is obvious, therefore, that if the Merchant Navy is to be of any value as a defence factor it must be considerably increased and strengthened.

* * * * *

38. 1938

British Naval Air Progress: The Fleet Air Arm Decision

H. G. Thursfield

In the 1937 issue of "Brassey," an account was given of the controversy which had long been going on between the Navy and the Air Force on the question of the organisation of the Fleet Air Arm and of other air units which may operate at sea in war. The controversy is now at an end. In March, 1937, Sir Thomas Inskip, Minister for the Co-ordination of Defence, announced that he was undertaking a systematic enquiry into all the factors involved in the question at issue between the two Services, with the assistance of the Chief of Staffs' Sub-committee of the Committee of Imperial Defence and the other authorities concerned. The enquiry lasted some months, and as its result the decision of the Cabinet was announced by the Prime Minister in the House of Commons on July 30 in the following words:

> The proposals the Government have had under consideration refer to two classes of aircraft. The first class includes all aircraft borne in ships of the Royal Navy. These are known as the Fleet Air Arm. They are under the operational control of the Admiralty, but as part of the Royal Air Force they are under the administrative control of the Air Ministry. The second class includes shore based aircraft employed in co-operation with naval forces. These are under the operational as well as under the administrative control of the Air Ministry.
>
> Under one proposal which has been before the Government the Admiralty would in each case have been given both the administrative and the operational control, and the whole of the personnel would be naval. The Government have, however, decided that in the case of the second class, namely, shore-based aircraft, which term includes flying-boats, there shall be no alteration in the present system. In the case of the Fleet Air Arm the Government consider that these ship-borne aircraft should be placed under the administrative control of the Admiralty. The necessary steps to give effect to this decision will be taken.
>
> The change can only be carried out gradually, and with the fullest co-operation between the two Services. The same close co-operation between the Services is indeed vital in the whole strategic field where both ships and aircraft are concerned, and I am happy to give the assurance that this co-operation will be forthcoming without reserve.
>
> I wish to make it plain that the decisions which have been reached do not reflect upon the present condition of the Fleet Air Arm, where a keen and efficient service has been built up, but have been reached because the Government believe that the lines now laid down will be the most satisfactory arrangement for the future. I also desire to express the

Government's appreciation of the untiring efforts of the Air Ministry to make a success of the system for which they have been responsible. I hope that these decisions which the Government have reached after full enquiry will be accepted in every quarter as a final and satisfactory settlement of a prolonged controversy which it is in the public interest to close.

In reply to questions, the Prime Minister also stated that, in principle, the Fleet Air Arm would have the shore establishments it needed, which would be under the Admiralty.

Discussion between the Admiralty and Air Ministry in order to evolve the measures necessary to give effect to the Government decision has been proceeding ever since the latter was given. Up to the end of the year, no measures had been made public, and in the meanwhile all former orders and organisations remain in force. It is fairly obvious, however, what the arrangements to be adopted must be.

Certain aerodromes and air stations, sufficient to accommodate the squadrons of the Fleet Air Arm when disembarked and to provide for Fleet Air Arm training will be transferred to naval control, so that the Fleet Air Arm will remain at all times under the same authority.

The preliminary flying training of naval officers of the Fleet Air Arm—and, presumably, ratings—is expected to remain under the Air Force, but it is likely to be combined with more "naval air" training than has been possible up to now, and the liaison between the two Services in this respect should be improved. The dual control of the Fleet Air Arm will come to an end, and with it, presumably, the system whereby naval officers are given rank in the Air Force, since there will be no further need for the latter.

For some years, however, until the Fleet Air Arm can overtake arrears of expansion, it seems probable that the services of a number of Air Force officers will be needed by it, just as they are needed now. There should, however, be no difficulty in arranging for this. The officers are serving there now, embarked in H.M. ships and, like every other person, subject to naval discipline while so embarked. The only difference will be, presumably, that when the dual control comes to an end, any Air Force personnel lent to the Navy will remain under naval discipline at all times when so seconded, instead of passing completely out of it when they land their machines on an aerodrome instead of a carrier.

Similarly, the services of aircraft mechanics and riggers from the Air Force will continue to be needed in the Fleet Air Arm until such time as the Navy can recruit and train ratings to replace them. No doubt the assistance of the Air Force in the training of these ratings will be available, though the establishment of the Fleet Air Arm will eventually be large enough to provide for its own training; but even then it would be of advantage, from the point of view not only of efficiency but also of inter-Service co-operation, that there should be an interchange of skilled ratings between the Services.

Policy

Various questions of policy consequent upon the change in the status of the Fleet Air Arm remain to be settled. The most important perhaps, is that of personnel. It is perfectly obvious that a large increase in the number of pilots in the Fleet Air Arm must be provided in the not far distant future. The number, given in last years "Brassey," of aircraft embarked in H.M. ships amounted to 176, of which 145 were in carriers and 31 were catapult aircraft carried in fighting ships. There are five aircraft carriers now building, one of which, the Ark Royal, is due to complete in the current year and the remainder in the course of the next two years. Each of the new cruisers to be passed into the Service carries more aircraft than the ship she replaces, while some of them instead of replacing earlier ships are allocated to bring cruiser squadrons up to their earlier strength—as the 2nd Cruiser Squadron of the Home Fleet now comprises five ships instead of four. Every battleship or heavy cruiser reconstructed carries more aircraft than she did before.

With the completion of the new aircraft carriers, some of those now in service will no doubt be placed in reserve; but aircraft carriers in reserve are of no use unless the aircraft which they are to operate on mobilisation are in existence, together with the pilots who are to fly them, the observers who are to be carried in them, and the artificers and craftsmen who are to maintain them. It cannot be an exaggeration, therefore, to estimate the strength of the Fleet Air Arm three years ahead at the equivalent of some 400 machines against the 176 of to-day. Indeed, allowing for the numbers under training for the larger force and the numbers required for administration of the naval shore aerodromes which will then be in existence, the effective strength of naval air personnel may well be even greater. A large expansion in personnel in the near future is thus inevitable.

At present, naval officers provide only 70 per cent of the pilots of the Fleet Air Arm. As soon as they can be replaced, the R.A.F. pilots will presumably no longer be available. The increase in the number of pilots required is thus even greater than the proportionate material expansion of the Fleet Air Arm. It seems unlikely that this increase can be provided wholly by volunteers from the ordinary Lieutenants list; and if not, the Fleet Air Arm must look elsewhere, in addition, for its pilots.

Lower Deck Pilots

The first and most obvious source of supply is the lower deck. The R.A.F. has long utilised the services of men from the ranks as pilots, as have foreign Services. Many will remember the very efficient "quartermaster" pilots of the French Naval Air Service, who did much sterling work in collaboration with British forces in the Mediterranean during the Great War. It is no secret that the Admiralty have long desired to train Petty Officers and seamen, who volunteer and are found qualified, as pilots, but that that measure is believed to have been vetoed up to now by the Air Ministry, for reasons which have never been made public. In any case, presumably no such veto will be effective under the new organisation, and it is to be expected that the position of pilot will before long be thrown open to the naval ratings, as it has for years been open to non-commissioned officers and men of the R.A.F.*

* * * * *

---

*An official announcement to this effect was made on March 5, 1938.

39. 1939

The Interdependence of Home Defence and Commerce Protection

"Securus"

The immediate re-actions in this country to the crisis in September, 1938, have thrown the question of Home Defence into stronger relief than at any time since the Napoleonic wars. In the United Kingdom, in fact, we have never as a people had to study very carefully the question of home defence since the fall of Napoleon. Although there have been minor panics it has never been a burning question to us in the same way that it has to Continental countries; we have had the sea as a barrier, as a bulwark to the castle of the Englishman's home. It is clearly for this basic reason that we have appeared to the French, since the Great War, to be ridiculously altruistic towards our late enemies. It has been said by a careful student of international affairs, when speaking of the American attitude towards European questions, that the Eastern States are some 3,000 miles, and the Western States some 5,000 miles, more altruistic towards ex-enemy powers than is France, and we are clearly in a somewhat similar situation. The consequence of this lack of real appreciation as to the meaning of Home Defence has in the past encouraged us, as a nation of shopkeepers, to look towards commerce protection and control of sea communications as the principal object of our armed forces in war. Commerce protection was our conception of Home Defence.

* * * * *

It may be stated categorically, indeed it has been so stated, that the foundation of our edifice of imperial defence is the maintenance of the security of the United Kingdom. It is recognised that our main military strength lies in the resources of man power, in the powers of endurance, and in the industrial capacity of the United Kingdom. It is also appreciated that with the advent of aircraft these fundamental bases of our power of national resistance can no longer shelter behind the steel fence of naval power and permit the slow, steady, and immense development of our national war effort to go on. They can be attacked from the first moment, and an umbrella of air security is now our most vital defence requirement.

Before going on to discuss how this umbrella is to be provided, it is important to recognise how closely connected are the problems of the security of the United Kingdom, so that reasonably normal life can continue, and the provision of those vast quantities of foodstuffs and raw materials that we require to be brought to us in ships from the four corners of the world. Efforts have lately been made to make us less dependent upon imported goods, but at the best they can be only a palliative to tide over the first few vital months. We can neither eat, nor manufacture and export the goods we must produce to pay for our food requirements, without the ability to use the sea.

This preliminary review enables us to define in quite simple terms the particular new factors that the development of the aeroplane has added to our problem of national security.

(i) It has introduced the new factor of air invasion.
(ii) Direct air attacks from shore bases upon ships at sea are now possible.
(iii) Considerable assistance is available to our organisation for controlling sea communications by greatly extending the area of vision of cruising units, and the power to attack isolated raiders, surface or submerged, hundreds of miles away from the parent naval unit.

At first sight it looks as if it would be a simple matter to divide responsibility and organisation as between the factors which quite clearly affect ships and their operating organisation, and air invasion of the country as a whole. Closer examination, however, shows this to be false. To begin with the same bomber aircraft can attack any target at will. They may attack ships in narrow waters, shipping in docks, naval bases, centres of land transport, Army concentration areas, towns, factories, or aerodromes. From the operational point of view the sole criterion is adequate range and refuelling and rearming facilities. The organisation for the offensive attack upon the hostile air organisation must, therefore, clearly be centrally controlled. The complementary organisation is close defence, consisting of A.R.P. organisation, guns, lights, balloons, observers and fighter aircraft. For similar reasons this also must be centrally controlled or else no sort of reasonable economy could be effected. There is clearly then some fundamental interdependence, but is it so fundamental that only one organisation is required to provide security in both directions? The principal difficulty is that in considering the protection of shipping one immediately begins to turn away from the United Kingdom and to pay attention to the ocean routes, the naval bases, the sources of supply and the air, and other threats that may be present. Here then is a distinction.

At home one central organisation seems clearly to be required to provide that umbrella of security without which we cannot live. To a considerable extent the same aircraft can act as the eyes and the teeth as well. It can co-operate with the Navy on one day to find a hidden enemy, with an Army on the next to provide protection or offensive power, and on the third it can bombard an enemy's factory organisation, centre of transportation, or an air base.

Out of range of these shores, however, the problem is different in urgency and degree and seems to require even further subdivision. First there is the question of the local protection of Dominions and Colonial territories. Each will have its separate home defence problem showing in small scale the principal characteristics of the Home Country in greater or lesser degree. Systems of inter-reinforcement by land, air, and sea will be features of the organisation, and a common doctrine and common equipment and idea will mark them all. Secondly there is the question of co-operation with the Navy, not only in its war against the enemy naval power, the war of cover, but also in its far more exacting war of sweeping the sea free from all submarines, cruisers, and commerce and protecting our own commerce on the high seas. The main Fleet requires its quota of aircraft for reconnaissance, spotting and striking, and this is provided by Fleet carriers and aircraft embarked in the capital ships and cruisers, assisted by the operations of General Reconnaissance shore based aircraft. If the situation permits, indeed, assistance can also be rendered both by the offensive and defensive air units provided to produce the umbrella of air security now required in war wherever an air threat exists. The Fleet bases require this protection just as badly as other centres of national activity, including the manufacturing capacity which maintains the Fleet.

* * * * *

What is the problem to-day?

Does a paramount Fleet still protect our shipping and secure us from invasion? Are our bases upon which the successful action of our naval power depends properly sited and adequately secure? If not, what has altered the situation and what must we do to reduce the adverse balance? Unfortunately it no longer needs the air enthusiast to point out the potentialities of his wares. Development of man's conquest of the air had fundamentally changed the whole basis of our national defence.

* * * * *

From this brief review of this complicated problem it is apparent that the influence of air power has reoriented our views on home defence. The protection of the United Kingdom is the corner stone of our Imperial defensive system. It is now subject to an increasing scale of air invasion and our most pressing need is to produce an umbrella of air security beneath which the national ability to live, work, and develop the war potential can go on. Only second to this is the need to protect the flow of sea-borne trade from the four corners of the globe, which is a necessity for our national life and activities. Both ability to live and work and continuity of supply are essential ingredients in national security, and if ability to live is the more important in terms of time in the initial stages, ability to continue to operate and to extend industrial output is a vital necessity for the survival of the Empire. Home Defence against air attack can be considered as a self-contained problem at the centre calling for a central directing and co-ordinating authority. The air defence forces required will comprise both striking and security elements and will provide for the necessary co-operation with land and sea forces. Commerce protection out of range of the Home Defence forces is another problem, calling for a different organisation, more elastic systems of control. Defence of bases and overseas territories partakes closely of the character of Home Defence as such, most of the air defence element will be represented, and central co-ordination of the forces will be required. The inter-relation between the particular activities of Home Defence and the particular activities of commerce protection, however, is much more marked than in the United Kingdom. The two portions of national security Home Defence and commerce protection may be likened to the heart and the stomach. If the heart is struck life ceases abruptly; if the stomach is not filled life slowly drains away. A comprehensive air shield is now a necessary part of our national equipment and this shield is now in the process of being forged.

The Coming of the Aircraft Carrier

(*Imperial War Museum*)

H.M.S *Ark Royal*, the first carrier of the name, operated seaplanes in the early days of naval flying

(*Imperial War Museum*)

The famous *H.M.S. Ark Royal* of World War II, taken from *H.M.S. Sheffield* in May 1941, shortly before she was sunk by a U-boat

THE LATEST TYPE OF AIRCRAFT CARRIER FOR THE BRITISH NAVY: H.M.S. HERMES (BOW VIEW).

(Photo: Abrahams & Sons, Devonport.)

(Constructed by Messrs. Armstrong, Whitworth & Co., Ltd., Walker-on-Tyne.)

THE HAWKER NIMROD FLEET FIGHTER.
Rolls-Royce "Kestrel" Engine.

THE HAWKER OSPREY FLEET FIGHTER RECONNAISSANCE AIRCRAFT.
Rolls-Royce "Kestrel" Engine.
(By courtesy of the builders. Photos by "Flight" and "The Aeroplane.")

U.S. ELECTRICALLY-PROPELLED NAVAL AIRCRAFT CARRIER SARATOGA.
(Constructed by the Bethlehem Shipbuilding Corporation, Fore River, U.S.A.)

PART VI: 1940–1949
World War II
Interpreting the Lessons

Introduction	185
40. 1940	
The Merchant Navy in War and Peace	186
ARCHIBALD HURD	
41. 1945	
The Organisation of Fighting Forces	190
H. G. THURSFIELD	
42. 1946	
The Air War at Sea	199
OLIVER STEWART	
43. 1948	
The Future Employment of Naval Forces	204
FLEET ADMIRAL CHESTER W. NIMITZ USN	

Introduction

Brassey's was able to continue publication throughout the war and devoted a great amount of space to narrative material which, although rightly appreciated at the time, has now been supplanted by historical research and perspective. Much more interesting are the early attempts to analyse the novelties of maritime operations and to interpret their lessons for the future of war at sea.

It was already clear by 1940 that there was a shortage of merchant shipping and later research has established that this was one of the most important factors in delaying Allied victory on land. There was no disagreement that inter-service co-operation, or the lack of it, had been of vital importance but very strong differences as to whether this pointed to the creation of a single unified armed service. Most interesting of all were the first attempts to evaluate the effect of the atomic bomb and the missile on the nature of warfare and the future of mankind.

Recommended Reading

John Creswell, *Sea Warfare 1939–1945* (Revised Edition, 1967).

40. 1940

The Merchant Navy in War and Peace
Archibald Hurd

When the Royal Navy was mobilised on the outbreak of the war, the Government took control of merchant shipping and of the principal ports of the country as well as of all the shipyards, those concerned with the repair as well as the building of ships, both of war and of commerce. These simultaneous events, ensuring unity in direction of the affair at sea, were without precedent in our history.

The course which the struggle had since taken suggests that if these measures had not been adopted, the opening weeks of the war might have been marked by a disastrous loss of tonnage. The enemy, it subsequently became known, had U-boats stationed at the focal points of the trade routes in readiness for the outbreak of hostilities, expecting to sink ships which were without defence, and had prepared many thousands of mines, including a new type of magnetic mine, to be sown in the traffic lanes of commerce, and particularly in the approaches to the most frequented English ports. The Germans had, in effect, planned a blockade of this country, which was to be pressed home without regard for the lives or property of neutral seamen, shipowners, or merchants. The failure of that design was the initial British success in the war. The prompt action of the Board of Trade and the Admiralty not only saved many merchant ships from destruction, with the loss of their cargoes, but it established confidence in the administration of the sea services.

In order to appreciate the significance of the Government's action on the outbreak of war, it should be recalled that the Royal Navy belongs to the nation and is built, manned, and administered by the Admiralty on behalf of the nation, while the Merchant Navy, provided by private capital, is managed by business men (who are usually described as "shipowners" though they are for the most part only the trustees of investors), and is manned by officers and men, who, apart from naval reservists, are uncovenanted to the State. The Merchant Navy, as it is called, consists of upwards of 300 fleets; some of these are grouped, though, as a rule, preserving separate managements, such as the P. and

O., Furness, Ellerman, Alfred Holt, and other groups. These ships, to the number of approximately 2,500, usually go about their business in the great waters of the world unfettered by the Government, except in relation to matters affecting the safety of life and property and, even in that respect, some of the regulations are less strict than those of Lloyd's Register of Shipping, an organisation of shipowners, shipbuilders, and underwriters, which was created many years ago for the purpose of mutual protection.

The significance of the measure of control over the shipping industry assumed by the Government when the war opened cannot be appreciated unless these considerations are borne in mind. Though the Merchant Navy was, in fact, the mother of the Royal Navy, it is provided by private individuals and managed for profit to the advantage of the nation. In one year (1920) it contributed invisible exports of the value of £340,000,000 to the credit side of the National Trading Account. It provides employment in shipyards and engine shops, in the ports and in offices and at sea for about one million men. In the creation of the great volume of shipping on the Register of the United Kingdom, fully and efficiently manned at all times, the State has had no part since the Navigation Acts were repealed. And as the State can claim no credit in this respect, so it has had no hand in the equipping of the shipyards or the provision of ports round the coasts of Great Britain and Northern Ireland which are used by the merchant ships. The whole maritime effort has sprung from individualistic action springing from what is often described as "the acquisitive faculty"—the desire for profit. Business competition has resulted in a high standard of efficiency.

* * * * *

Transport of the Expeditionary Force

The lessons of the Four Years' War were not without their influence on the Government when the present struggle opened. It was feared that the enemy might at once resort to unrestricted submarine warfare, in spite of pledges which had been given to respect not only international law but also the dictates of humanity. In these circumstances, there was no saying how serious the shipping losses in relation to the seagoing tonnage—2,000,000 tons less than in August, 1914—might be. So it was determined that, on the one hand, shipping should be used to the greatest national advantage from the opening of the struggle, and that the law of supply and demand should not again raise freights to dizzy heights. Within a few days of the declaration of hostilities, the Marine Department of the Board of Trade, which had taken over the limited peace duties of the Ministry of Shipping of 1916–18, took virtual control of all tonnage afloat under the British flag as well as the shipyards, and administered the War Risks' Insurance Act, which had been passed in August, from an office which was immediately opened at Lloyd's. Many ships had already been requisitioned for the fighting services and the number was increased within a few days. Overseas voyages of British merchant ships were brought under control by means of licences and a committee was established to consider applications for licences. Four prominent members of the shipping industry, Sir Vernon Thomson, Sir John Niven, Mr. Harrison Hughes, and Mr. Philip Runciman, were included in this committee.

On the outbreak of war, the Government decided, in accordance with the principle that it is an error "to swop horses when crossing the stream," to leave the Marine Department to carry out its prearranged plans instead of immediately setting up a Ministry of Shipping. No praise can be too high for the manner in which the understaffed section of the Board of Trade, working in close association with the Admiralty and the War Office, carried on its work in face of many difficulties. It was responsible for requisitioning and equipping merchant ships which were required by the fighting services, whether as armed merchant cruisers, minesweepers, hospital ships, or troopships, etc. Sea transport officers were appointed at all the principal ports of the United Kingdom and abroad to look after the requirements of merchant vessels requisitioned for the defence services. In conjunction with the War Office and the Admiralty, plans were prepared for the transport of the British Expeditionary Force.

An organisation was at the same time set up for the purpose of providing shipping space to meet the requirements of the Government purchases of essential imports, and commercial agents were appointed in ports all over the world. Chartering committees, drawn largely from members of the Baltic Exchange, were appointed to act on behalf of the Government and a schedule of freight rates was instituted.

Before passing to later developments, a tribute must be paid to the success with which the British Expeditionary Force was transported to France, the movement beginning within four days of the beginning of the war. The co-operation of the railway and harbour authorities was of the most efficient character, and the officers and men of the merchant service earned the highest praise for their skill and resource. This was the greatest and most successful operation of its kind ever performed. In a period of five weeks no fewer than 158,000 men were embarked and conveyed safely—at the rate of three convoys a night on the average—to their ports of disembarkation. One hundred and seventy-four ships were employed and over four hundred voyages were made. The movement was on a greater scale than in the war of 1914–18. Not only was the force larger, but the army had in the meantime been further mechanised. Then the men marched on to the ships, the horses were led, and a light derrick could lift what the soldier could not carry. There were only eight hundred mechanised vehicles in all, and it was a rare load that exceeded two tons. The transportation which began immediately the present war opened involved the carriage of more than 25,000 vehicles, including tanks, some of them of enormous dimensions and weighing fifteen tons apiece or more. The ordinary shore cranes could not raise such weights; special ships were required to carry them and highly trained stevedores to manipulate them.

* * * * *

Exaggerated Estimates of Tonnage

Not until the war had been in progress several months—all available neutral tonnage having been purchased or chartered by the Government and a limited scheme of food rationing enforced—was it realised that the country had been short of ships from the first days of the war. Fortunately the losses inflicted by the enemy were smaller than might have been expected from the ruthlessness with which U-boat and mine were employed. Several Ministers had contributed to the widespread belief that the country had plenty of tonnage at its disposal. Time and again it was claimed that 21,000,000 tons of shipping were available when hostilities began. The Chairman of the Liverpool Steam Ship Owners' Association (Mr. F. Fletcher Hunt) subsequently showed that this was a complete misconception. He pointed out that the figure of 21,000,000 tons represented the total tonnage of all ships of 100 tons gross and upwards registered in the British Empire. It included, on the United Kingdom register, a large quantity of small craft—hoppers, salvage vessels, tugs, ferry boats, pilot boats, pleasure steamers, dredgers, and the like plying habitually within territorial waters. It also included, on registers outside the United Kingdom, vessels trading on the Canadian lakes and in the coastal trades of India, Australia, and other parts of the Empire. It had, consequently, no relationship to the volume of British merchant navy available for the overseas carrying trades of the United Kingdom.

* * * * *

41. 1845

The Organisation of Fighting Forces
H. G. Thursfield

It will generally be admitted on all sides that the most outstanding lesson to be learned from the course of the great war which seems, in Europe at least, as these words are being written, to be drawing to a close, is that victory is won, not by this arm or that, but by the full collaboration of all arms; by their employment, not independently, each in its own sphere or in its own element, but interdependently, all directed alike towards one end. That lesson has been driven home as much by our reverses as by our successes; for every reverse has been traceable ultimately either to a lack of collaboration between the separate arms on our part, or else to its having been achieved to a greater degree by our enemies. To demonstrate this proposition, it is not necessary to examine in detail every campaign or action that has been fought in the last five years, though the same conclusion is reached in the case of each of them; it will suffice for my purpose here to take a few examples.

In the Norwegian campaign the German success on land was by common consent largely the result of the close integration of their air forces with their armies; the whole of the energies of the *Luftwaffe* were directed to furthering the advance of the military forces with the aim, as quickly as possible, of completing the military occupation of the whole country. Moreover, the ultimate object of that occupation was the acquisition of the harbours of Norway whence U-boats and sea-going aircraft could prey on British Atlantic communications, which were vital to us, with greater freedom than was possible to them when compelled to work from German bases only. On the other hand, our great handicap was that we lacked at that time the air arms needed by sea and land forces operating against the German invaders of Norway; and that, moreover, British sea and land forces alike lacked both the equipment and the training needed for the successful conduct of an amphibious campaign overseas. The subsequent campaign in France told the same story on both sides. The enemy's chief object was the acquisition of the French Atlantic ports, needed for the same purposes as those of Norway—I leave out of account for the time being the question, about which two opposite views are held, of whether a serious invasion of this country was intended by Germany in 1940. German success in France was chiefly due again to the full integration of land and air forces under a common direction to the same end. Allied failure was largely due to the inadequacy—which arose from various prewar causes—of the air arm which made full collaboration, of the standard set by the Germans, an impossibility. Yet when circumstances made possible the full collaboration of all three arms, sea, land, and air, of the British Forces of Dunkirk, it became possible to snatch them from what had seemed inevitable and irretrievable disaster.

The same conclusions emerge from study of the whole course of events in the Mediterranean. As long as we were deficient in this arm or that we had the greatest difficulty in holding our own, despite the wonderful individual prowess of the fleet under Admiral Cunningham or the armies under General Wavell and his successors. The final success in the Mediterranean area, which started with the expulsion of the Axis forces altogether from North Africa, followed from the employment of all arms towards a common end. The armies and air forces were built up to the necessary strength in Egypt by virtue alone of control of the outer seas by the Navy and, in certain areas, by the air forces working with it; their famous final advance westwards was achieved through the intimate collaboration of all three arms. The armies and air forces which secured control of North-West Africa did so only by virtue of the naval control of the intervening sea across which they had to be conveyed to reach their theatre of operations. The military occupation successively of the whole of North Africa enabled air forces to be based within reach of the whole African shore of the Mediterranean, whereby control of the sea routes from end to end could once more be secured by sea and air forces, in proper proportion, working in collaboration. The control of the Mediterranean sea routes thus acquired enabled the military occupation to be extended through the intermediate islands—needed again as air bases—to Sicily, Italy, and eventually to Southern France. The same story of success following full collaboration, the provision of all arms in adequate proportion, and their common direction to one end, is told by the course of events in the English Channel theatre from June, 1944 onwards.

On the other hand, failure in collaboration has invariably been the prelude to defeat. An example of this on the small scale is to be found in the ill-starred campaign in Greece in 1941. The British forces, both land and air, but especially the latter, were all too weak to accomplish the task, for which they had been sent to Greece, of holding up the advance of Axis forces southward through that unhappy country; but although at first, when troops and air forces pursued different tactical objects, they produced little effect against an enemy employing great and integrated forces, as soon as tactical collaboration was established, the improvement in results obtained became very marked. The same thing on a much larger scale was demonstrated at sea. The enemy, realising what it would almost seem that many people in this country had forgotten, that British strength depended fundamentally on our being able to retain command of the sea and assure the continuity of our sea communications, made one of his greatest efforts in the attack which had nearly brought him victory in 1917, on merchant shipping by U-boats. British strength in the destroyers, corvettes, and other classes of the small ships needed to defend shipping against U-boat attack, had been allowed, in the years of treaty-making between the wars, to fall far below strategic needs, on the mistaken theory—the fallacy of which was constantly pointed out by Admiral Sir Herbert Richmond and other writers—that the numbers of such ships that we should need in war would be governed by the strength of foreign navies in submarines, rather than by the volume of British traffic needing protection. At the same time, the air forces needed for work at sea in the defence of shipping, in collaboration with the sea forces performing the same duty, simply did not exist. Even the squadrons which had inherited the functions of the Royal Naval Air Service in the 1914–18 war were no longer called "Naval Co-operation Squadrons" but had been renamed "General Reconnaissance Squadrons".

Taking advantage of this lack of the forces needed for the most fundamental of British tasks in time of war, the enemy put vast resources into the multiplication of U-boats, standardised both their construction and tactics so as to obviate the need for intensive training and great skill, and devised methods of attack that could exploit the British shortage in defending forces. Those methods could only be countered by intimate collaboration between sea and air forces. Sea forces alone could not have done it, unless in numbers so vast as to be quite impracticable to provide; air forces alone could not do it at all, since an aircraft is completely impotent against a U-boat once the latter has dived below the surface, moreover, in the early months of the war, they were also impotent at night, when the most damaging attacks by the U-boats were made. But ships and aircraft in collaboration—as was mentioned by Mr. Churchill in a speech in Parliament on February 11, 1943—proved deadly, once both were available in sufficient numbers.

Yet it was not until April, 1943 that they were available in sufficient numbers. It was only then that long-range aircraft, working from both sides of the Atlantic were provided in sufficient force to give adequate protection to trans-Atlantic shipping, while the gap in the middle, that neither of them could reach, was bridged by means of the small so-called "escort carriers". Yet all through 1940, 1941, and 1942, great resources in the air were being devoted to the bombing of German shipyards and German cities; the first, on the theory that to do so would prevent U-boats being built at all, and the second, presumably on the theory, so often advanced by air enthusiasts before 1939 yet now fully disproved by the hard experience of six years of war, that it would induce the enemy people to give up the struggle and make peace. Be that as it may, and whatever the theories that were the basis of our methods, the two arms were pursuing different aims and objects, neither of which were being attained. The building of U-boats was not being held up by the bombing of shipyards but on the contrary went on steadily, on a uniformly rising curve. The German people were not being induced to force Hitler to make peace, but on the contrary, fought on until their armies were beaten in the field in their own country, despite an unthought-of intensification in the weight of the bombing attack. Atlantic convoys were not getting through, but on the contrary, so heavy were their losses, especially in ships carrying the oil fuel on which all arms relied, that it began to be doubtful whether the bombers would be able to get off the ground. As the Prime Minister and the President of the United States remarked, in summing up the whole course of the U-boat war, in 1941 and 1942 the issue hung in the balance. But in 1943, when full collaboration between sea and air forces in the Atlantic was at last established, the upper hand was gained which was never again lost.

Pacific operations have told the same story. The initial Japanese attack on Pearl Harbour was an outstanding example of intimate and effective collaboration between sea and air forces, rendered easier in that case by the fact that they were all drawn from the Japanese Navy and not from separate Services. The American defence against that surprise attack largely failed because of failure in collaboration between the Army and the Navy—*vide* the Report of the Commission of Enquiry held immediately afterwards.* All the Japanese operations by which they overran the Philippines, the Dutch East Indies, and the South Sea Islands were little masterpieces of full collaboration between all arms; and each move forward was preceded by careful preparation so that full collaboration could be achieved for the next step. On the British side, the disaster of the loss of the Prince of Wales and Repulse was the direct result, not only of lack of provision of the different arms in due proportion, but of the failure to use in collaboration even those that were available. And the disaster of the loss of all Malaya, culminating in that of Singapore, was traceable largely, of course, to the numerical weakness of the forces defending those

*"Brassey", 1942, page 111.

possessions, but even more to the fact that the sea and air arms necessary to an adequate resistance to an invasion on the scale to be expected were not provided to collaborate with the military garrison of that territory. The fact was, of course, that in the military situation then existing in Europe, they simply did not exist; but I am not here concerned with the causes of the failure, or omission, to supply properly balanced forces of all arms for the defence of Allied possessions in the East Indies, but merely to point out the inevitable result of their lack. It is not a question of numbers alone, as can be seen from the final phases of the campaign in Malaya, culminating in the fall of Singapore.

* * * * *

I have pointed out earlier how this lesson was brought home very forcibly to our American Allies by the events at Pearl Harbour on December 7, 1941, with the result that their Service leaders quickly recognised the paramount necessity of precautions against any survival of the lack of inter-Service collaboration then manifested. One of their most successful commanders, Admiral Halsey, who has held seagoing commands throughout the victorious advance westwards across the Pacific, in a series of amphibious operations by land, sea, and air forces in collaboration, made it a principle that there should be no distinction of Service amongst the officers employed on his staff. He appointed to each position the officer best fitted for the work attached to it, regardless of whether he belonged to the Navy, to the Marines, or to the Army; and moreover he ordered that they should all wear the same uniform. A visitor to his headquarters had no means of knowing from which Service any staff officer with whom he might come in contact had originally been drawn. Admiral Halsey's operations were so uniformly successful, and the spirit of collaboration, and absence of any inter-Service jealousy or disagreement, in the forces under his command were so very marked, that many people have urged that his example should be followed and that the same principle should be carried even farther to the point of amalgamation of all fighting services, land, sea, or air, into one fighting force. The same conclusion has been reached, and the same action has been urged in this country, independently by other observers, also impressed, as a result of analysis of war experience in the North African, Mediterranean, and Atlantic theatres, with the supreme need to insure full collaboration of all arms, not only in the field but also on all planes up to the highest, and by the lack of collaboration which the system of separate Services seems to some extent to have bred. They would amalgamate Army, Navy, and Air Force into one service. Such an amalgamation would be so far-reaching a change that it seems desirable to examine it from all aspects in order to make sure that of the results that would flow from it, not all of which could be clearly foreseen, the good should unmistakedly outweigh the bad.

* * * * *

Amalgamation, however, would undoubtedly involve abandoning many things by which each of the Services at present sets much store, and which undoubtedly have very considerable moral value, in the shape of tradition and *esprit de corps*. It is thus clearly desirable, before these things are thrown overboard, to be quite sure that that is the only way of achieving the desired end. Moreover, there are solid reasons for the differentiation between soldiers and sailors, based on severely practical considerations; and it would clearly be highly unwise to ignore experience and practical necessities in favour of a form of organisation built up *de novo* from pure theory—although in the political sphere that sort of thing seems to be very much in the fashion these days. The desired end is full collaboration between different Services and their common direction towards the same end. If that end can be achieved under the existing organisation—which, in part at least, has stood the test of time and has a great deal to recommend it— it is surely preferable to do so than to scrap it and adopt a wholly new and untried form of organisation which may prove, as such things often do, to possess all sorts of unforeseen defects and disadvantages. Whether or not that is possible is a point upon which history can perhaps give some guidance, in spite of the fact that in the wars of the past only two Services were concerned instead of the three which we— though not the Americans—have now adopted.

* * * * *

It seems therefore desirable to examine the causes which have brought differentiation between the Services. I leave out of account for the time being the question whether the British system of constituting air forces into a third Service, or the American system of adhering to the time-honoured organisation which differentiates between land and sea forces only, is the sounder. It will suffice to note that the three-Service system has been adopted in this country and there seems little likelihood, or even possibility, of its being modified in favour of any return to a two-Service system; and that the American Forces, having had war experience both of their own organisation and of ours, are quite determined never to adopt the latter. But history again is illuminating on the causes of differentiation between Army and Navy.

The early history of navies shows the soldier afloat, maintaining his traditional position as the fighting man, whatever the place and conditions in which fighting had to be done, the seaman at first merely providing the vehicle for carrying him to battle. It was Drake who was the first to recognise that this sytem was unworkable and insisted that all afloat in his ships should be, "one Company", abolishing the distinction between the seaman and the fighting man. "I will have the gentlemen haul and draw with the mariners," was his method of laying down the great principle that all in a ship-of-war, as well as any other ship, must be professional seamen, live by the sea and hold the true faith of the sea. Seamen indeed do not form a race apart but they do form a profession apart, for those unaccustomed to life afloat are frequently incapacitated for long periods when they find themselves on the sea. Most landsmen, in fact, are miserable afloat, except in the comparatively rare periods of calm weather; not everybody has a taste for seafaring—though the proportion of Englishmen who have is probably higher than that of most nationalities—so that the seafaring branch of the fighting forces must, in practice, inevitably be differentiated from the landsmen. On the other hand, the profession of soldier too, especially in these days, is a whole-time job in itself, just as much as that of seaman. It is not physically possible for any man to be fully efficient in two separate professions as different from one another, and each as exacting, as that of the modern soldier and the naval seaman of to-day—or indeed the naval seaman of any past age.

Thus it seems clear that, even if on the grounds of logical organisation and the need for "integration of the Services" in order to ensure that they shall all direct their efforts towards the same end, the amalgamation of Army, Navy, and Air Force into one Service were decreed to-morrow by a courageous stroke of the pen, actually the existing differentiation would in practice have to persist. The seaman would have to remain a seaman, and the soldier to remain a soldier if they were to maintain, as clearly they would be required to, their present standard of professional efficiency. Any step in the direction of assimilating the duties of soldier and seaman could hardly have any result other than to reduce efficiency in one capacity or the other; for there is much truth in the old saw that Jack of all trades is master of none. The only actual difference would be that the traditions of both Services, by which they set so much store, would have to be scrapped, and it is difficult to see how the great object of collaboration of different arms would be served by that change.

* * * * *

The strongest argument against amalgamation, on the other hand, is that derived from tradition. Officers and men of the Royal Navy, though they do not talk much about it, are immensely proud of belonging to "The Navy, whereon, under the good providence of God, the wealth, safety and, strength of the kingdom do chiefly depend". They are immensely proud of their status as members of the Senior Service, and very jealous of it, though of this too, they make no parade. Naval traditions go back to Drake, a very real figure even to-day in the West Country that has always furnished so large a proportion of naval strength, as are the other great names in our naval history. The seaman of to-day, if he grouses at the rate of pay granted by a niggardly Treasury, remembers that at the time of the Mutiny at Spithead, which arose out of the same cause, it was Lord Howe who took the seamen's part and saw that their just grievances were redressed. The seamen of to-day has an immense admiration for the fighting qualities of the British soldier, and there is nothing in his power that he will not do for him—there was never a complaint heard in the destroyers of the Mediterranean Fleet when in 1941 they were called on to go back, time after time without air support, into the hell that the *Luftwaffe* were creating in the waters round Crete, to make sure as far as was humanly possible that no British soldier should be left behind that could by any means be brought away. But in his heart of hearts, despite his admiration, in a way he looks down on him as really only a poor landsman, dependent in the last resort on the sea service to care for him—an attitude which the soldier to some extent shares, in his confidence in the Navy's capacity to look after him when he has to be afloat, or supplied from the sea. But the sailor's pride is in being of the Navy, to which he owes allegiance, and of the traditions of which he is the guardian; he does not owe allegiance to an abstraction which contains all sorts of other men, not seamen at all, newly invented by some theorist who does not understand the sea or ships, and thinks that there is no difference between a seaman and the landlubbers who depend on him. It is not going to add strength to the Navy or those who compose it to cut them off from the long tradition of which they are the conscious heirs, or to destroy the separate identity of the Service which is their pride; nor is it going to make him do more in the way of co-operation with the other Services than he has done in all the wars of our history.

It is the same thing in the Army, of which the great strength lies in regimental tradition, embodied in the battle honours borne by the Regimental colours. No good could possibly come of a clean break with those traditions; yet there can be no real amalgamation of the Services otherwise. For if regimental traditions are to be retained in the infantry units of the new integrated Service, or whatever it is to be called, they can be nothing but the old Regiments under another name, though no longer allowed to be parts of the Army to which they were proud to belong. If amalgamation were to be real, and not a mere paper abstraction without practical effect, tradition and *esprit de corps* must go. We should have to be very sure that the change was going to create a more efficient fighting organisation before that loss could be accepted.

The Royal Air Force's traditions do not go back as far as those of the Navy or Army—naturally enough, since it was only created in 1917. But they are very real and individual, none the less, and much value is attached to them within the Service. They, too, would have to go in amalgamation, and formations formerly of the R.A.F. could not even absorb the traditions of the older Services in that process, for they would have disappeared too.

The conclusion that emerges from these considerations is that a real amalgamation of the three fighting Services into one is neither a practicable proposition, nor is it necessary in order to ensure that, in future, the full and cordial collaboration that was achieved by the forces under General Eisenhower shall always be forthcoming. Unity of effort and direction, and full collaboration of all arms to the same end, can already be achieved, once the paramount necessity for achieving them is recognised, and kept firmly in view, in all the Services; and once the mischief-makers, intent on stirring up jealously between the Services, and keeping it alive by continual exaltation of one and depreciation of the others, are exposed and suppressed. It is perhaps unfortunately true that we are unlikely ever to be free from some of that kidney, even in high and influential positions. But in spite of them, in the present war in the Services themselves have broken away from the attitude they maintain, and have achieved full collaboration in the face of the enemy. They have done it once, and they can do it again, provided we do not, at the bidding of visionary theorists, scrap well-tried institutions in favour of a brand-new product of the armchair planner's brain.

42. 1946

The Air War at Sea

Oliver Stewart

*   *   *   *   *

It is now possible to turn to the more general survey of the war developments. The Pacific war was notable for the way in which it stimulated the employment of aircraft carriers. Aircraft carriers were dominant in many of the actions. Progress in their design, equipment, and operation was greatly accelerated by experiences in the Pacific. In fact the lesson of the war in the Pacific was that the aircraft carrier, instead of being near the end of its useful life, was only at the beginning of it. This lesson would have remained firm had it not been for the coming of the atomic bomb. There can be no doubt that the successful production of this kind of bomb casts doubt upon most of the direct inferences that can be drawn from the war.

That is the great problem of the present time: that a large amount of military experience was bought at an enormous cost during the war and that just at the end of the war a new weapon arrived which was of such power that it affected all foregoing lessons. Before the attack on Hiroshima on August 6, the lessons of the Pacific war were plain for all to see. They emphasised the indivisibility of the three Services; they emphasised that three Services working in really close collaboration are more powerful than the sum of their individual strengths; they emphasised that aircraft can be made more effective in a war waged over great distances if as many of them as possible can be given mobile bases.

But that last and most fundamental lesson, as it seemed at the time, is disturbed by the coming of the atomic bomb. The aircraft carrier bases its value on the fact that it is a mobile base and that in consequence it allows aircraft to operate at higher battleworthiness. That is to say it enables aircraft to go into action, stripped for action, not lugging round with them hundreds of pounds of fuel which is of no value in the combat, but which must be carried in order that they may be able, if they win, to return to base.

Here is the foundation of the value of the aircraft carrier. It sends up an aircraft, be it fighter or bomber, with a minimum fuel load. In consequence that aircraft can carry either heavier arms or else it can enjoy the advantages of improved performance. Theoretically one ounce of petrol more or less, reduces or increases the air performance in speed and climb of that aircraft. And in actual fact the petrol load does affect performance. The aircraft working at long range is—other things being equal—always less effective in combat than the aircraft working at short range. It follows that if the aerodrome can be moved about so that the aircraft take off fairly near the scene of battle they will enjoy a notable advantage. The aircraft carrier does just that. Its fighter is stripped for combat, carrying as little dead load as possible. Its bomber is using as much of its lift as possible for lifting bombs—not petrol to get it home.

All this was known and was the basis of the large carrier programme launched by the progressive navies. The Midway Island battle of June, 1942, rubbed in the point. The Solomons battle of April, 1943, taught the same lesson, and there were hosts of other examples which all tended to make the lesson look as if it were a permanent one. It seemed to say that the aircraft carrier was one of the kinds of ship that had a long period of intensive development before it.

How is this conclusion affected by the coming of the atomic bomb? The answer is first that the atomic bomb is not heavy. A large bomber was used to launch the atomic bombs that were dropped on Japan; but a small bomber could have carried them. Moreover the atomic bomb could be made as the war-head of a rocket. And General Arnold has said that the means are already in sight for increasing the range of rockets to 1,000 miles and more. And as the speed of rockets and the height to which they go are both so great, the difficulties of interception are correspondingly greater.

Surveys intended to sort out these matters must have an enormous scope and range. It appears that the main tendency is for striking weapons to change drastically; but for carrying vehicles to change more slowly. It appears to me that in this differential in the rates of change there lies the secret to future weapon policy.

Let it be supposed that the atomic bomb is so developed, and with it the rocket, that strikes can be conducted at great ranges and with all the explosive power needed to destroy large areas. Then it will follow that the ordinary bombing aircraft would play a negligible rôle if there were another war. The means of striking at the enemy would be the rocket with atomic war head. The bomber could justify its existence only on the grounds that there might be some targets which one side would want to destroy while leaving contiguous buildings or installations undamaged.

* * * * *

For during the war no effective means of countering the rocket assault was found. Owing to the extremely high speeds of the rocket weapon and to the great heights to which it ascends in its trajectory, it is difficult to find any means of dealing with it. Means of some kind will be found no doubt; but it may confidently be expected that the balance between defence and offence will, as in the past, swing a little this way and that, but never tip finally and completely in either direction. So the rocket with atomic war-head will get through. Means may be found for destroying some rockets before they reach their targets; but not many would have to arrive for the destruction to be massive.

The method of striking at an enemy, therefore, has undergone fundamental change. There will be a leaning towards massive destruction as opposed to the attack upon precisely defined targets. And the atomic bombs will be conveyed not by manned aircraft, but by guided missiles working at speed far in excess of the speed of sound and attaining heights of perhaps 80 or 100 miles above the earth at the peak of their flight. If this prediction be correct, the need for a large number of aircraft disappears. That applies whether the aircraft are land-based or ship-borne. They will not be required. The strike of the future will be massive and will not be selective. Instead of choosing a precise target and then selecting the precise weapon which fits that target—torpedo, bomb, rocket, or other—the whole of a large area in which the target may be found will be wiped out. With the passing of the bomber and of the torpedo-carrying aircraft, there must also be the passing of the interceptor fighter. And even if the change is gradual, it must affect both the bomber and the interceptor fighter simultaneously.

* * * * *

But wars are not won by striking, however massive the destruction wrought. Striking is only the first stage. There must always be the ultimate stage of occupation. It is in considering the means of achieving occupation that we come back to the carrying vehicles; to the ships and the aircraft. To wreck a city and so to destroy the ground on which it stood as if to sterilise it, is not victory in any sense accepted now or likely to be accepted in the future. Victory is when that ground is taken over and made fruitful by or partly for the benefit of the victors. In other words occupation is not only the outcome of victory, it is the purpose of war. And for occupation there must be the whole process of carrying.

We come, therefore, to the central fact that the final stages of the war, which were carrying operations, emphasise the real purpose of all wars and act as a reminder that whatever the weapons that may be used, there must come a stage when vast numbers of men and huge quantities of equipment must be moved from one place to another and must be safely landed at that new place and then fed and supplied while they remain there.

Apart from striking, therefore, there is still, and is always likely to be, carrying. Carrying has so far been done by a combination of land, sea, and air vehicles, and there is nothing to suggest that any drastic change will come. I believe that aircraft will be able in the future to take over increased carrying responsibilities; but I do not think that they will oust other methods of carrying. It follows that the operations in any future war which had to do with carrying would employ ships and transport aircraft. The aircraft would probably be much faster than the existing ones and they would probably be much bigger. They would fly higher and their rate of climb with full load would be better. But they would essentially be transport aircraft. There would likewise be transport ships, ships devoted to the carriage of men and supplies. Whether there would be aircraft carriers is more difficult to say. It may be that if there were another world war, there would be no place for the aircraft carrier. It would not be needed for the combat duties it performed in the recent war, and it would be unsuitable for aiding in any of the carrying duties. That applies to a world war.

* * * * *

Let the inferences to be drawn from the war of 1939–45 be summed up in this way: the operations of the war laid emphasis throughout upon the interdependence of the three Services; they enabled especially large strides to be taken in the development of the design and use of shipborne aircraft and of aircraft carriers; they showed that carrying must in the future form a large part of the aviation side of any operation. That was the first set of inferences. All of them lead to the stressing of collaboration and of the employment to the full of the advantages of the mobile base for improving the performance in combat of aircraft of all types.

The second set of inferences were all the result of the successful employment of the atomic bomb against Japan, and of the development by the Germans of guided missiles, especially of the rocket type, which are hard if not impossible to intercept. This set of inferences indicates that the ordinary heavy bomber has had its day and that major blows will be struck by guided missiles with atomic war heads. Therefore the weapons appropriate to major war will be these guided missiles and the apparatus of transport and supply which will provide the necessary sequel to the destruction of any part.

There are even wider problems to be faced. For instance the very nature of the atomic bomb makes it better suited for use against countries with high population densities, large industrial and built-up areas. Against primarily agricultural communities it would obviously be far less effective. If Great Britain were ever to take the offensive she would be at a disadvantage in comparison with any other country in the world if the atomic bomb were the main weapon. It follows that Britain's defences against the atomic bomb, launched by rocket, are more vital to her future existence than to the future existence of any other country.

But these are wider speculations. Enough has been said to show the complexity of the problems which now face those charged with shaping weapon policy. Let us look carefully and clearly at the new weapons; but let us not forget the special demands of carrying or neglect to continue the appropriate development of the old weapons.

43. 1948

The Future Employment of Naval Forces
Fleet Admiral Chester W. Nimitz USN

U.S.N. The Nimitz Report

On the day he relinquished office as Chief of Naval Operations of the U.S. Navy in December 1947, Fleet Admiral Chester W. Nimitz submitted to the Secretary of the Navy the subjoined report, giving his views on the function of the Navy in maintaining the future security of the United States.

* * * * *

Employment of Naval Forces in the Future

In addition to the weapons of World War II the Navy of the future will be capable of launching missiles from surface vessels and submarines, and of delivering atomic bombs from carrier-based planes. Vigilant naval administration and research is constantly developing and adding to these means. In the event of war within the foreseeable future it is probable that there will be little need to destroy combatant ships other than submarines. Consequently, in the fulfilment of long accepted naval functions and in conformity with the well known principles of warfare, the Navy should be used in the initial stages of such a war to project its weapons against vital enemy targets on land, the reduction of which is the basic objective of warfare.

11. For any future war to be of sufficient magnitude to affect us seriously it must be compounded of two primary ingredients: vast man-power and tremendous industrial capacity. The conditions exist today in the great land mass of Central Asia, in East Asia and in Western Europe. The two latter areas will not be in a position to endanger us for decades to come unless they pass under unified totalitarian control. In the event of war with any of the three we would be relatively deficient in man-power. We should, therefore, direct our thinking towards realistic and highly specialised operations. We should plan to inflict

unacceptable damage through maximum use of our technological weapons and our ability to produce them in great quantities.

12. Initial devastating air attack in the future may come across our bordering oceans from points on the continents of Europe and Asia as well as from across the polar region. Consequently our plans must include the development of specialised forces of fighter and interceptor planes for pure defence, as well as the continued development of long-range bombers. Offensively our initial plans should provide for the co-ordinated employment of military and naval air power launched from land and carrier bases, and of guided missiles against important enemy targets. For the present, until long-range bombers are developed capable of spanning our bordering oceans and returning to our North American bases, naval air power launched from carriers may be the only practicable means of bombing vital enemy centres in the early stages of a war.

13. In summary it is visualised that our early combat operations in the event of a war within the next decade would consist of:

Defensively:

(*a*) Protection of our vital centres from devastating attacks by air and from missile-launching submarines.

(*b*) Protection of areas of vital strategic importance, such as sources of raw materials, our advanced bases, etc.

(*c*) Protection of our essential lines of communication and those of our allies.

(*d*) Protection of our occupation forces during re-enforcement or evacuation.

Offensively:

(*a*) Devastating bombing attacks from land and carrier bases on vital enemy installations.

(*b*) Destruction of enemy lines of communication accessible to our naval and air forces.

(*c*) Occupation of selected advanced bases on enemy territory and the denial of advance bases to the enemy through the co-ordinated employment of naval, air, and amphibious forces.

14. Of the above activities or functions there are certain ones which can be performed best by the Air Force and certain others which can be performed best by the Navy—it is these two services which will play the major roles in the initial stages of a future war. The 80th Congress took cognisance of this fact when, in the National Security Act of 1947, it specifically prescribed certain functions to the Navy, its naval aviation, and its Marine Corps. In so doing the Congress gave emphasis to the fact that the organisational framework of the military services

should be built around the functions assigned to each service. This is a principle which the Navy has constantly followed and is now organised and trained to implement.

15. Defensively, the Navy is still the first line the enemy must hurdle either in the air or on the sea in approaching our coasts across any ocean. The earliest warning of enemy air attack against our vital centres should be provided by naval air, surface, and submarine radar pickets deployed in the vast ocean spaces which surround the continent. This is part of the radar screen which should surround the continental United States and its possessions. The first attrition to enemy air power might be by short-range naval fighter planes carried by carrier task forces. Protection of our cities against missile-launching submarines can best be effected by naval hunter-killer groups composed of small aircraft carriers and modern destroyers operating a team with naval land-based aircraft.

16. The safety of our essential trade routes and ocean lines of communication and those of our allies, the protection of areas of vital strategic importance such as the sources of raw material, advanced base locations, etc., are but matters of course if we control the seas. Only naval air-sea power can ensure this.

17. Offensively, it is the function of the Navy to carry the war to the enemy so that it will not be fought on United States soil. The Navy can at present best fulfil the vital functions of devastating enemy vital areas by the projection of bombs and missiles. It is improbable that bomber fleets will be capable, for several years to come of making two-way trips between continents, even over the polar routes, with heavy loads of bombs. It is apparent then that in the event of war within this period, if we are to project our power against the vital areas of any enemy across the ocean before beachheads on enemy territory are captured, it must be by air-sea power; by aircraft launched from carriers; and by heavy surface ships and submarines projecting guided missiles and rockets. If present promise is developed by research, test, and production, these three types of air-sea power operating in concert will be able within the next ten years critically to damage enemy vital areas many hundreds of miles inland.

18. Naval task forces including these types are capable of remaining at sea for months. This capability has raised to a high point the art of concentrating air power within effective range of enemy objectives. It is achieved by refueling and re-arming task forces at sea. Not only may the necessary supplies, ammunition, and fuel be replenished in this way but the air groups themselves may be changed. The net result is that naval forces are able, without resorting to diplomatic channels, to establish offshore anywhere in the world, airfields completely equipped with machine shops, ammunition dumps, warehouses, together with

quarters and all types of accommodation for personnel. Such task forces are virtually as complete as any air base ever established. They constitute the only air bases that can be made available near enemy territory without assault and conquest; and furthermore, they are mobile offensive bases that can be employed with the unique attributes of secrecy and surprise—which attributes contribute equally to their defensive as well as offensive effectiveness. Regarding the pure defence of these mobile air bases the same power projected destructively from them against the enemy is being applied to their defence in the form of propulsion, armament, and new aircraft weapons whose development is well abreast the supersonic weapons reputed to threaten their existence.

19. It is clear, therefore, that the Navy and Air Force will play the leading roles in the initial stages of a future war. Eventually, reduction and occupation of certain strategic areas will require the utmost from our Army, Navy and Air Force. Each should be assigned broad functions compatible with its capabilities and limitations and should develop the weapons it needs to fulfil these functions, and no potentiality of any of the three services of the Military Establishment should be neglected in our scheme of National Defence. At the same time each service must vigorously develop, in that area where their functions meet, that flexibility and teamwork essential to operational success. It should also be clear that the Navy's ability to exert from its floating bases its unique pressure against the enemy wherever he can be reached—in the air, on sea or land—is now, as it has been, compatible with the fundamental principles of warfare. That our naval forces can be equipped defensively as well as offensively to project pressure against enemy objectives in the future is as incontrovertible as the principle that every action has an equal and opposite reaction.

20. In measuring capabilities against a potential enemy, due appreciation must be taken of the factors of relative strength and weakness. We may find ourselves comparatively weak in manpower and in certain elements of aircraft strength. On the other hand we are superior in our naval air-sea strength. It is an axiom that in preparing for any contest, it is wisest to exploit—not neglect—the element of strength. Hence a policy which provides for balanced development and co-ordinated use of strong naval forces should be vigorously prosecuted in order to meet and successfully counter a sudden war in the foreseeable future.

World War II—Men in Action

SUNDERLAND FLYING-BOAT OF THE COASTAL COMMAND.
The control cabin, showing pilot in port seat, second pilot, using Aldis lamp, in starboard seat.
(*Official R.A.F. photograph.*)

Life in a British submarine. 1. The engine room.
(*The Times* photograph.)

Life in a British submarine. 2. The ward-room.
(*The Times* photograph.)

With an arctic convoy. Destroyer's forecastle at dawn.
(British official photograph. Crown copyright reserved.)

Vehicles landing at Salerno from an L.S.T.
(*Official photograph. Crown copyright reserved.*)

The first men ashore in Normandy. A British Naval Commando.
(British official photograph. Crown copyright reserved.)

World War II—Merchant Ships at War

ANCHOR-DONALDSON LINER ATHENIA.

(Photo: Maclure, Macdonald & Co., Glasgow.)

(Constructed by The Fairfield Shipbuilding and Engineering Co., Ltd., Govan.)

(From a drawing by Arthur J. W. Burgess.)

S.S. RAWALPINDI FOR THE PENINSULAR AND ORIENTAL STEAM NAVIGATION CO.

(Under construction by Messrs. Harland and Wolff, Ltd.)

THE CUNARD WHITE STAR TURBINE LINER QUEEN MARY.
[FROM A PICTURE.]
(*Builders, Messrs. John Brown and Co., Clydebank.*)

PART VII: 1950–1956
Infantry Wars in a Nuclear Age

Introduction	211
44. 1951	
Western Defence	212
COLONEL E. H. WYNDHAM	
45. 1954	
The Role of Modern Infantry in Battle	217
MAJOR-GENERAL SIR HAROLD E. FRANKLYN	
46. 1955	
The Strategic Air Command, U.S. Air Force	221
AIR VICE-MARSHAL W. M. YOOL	
47. 1955	
Operations of Regular Troops Against a Guerrilla Enemy	226
MAJOR-GENERAL E. K. G. SIXSMITH	

Introduction

Brassey's responded immediately to its new title of *The Armed Forces Yearboook* (see p. xix) by devoting considerable space to the Army and Air Force. The latter had, of course, figured considerably in earlier years because of the impact of aircraft on maritime war and the bitter struggle for control between the two services. The Army, however, had had little consideration and this was now corrected.

The British Army found itself in a paradoxical world. On the one hand, it had to adjust itself to a permanent role as a component of NATO forces in Europe and, in that context, to work out the effect of nuclear weapons on the battlefield. But in contast to the theorising and training involved in this, the Army in fact spent a great deal of its time in fighting conventional wars in Kenya, Korea and Malaya and elsewhere. Although these were predominantly infantry wars, with the other services providing fire and logistical support, they were also wars of ideas and social phychology because of the ideological inspiration of the enemy. These were the wars for "hearts and minds" as well as of close combat.

In startling contrast to all this was the emergence on the scene of the United States Strategic Airforce, the first organisation designed specifically for nuclear deterrence—an issue which was to dominate the coming decades.

Recommended Reading

Sir Robert Thompson, *Defeating Communist Insurgency* (1966).

44. 1951

Western Defence

Colonel E. H. Wyndham

This Chapter was written in the spring of 1951 and therefore only deals with the situation as it was at that time. No doubt great advances will have been made by the time of publication. We are, however, concerned with the erection of an edifice upon foundations already well and truly laid. Between spring and autumn the walls will grow higher, but the plan and elevation remain constant. It would indeed be a disaster if they were to be changed and building to start again on fresh foundations. The measure of the progress made in the spring of 1950 is at once apparent if we look back at Chapter IV of "Brassey's Annual" for that year and compare the situation as it was then with the present organisation as revealed in the White Paper published in April, 1951 (see Reference Section). The main weakness in the organisation is perhaps a surfeit of committees.

Now we have General Eisenhower once more in Supreme Command with Lord Montgomery as his Deputy. There had been no change of plan, there has merely been an extension and development of the previously existing plan. The new Supreme Commander has started off from the point on the road already marched by the organisation set up by the Treaty of Brussels and he marches in the same direction.

The terrain to be dealt with is so huge that decentralisation was clearly necessary. The division into three areas, northern, central and southern, suffices for the present. The central area, consisting of Western Germany, the Benelux countries, and France is the keep, the loss of which would be disastrous, though, as the last war proved, not necessarily decisive. Here will be the greatest concentration of forces under the direct control of the Supreme Commander himself. In view of the geographical position of France as the backbone of the keep, it is reasonable that the command of the land forces of the central area should be in the hands of a Frenchman.

The northern area, consisting of Scandinavia, the Baltic, and the North Sea, is equally reasonably to be commanded by a British Admiral, Sir Patrick Brind, with an American in command of the Air, and the Danish and Norwegian Armies under their own commanders.

At the time of writing the set-up of the southern area is still undetermined. It is clearly a much more complex problem than is the case with the other two. Presumably it will comprise Austria, Italy, and North Africa. Clearly it is inextricably bound up with the question of the defence of the Mediterranean, mainly a Naval and air problem. The Mediterranean must surely be treated as a whole.

What is the role of Portugal in Western defence? Will she come under the central or the southern area? Presumably her main contribution will be to supply Naval and air bases for forces engaged in the Battle of the Atlantic, and in this to some extent making good the gap left by the absence of Eire from the Atlantic Treaty. If Spain was a participant in the plan, both she and Portugal would clearly fall into the central group. In her present isolated position Portugal seems to fall more naturally into the orbit of the Atlantic Command than into any of the three areas under General Eisenhower.

The strategic importance of Great Britain in any system of defence for Western Europe is obvious. The facts of history have proved that for many centuries this island was rendered inviolable thanks to "the silver sea, which serves it in the office of a wall." Its security has been lessened in modern times by man's conquest of the air and the development of long-range missiles. Its importance as a base for sea, land, and air operations for the defence of Western Europe remains, however, unimpaired. The problem is to give it the highest possible degree of security against air and long-range missile attack. It forms a link between the northern and central areas, and its defence falls within the orbit of both. It is in effect a second keep. The northern and central areas form its outer defences, but clearly its last ditch defence must be a purely British responsibility. So long as the command of the sea rests in Western hands, this will be mainly a question of anti-aircraft defence, long-range missiles being in this context the equivalent of aircraft. It will be liable also to airborne attack. Defence against this may be primarily based on a Home Guard, but a mobile striking force of well-trained troops will also be necessary. Finally, it is of course clear that it is not Great Britain alone but the British Isles as a whole which are important to Western European defence. The gap left by the neutrality of Eire seriously complicates the problem of protecting the Western Approaches.

The northern area is likewise weakened by the regrettable decision of Sweden not to sign the Atlantic Treaty. Her particular difficulty is dealt with later in this article. At the same time there is no doubt as to what she would do if attacked and she has set her defences in order. But, should the course of a future war drive her into the arms of N.A.T.O., she would come as a stranger without previous co-ordination of plan or tactical doctrine. But, whatever line Sweden may eventually take, it is clear that the security of Scandinavia as a whole is of great importance. If Norway and Denmark can be held the Baltic is sealed off, submarine and other warships have a long way to go from Archangel and Murmansk before they can seriously threaten the passage of the North Sea and the Western Approaches, and Great Britain is much more secure against air attack. Norway and Sweden have another asset; their topography is difficult for an invader to cope with and the communications leading into them from the east are very poor. On the other hand, Denmark is inevitably lost if the forces of the central area are compelled to make any withdrawal from their forward positions at the start of the campaign. Norway would then be in a very isolated position and threatened from the direction whence she was overrun in 1940.

The proximity of Denmark to the forward Western positions on the Elbe brings out one of the weaknesses of the central area. N.A.T.O. will never embark upon a preventive war and hence will not be the aggressor if war comes. This means that the opening battle would be fought in the place and at the time selected by the enemy with whom the initiative would therefore lie. An initial withdrawal by the N.A.T.O. troops in the central area is therefore almost inevitable. As already mentioned, any withdrawal whatever on the northern flank puts Denmark in very serious jeopardy; while from every point of view it would be important to bring the withdrawal to a halt on the Rhine at latest and this only leaves a distance of 250 miles to play with. Not much in these days of rapid movement. The other chief weakness of Western Europe, including Great Britain, is that it presents a far more concentrated target to air attack than any to be found beyond the Iron Curtain. On the other hand, the enemy's communications will be very long and contain a number of vulnerable bottlenecks.

* * * * *

The problem of the southern area resembles that of the northern in that it seems to be inextricably bound up with the sea. If Austria and northern Italy were overrun, the security of the central area would not necessarily be endangered provided forces were available to defend the French Alpine frontier. But if the Italian peninsula falls into the hands of the enemy, the security of the Mediterranean is seriously weakened to say the least. Surely the southern area cannot be considered merely from the point of view of the land and air defence of Austria and Italy. Turkey and Greece, though not signatories of the Atlantic Treaty, are associated with N.A.T.O. plannings in the Mediterranean and their problems need to be taken fully into account. It is also impossible to divorce the defence of Western Europe from the question of the defence of the Middle East, and the importance of the sea passage of the Mediterranean in this connection is obvious. From this thought the mind naturally turns to the question of Yugoslavia, ideologically communist but independent in outlook and determined to resist aggression from any quarter. There can of course be no question of it falling upon her from any direction but the east. Therefore she surely must be taken into account in considering the defence of the southern area. While it is highly improbable that she will ever adhere to the Atlantic Treaty, at the time of writing she is seeking assistance from the West in the matter of the supply of armaments. Should this come to pass, clearly they should only be supplied if certain conditions are fulfilled. The first condition must certainly be that she would cooperate in the fullest degree in the important point of keeping the enemy out of Greece and eliminating Albania as an isolated outpost of the eastern bloc, as she is at present. Obviously the whole position would be greatly strengthened if the enemy could be kept out of Yugoslavia, Greece and Turkey. The passage of the Mediterranean would then be almost completely secure. But from the North Cape through Norway and Denmark, thence across Western Germany to Switzerland, on again through Austria and Yugoslavia to Greece and finally to the frontiers of Turkey is an immense length of ground to be made secure against an enemy who would be on interior lines and able to choose his point of attack. Penetration somewhere would be certain to occur.

We are told that the southern area of General Eisenhower's Command is to be commanded by an Italian general. This must clearly mean that its main purpose is the defence of the Italian peninsula. Therefore, if forced back from Italy's north-eastern frontier, the line of withdrawal will be towards the neck of the Peninsula and away from the central area. It would hardly be good for the morale of the Italian forces to expect them to abandon their native land to the enemy and turn aside to defend the Alpine frontier of France. It is difficult to avoid the conclusion that a better set up would be to have the northern and central areas under General Eisenhower and a separate Mediterranean Command, as we had in the last war. This Command to control the whole Mediterranean from the naval point of view and Italy and any possible Balkan operations from the land point of view. The air, of course, covering both under one commander. We are quite rightly advancing step by step; surely this will be the next step.

There is another most important point to be considered which we in this country find it hard to understand because we have been more fortunate in our experience of European war than have other Western European countries. We have been defeated in the past on the Continent, at Dunkirk for example, but we have never experienced the ultimate consequence of defeat; we have never suffered enemy occupation. Other countries have, some once, some twice, within living memory. Some many times in their history. The people of these countries very naturally and rightly now tend to consider the problem with their heads rather than with their hearts. They are ready to die for their country, but they do not wish to die for it in vain. They envisage the possibility of resistance, if it be not backed with sufficient strength, merely leading to total destruction of the whole scheme of things in their native land. It is therefore quite plausible for them to argue that, rather than suffer total destruction with those who survive shut up in concentration camps, it may be better to let the enemy in and await liberation from the West, as they had to do last time. They know, from their previous experience of enemy occupations, that they are past masters of the art of making the best for themselves of such a situation and at the same time making the worst of it for the occupying force. Then, when liberation comes, there may be something left in the country to carry on with. All this is perfectly logical and we should try to understand and appreciate it. None of these countries can hope to survive by their own unaided efforts. They must have help and they must see beforehand that the help is there. The lesson for N.A.T.O. from all this is simply that men, guns, tanks and aircraft must be on the spot in Western Europe in such quantities as to engender a general feeling of security as soon as possible. We still have a long way to go before this will be the case.

* * * * *

45. 1954

The Role of Modern Infantry in Battle
Major-General Sir Harold E. Franklyn

Experience in World War II shows that although Armoured Forces often play an important part in the opening and concluding stages of campaigns, the Infantryman is usually in greatest demand during the middle period when the most stubborn fighting takes place. As we know this period may last for years. It is also interesting to note that in the most recent campaigns on any considerable scale—in Korea and in Indo-China—the foot soldier has been supreme on both sides. Only he could climb the hills of Korea, wade the numerous rivers and streams and negotiate the swampy rice fields of Indo-China. This is not due entirely to the peculiarities of terrain in either country, as they are in many respects quite different. It seems likely, therefore, that the Infantryman will still assume the major role on land in any future war, and if this is so it will be profitable to discuss the organisation, equipment, and methods of the modern Infantry soldier.

Before attempting to estimate the part which infantry will be called upon to play in wars of the future it is first necessary to examine whether there is likely to be a radical change in the way in which these wars will be fought. There is a temptation to attribute exaggerated importance to every new invention designed to increase the power to wage war. Some of these inventions seem to be of such a revolutionary character that there is a demand, usually among those without much experience or balance, that previously accepted methods should be discarded. The advent of gunpowder, rifles, automatic weapons, modern artillery, gas, aeroplanes, and tanks, each in their turn, had this effect to a greater or lesser degree, and yet sooner or later antidotes were found. The upshot has been that, although tactical ideas have had to be changed materially, the clash of armed forces in battle has not been fundamentally different from what it was centuries ago.

At the moment atomic power, whether used from the air or eventually from the ground, appears so overwhelming that it is indeed difficult to assess its effect on wars of the future. Yet surely and inevitably the balance will be restored, as it always has been, by antidotes or at least palliatives. Until atomic missiles have been used on the battlefield their effect on tactics must remain a matter of conjecture. Although it is of vital importance that military thinkers should try to peer into the future, there is no justification for anyone to be dogmatic on the matter yet. Some would have us believe that such will be the effect of atomic power that campaigns will be fought out almost entirely between swiftly moving armoured forces supported from the air. The offensive power of such forces is almost unlimited and their mobility makes them a poor atomic target. Their defensive power, however, is small, yet base installations, lines of communication, capital cities, industrial areas, and all the varied means of production will have to be protected from assault by land as well as from the air. No amount of manœuvring by armoured forces will suffice to protect these and other vital areas; they must be defended by troops of all arms. If this is so the attack on positions defended in this way cannot be made either effectively or economically by armoured divisions, the attack must in fact be made also by a force of all arms. In such battles infantry must continue to play a predominant part and it is therefore important to see how this part can best be played, making due allowance for the effect of atomic weapons as far as this effect can be estimated but without exaggerating it.

* * * * *

The Influence of Atomic Weapons on Infantry Tactics

Until some form of satisfactory antidote has been found it is clear that atomic power must be such an important factor that both strategy and tactics will be profoundly affected. It would appear that any dense concentration of troops, of transport or of installations will form such a vulnerable target that survival is unlikely. Dispersion and mobility would seem to be the best safeguards and these safeguards must be observed if infantry are to play a useful part in battles of the future.

It is apparent that any form of static defence, particularly in well defined localities, will be highly dangerous. This fact is so obvious that it has already been suggested that in future attackers will depend for success on blasting a passage in a defensive zone by atomic missiles, after which armoured divisions will pour through to exploit success. If this is to be the plan, then the defence must be so arranged as to minimise such a danger. The more passive the defence the greater the risk and so resistance must be as active and mobile as possible. As long as space is available a manœuvre battle may be the best method of protecting vital objectives, but sooner or later space will dwindle and the side which strategically is on the defensive will be forced to stand and fight. Once an army has to occupy a defensive position there are three main considerations which should govern the deployment of the troops. First, the actual position should be lightly held, otherwise the casualties in the area when the enemy decides to make his main thrust will be shattering. Secondly, reserves for counter attack should be strong, be well hidden, be far enough back to escape the initial atomic attack, and yet mobile enough to act quickly when needed. Thirdly, there should be an anti-tank obstacle behind the defensive zone, so that after the enemy has torn a rent in this zone with his atomic missiles his armoured troops cannot immediately seize the opportunity which has been made for them. This rather novel form of defensive plan must now be examined in more detail.

The first object must be to prevent the attack, by using ordinary pre-atomic methods, from brushing aside opposition. This object can only be accomplished by holding a sufficiency of defended posts. The continued survival of these posts will depend as much on counter attack as on the tenacity of their garrisons. It is presumed that attackers, once they have established close contact with the defence and find further progress difficult, will resort to atomic assault. Assuming that the garrisons in the area of atomic attack are wiped out, although this may not be inevitable, it is now necessary for the attackers to establish bridgeheads beyond the anti-tank obstacle before the armoured divisions can be unleashed. It would seem that at this point of the battle atomic advantage may well swing towards the defence. Infantry moving forward to establish bridgeheads will be confined by the anti-tank obstacle in front, and on the flanks by the posts of the original position, the garrisons of which should still be holding out. These infantry thus confined and partially static should themselves be a prey to atomic attack.

This picture only intends to portray a possible form of defence against an army using atomic weapons; but it serves to show that even so an attack will not necessarily be a smooth uninterrupted process and that during the pauses when the attack is temporarily halted there will be opportunities for atomic counter attack. It is of course possible that the attack will take advantage of the fact that the defence is widely dispersed with a view to meeting atomic bombardment and that these dispositions will be less well suited to meet an ordinary attack.

Theoretically this brief examination of the effect of atomic power on infantry tactics may be correct as far as it goes. It is difficult, however, to estimate the moral effect on those battalions dispersed even thinly in a defensive position and at the apparent mercy of atomic attack. Still more doubtful will be the fighting value of battalions which have already undergone such an experience. There is, however, a likelihood that no nation will for some time to come dispose of sufficient nuclear material to be able to use it on any but the most profitable targets. Dispersed infantry do not present a target comparable with the many installations which are vital to the needs of an army. Inevitably less accuracy in bombing will be needed and more telling results obtained by attacking these installations rather than the army itself.

Morale

So far attention has been almost entirely concentrated on weapons, equipment, and the various tactical methods applicable to modern infantry. To whatever degree of perfection these may be developed it is the quality of the infantry as fighting soldiers, however, that will be the deciding factor between success and failure. Fortitude and self-reliance in the individual and discipline and *esprit de corps* in the battalion count for much more than skill with weapons, efficiency with equipment, or tactical ability. Similarly, dash and self-confidence are more valuable qualities for the junior leader than a high degree of military knowledge. Above all the excellence of British infantry has depended, and will surely continue to depend, on complete loyalty to the regiment. It can be argued that other arms of the service and other nations can achieve results with only a much wider form of loyalty: it can be proved that administrative difficulties over reinforcements in time of war make the working of the regimental system well nigh impossible; yet it still remains true that any action calculated to undermine this loyalty to the regiment, however plausible it may appear, is a blow at what has made British infantry incomparable.

46. 1955

The Strategic Air Command, U.S. Air Force*

Air Vice-Marshal W. M. Yool

The Background

The advent of the thermo-nuclear or hydrogen bomb as a practical weapon of war has resulted in the atomic bomb being regarded already as a largely outmoded weapon. Whilst there is no doubt as to the enormous destructive power of the hydrogen bomb, there has been considerable discussion as to whether the existence of nuclear weapons in the hands of the free world is in fact the main deterrent to aggression, and on whether they should be regarded as the primary war-winning weapon or their use confined to supplementing the more conventional weapons.

The position of the N.A.T.O. powers in regard to the use of nuclear weapons was made clear at the meeting of the Council of Ministers in Paris in December, 1954, when it was agreed that whilst N.A.T.O. defence planning and preparation is to be based on the assumption that nuclear weapons will be used, the actual decision to use them is specifically reserved to the Governments concerned. As the whole structure of N.A.T.O. is thus based on the use of nuclear weapons, there seems little doubt that if war should come permission to use these weapons would follow almost automatically. It does not necessarily follow that they would also be used in a world-wide strategic role, as the forces capable of using these weapons are outside the control of N.A.T.O.

Mr. Dulles, on his return from the N.A.T.O. meeting in Paris, stated that the reductions proposed in the U.S. Army were justified by the fact that the employment of atomic weapons would enable Western Europe to be successfully defended without having to look to its eventual liberation. Mr. Dulles then described the current military policy of the United States as the consummation of the policy of "continental

---

*The information on which this article is largely based was kindly supplied by the Command.

defence" (a new term which has the same meaning as "massive retaliation" and refers to the ability of strategic bombers to reach any part of the world), based essentially on long-range striking power which would serve as a deterrent to aggression and in the last resort ensure victory.

This policy has not gone unchallenged and many writers both here and in the United States have queried its soundness, on the grounds that "massive retaliation" will only result in the simultaneous destruction of the major cities and production centres of both sides. They have therefore argued that this policy is largely bluff, as neither side would dare to use the weapon because of the possible consequences. This feeling is probably more vocal in the United States than in this country, although a letter to the *Times* by Captain Liddell-Hart in January, 1955, revealed that there are doubters also on this side of the Atlantic. In the United States these doubts were well summed up in a report issued in December, 1954, by the Centre of International Studies attached to Princeton University, which stated that if an antagonist challenged the policy of "massive retaliation" the United States would have to accept the consequences of executing its threatened action, and that this meant that they would "either have to put up or shut up." If they "put up" they would be plunged into the immeasurable horror of an atomic war. If they "shut up" they would suffer serious loss of prestige and would have damaged their capacity to establish deterrents against further Communist expansion.

The policy of defence through the deterrent effect of nuclear weapons is based on the fact that such is the power of the hydrogen bomb that only a few of these missiles would be required practically to obliterate the main centres of population and production, even in countries of the size of Russia or the United States. This is assuming that the bombs could be delivered at the correct points, and there are writers who have suggested that the defence should prove sufficiently effective to ensure that the threat will not prove quite so devastating in practice as is feared.

Be that as it may, there is no doubt that even one hydrogen bomb successfully exploded on an important target would have a profound influence on the course of a war, and this supports the views of those who consider that the existence of such weapons in itself provides the best safeguard against war. It is certainly the declared policy of the British Government, as of the United States, and is confirmed in the Statement on Defence, 1955, which states, "In the hands of the free world, which at present has a marked superiority both in the weapon itself and in the means of delivering it, and which has no thought of aggression, it is a most powerful deterrent. In the Government's considered view this deterrent has significantly reduced the risk of war on a major scale."

Role of the Command

In all these discussions comparatively little is heard of the only force in the hands of the Western Powers that is actually capable at present of implementing the "massive retaliation" or "continental defence" policy, namely the Strategic Air Command of the United States Air Force. The Command is outside the control of N.A.T.O. and any decision to use this force in war lies with the United States Government, although there are agreements limiting the operation of the aircraft from certain advanced airfields without the permission of the Government concerned, e.g. the strategic bomber aircraft cannot operate from the advanced bases in the United Kingdom without the agreement of H.M. Government.

What then is the Strategic Air Command, on which the Western Powers at present depend both for the deterrent effect of nuclear weapons and to implement the policy of "continental defence" should this prove necessary in war? The Command is the long-range nuclear striking arm of the U.S. Air Force. As such it is organised so as to be able to conduct strategic air operations on a world-wide scale, and to do so at a moment's notice at any time. For this purpose it is maintained on a war footing in a constant state of readiness for instant action. A good indication of the importance attached to the Command can be gained from the fact that out of some £2,500 million spent on the Air Force by the United States since 1947, probably 75 per cent of this sum has been spent on the Strategic Air Command.

In the event of sudden aggression against the United States or her allies, the strategic bombers of the Command, operating from their home bases and from advanced bases in other parts of the world, would immediately mount simultaneous nuclear attacks against a number of selected vital targets located over a wide geographical area of the enemy's homeland. These widespread attacks, aimed at such objectives as industrial and communications centres, sources of power, and stockpiles of strategic materials, would be designed to destroy the vital elements of the enemy's war-making capacity to such an extent that he would no longer have the will or the ability to wage war. The Command is also organised to give tactical air support if necessary to army commanders by attacks on suitable targets, and to prevent the operation of the enemy's long-range bomber force.

As at present organised and equipped there is probably no part of the globe that is not within reach of the bomber aircraft of the Command,

and new aircraft now being developed will not only substantially increase the striking power of the force, but bring potential targets nearer in terms of flying time.

As an example of its meticulous planning the Command has strike plans drawn up to meet all foreseeable contingencies, and hundreds of these plans have been prepared in detail. Each plan is based on the possibility of different world conditions, different availability of forces, and different objectives. If the Command should be ordered by the U.S. Joint Chiefs of Staff to execute its primary task, the Commander may select any plan, or any combination of plans designed to meet the particular situation. The relevant portions of the various plans are disseminated down to the most junior member of the unit who needs to have any of the information. Combat crews know their wartime targets, the routes to the targets and the bases from which they will operate, and have been studying their possible operational missions for years.

* * * * *

Aircraft

The bomber wings are at present equipped with the 600 mile-an-hour B-47, each capable of delivering a nuclear weapon with a far greater destructive force than hundreds of the bombers used in the last war. Each aircraft has sufficient range to take off from its home base, fly thousands of miles to attack a target in the enemy's homeland, and return to a friendly base without refuelling. The B-47 weighs 176 tons, has six piston and four jet engines, and over a thousand of these aircraft have been built.

The eight-jet Boeing B-52 Stratofortress, described as a heavy intercontinental bomber, will soon be replacing the B-47, and will have a greatly improved performance. It was announced by the Secretary of the Air Force, Mr. Talbott, on March 2nd, 1955, that the B-52 is a 185-ton swept-wing bomber with eight jet engines mounted in pairs, and that its speed is said to be close to the speed of sound. Each aircraft costs 8 million dollars, can carry any kind of bomb, including the hydrogen bomb, and with full load will have a range of between 5,000 and 6,000 miles. The first wing to be equipped will be the 93rd Bombardment Wing at Merced, California.

* * * * *

Conclusion

The United States has not relied only on the Strategic Air Command for the operation of nuclear weapons for strategic purposes, but after some controversy decided to proceed with the construction of the giant carrier Forrestal, of 60,000 tons, and to build two more similar vessels. The Saratoga is expected to be completed in 1956 and another was to be laid down in 1954. Mr. Thomas, the Secretary of the Navy, speaking in New York on February 16th, 1955, said that the U.S. Fleet now had aircraft capable of delivering nuclear weapons on targets many thousands of miles away. He went on to say that the American naval forces are complementing their fixed land bases by a new concept of mobile sea bases—carriers, jet aircraft, and guided missiles—thus giving the American reprisal system the flexibility and dispersal it must have in these times when sudden atomic attack might overwhelm land bases. Whether the suggestion that, as these giant carriers can hide themselves in the oceans of the world, they are less vulnerable to surprise attack than the strategic bomber airfields is a sound argument for their construction is not relevant to this article; but the existence of these carriers does increase the strategic bombing potential of the United States.

Whether or not one is in agreement with the policy of the deterrent effect of the hydrogen bomb, it is evident from this survey of the role and organisation of the Strategic Air Command that the United States possesses in the Command and in its stock of nuclear weapons the most powerful force for waging war yet known in the history of mankind.

47. 1955

Operations of Regular Troops Against a Guerrilla Enemy

Major-General E. K. G. Sixsmith

The purpose of this article is to examine the nature of guerrilla warfare and the problems involved in dealing with an irregular enemy. Since the end of World War II our regular forces have been involved almost continuously in such campaigns, in particular in Greece, in Palestine, in Malaya, and in Kenya.

Those not familiar with the problems inherent in such operations are often surprised and critical of the difficulties experienced by regular troops in dealing with the enemy effectively enough to restore peace. The short campaigns of the past on the North-West Frontier—possibly only a winter's season—appear a pleasing contrast, showing the way these affairs should be conducted.

The essential factor to remember is that these campaigns are completely political in character. It is true that all war is a continuation of policy, but in guerrilla warfare the enemy's activities are aimed directly at the political machine, and every measure which the government takes which upsets the normal tenor of civilian life—however essential to deal with the enemy—is to some extent a score for the insurgents.

Guerrilla campaigns are thus a matter of great political importance and are subject to decisions at the highest level. Yet the agent by whom these decisions are forced home is anything but a high political personage; he is the humble company commander, attempting to control an area the size of an English county, in co-operation with an equally humble police officer and a handful of police. For this reason this article is divided into two parts, the first the discussion of the general principles, and the second an examination, by way of example, of the way these principles present themselves to the company commander in action.*

*For the production of Part II the author is indebted to Major A. R. Kettles, M.B.E., M.C., The Camcronians (Scottish Rifles), who commanded a company in Malaya for a considerable period.

PART I
General Principles

It has already been indicated that guerrilla operations are almost always operations in aid of the civil power. There is in existence a standard work on such duties known as *Imperial Policing*, by Major General Sir Charles Gwynn.† Although this work was first published in 1934 and in general it considered operations less extensive in scope than those that have taken place since the war, it is most interesting to see how his chapter on "Principles and Doctrine" stands the test of judgment in the light of recent experience. The four principles enunciated in this work are:

(1) Questions of policy remain vested in the civil government, and even when the military authorities are in full executive control, the policy of the government must be loyally carried out.

(2) The amount of military force employed must be the minimum the situation demands, since the military aims to re-establish the control of the civil power and secure its acceptance without an aftermath of bitterness.

(3) The necessity for firm and timely action. Delay in the use of force and hesitation to accept responsibility for its employment when the situation clearly demands it will always be interpreted as weakness, and eventually necessitate measures more severe than those which would have sufficed in the first instance.

(4) Co-operation between Army and police is essential. When unity of control is not provided, the necessity for close co-operation and for mutual understanding is all the more important. Anything in the nature of jealousy or competition to secure credit is certain to lead to lack of co-ordination in the course of action. Not the least important aspect of co-operation is a system of military and police intelligence, working in unison but each making the special type of contribution which its normal function and practice demand and allow.

The author develops the theme of co-operation by discussing the merits and demerits of martial law. In Malaya the same unity of control as is given by martial law was established by appointing a Director of Operations in command of all the security forces. During the time in which the initiative over the terrorist was gained for the first time, the appointment of Director of Operations was held by General Sir Gerald Templer, who was at the same time High Commissioner of the Federation of Malaya. Since then it has been found possible and politically desirable to separate the appointments, but the Director of Operations

† *Imperial Policing*, by Major-General Sir Charles Gwynn, is published by Messrs. Macmillan & Co. Ltd. of London.

still exercises command over all the Security Forces, civil and military. It has not so far been found politically desirable to follow the same practice in Kenya.

Political Factors

It will probably be agreed that the above principles are still applicable, and as general guidance for senior commanders and officials it would be difficult to improve upon them. Certainly no military principle has been more clearly underlined by experience than the fourth: success has followed the real integration of army and police, and when this integration—and the mutual confidence which goes with it—has been lacking, there has been failure and confusion. It is prudent to remember, however, that although failure to obey a principle is usually followed by difficulties or disaster, obedience to principles is not always followed by quick and overwhelming success. As Shakespeare says: "If to do were as easy as to know what were good to do then chapels had been churches and poor men's cottage princes' castles." The principles of minimum force and effective action, important though they be, do not solve the problem for a commander faced with a situation which stretches his resources to the uttermost. The difficulties of the commander on the spot have recently been rather of finding sufficient forces to maintain a reserve and at the same time maintaining control in areas where the initiative has been regained.

Palestine had and Malaya and Kenya both have the common factor that in the mass friend and foe are indistinguishable. More than that, the friend of today can be the terrorist of tomorrow, and vice versa. The question of which side a man is on can turn easily on the fears, hopes, prejudices, or beliefs of the moment. The thoughtless action or attitude of a soldier or policeman in the conscientious pursuit of his duty may make enemies of a village, while weakness or hesitation will be measured against the ruthlessness of the terrorist. It is easy for the terrorists, by the most ruthless methods or alternatively by arguments of racial loyalty, to ensure considerable co-operation for themselves, or at least non-co-operation with forces of law and order. The answer to such a situation is largely a political matter, but political remedies work slowly and in the meantime security forces must act. There are only two alternatives before the security forces, either to maintain such a state of law and order that the local population sees that law and order are to its advantage and does not fear to support it, or to impose such restrictions and force against the local population that it is impossible or dangerous to help the terrorists. The first is obviously the better course, but demands so large a force that it cannot be used except in the main centres of population. The second course is often the only practicable one in outlying districts. In Malaya the two most effective measures have been the concentration of local inhabitants into specified centres and villages and the control of food supplies. The former has in the main enabled the local population to be defended and it has separated friend from foe in so far as it has enabled the security forces to know who was about his lawful business and to identify those at large in the jungle as terrorists. Control of food is obvious in its application, since unless they can live on the local inhabitants the terrorists cannot operate near the centres of population or the main routes which give them their targets. It can be imagined that concentration of the population and stringent control of food are not measures that sit lightly on the local inhabitants, any more than are summary powers of arrest and interrogation. Such measures, moreover, give ample opportunity to the misguided as well as the malevolent to criticise or attack the political authorities or to make propaganda which assists the terrorists' cause.

It is not only in the countries where operations are taking place that political susceptibilities have to be watched. Perhaps to our credit as a nation, there are always among us people ready to believe that any rebel anywhere is a fighter for freedom. The political advances which, with our assistance, have been made in Asia and Africa make it fairly easy at times for some well intentioned and all malicious intellectuals to see the activities of terrorists as if they were those of true nationalists. Indian opinion and that of other Asian countries works in the same direction. United States opinion is by tradition much opposed to what she describes as "colonialism," but where the enemy is identified in any way with communism it is not likely she will lack understanding of the real issues at stake. The Government of the day may therefore be called to account for an action which appears to go further than is absolutely necessary and military commanders would be unwise to neglect political scruples at home. To this extent the principle of minimum necessary force is paramount.

A minor example of the way political susceptibilities may be aroused lies in the publication of individual terrorist losses and the kills ascribed to specific units or forces. It can be said that to do so is to reduce a matter of life and death to a sporting competition between units and to represent the stern task of a fighting unit to terms of the chase. Nevertheless, it is necessary to remark that the penalty of being a terrorist must be brought home to the population, and that a terrorist killed or captured is often the result of weeks or delays of patient and efficient work by a unit justly proud of its achievement.

The British soldier has always been considered Britain's best ambassador, and there is overwhelming evidence in the countries in which he now operates that the national service man of today is living up to the old tradition. Nevertheless, this guerrilla warfare is one in which he has to be constantly on his guard against his anti-terrorist activities being confused with anti-racial activities. In the long run our success in Malaya and Kenya depends upon the co-operation of the Malayian Chinese and the Kikuyu respectively, and our best ambassador has never had a more difficult task.

Hard as the political considerations are to meet, it is freely admitted that in recent experience military commanders have not lacked support for those measures which they consider expedient and right. But it must not be thought that once political difficulties have been resolved the rest is easy. There are still a host of military difficulties to overcome.

Military Factors

When the enemy has been separated from the local population he is by no means brought to battle. In countries like Kenya and Malaya he can find secure sanctuary deep in jungle or mountain forests. He must then not only be found but fixed and brought to battle by superior forces.

This problem of fixing the enemy is perhaps the most difficult of all. Military forces are organised and armed primarily for modern war. The organisation of an infantry battalion can be adapted for imperial policing, which is one of the traditional rôles of British infantry; nevertheless, it must be realised that the British soldier is trained for average conditions and that hunting a skilful terrorist in the jungle is far from average. National service creates special problems since, if you deduct the time for basic military training and for the journey each way, less than eighteen months is left for service in the jungle with a unit. Eighteen months is admittedly a long time to learn quite a lot about jungle life, but battalions themselves alternate every three years between home and overseas, so, looked at from a unit point of view, life is a succession of losing your known and trusted men and teaching a new lot. If the regular army were well up to strength with a cadre of experienced non-commissioned officers, all this could be taken easily in its stride, but many of the subalterns and most of the non-commissioned officers are themselves national service men. All these difficulties are cheerfully tackled and in the result any commanding officer will agree that the national service man comes out of it magnificently, providing a proportion of really first-class patrol and section leaders in the jungle. There are besides British infantry battalions the Brigade of Gurkhas, the only long-service units now left in the British Army, and such units with special aptitude for jungle operations as the Fiji Infantry Regiment and the Malayan and African units operating more or less in their own country. Nevertheless, the infantry battalion at work in the jungle—as will be illustrated below—presents a very different picture from the methodical methods and known hazards of an Indian infantry battalion at work on a punitive expedition against some erring village on the North-West frontier in the 1920s and 1930s.

Much has been talked of the infantry and little of other arms or services. It is true that the Royal Air Force, armour, artillery, engineers, signals, and administrative services all play an essential part in operations against the terrorists. On the other hand, one of the difficulties of guerrilla warfare is that operations do not take place on the scale that allows these arms to exert the same weight as they do in normal warfare.

Perhaps the most valuable air support has been in roles which might be considered subsidiary in modern war. The light aircraft has been invaluable in reconnaissance, for example, discovering enemy food areas and camps. Supply dropping and modern methods of parachuting make possible deep patrolling in the jungle. The helicopter has immensely increased the speed and range of operations and may—as reliability and ease of control increase—revolutionise deep jungle operations. Nevertheless, at the moment is is infantry that bear the main burden of operations, and it is on their performance that military success depends.

Summary

In order to summarise the general problem it may be profitable to examine the requirements of the terrorist forces and to show in what direction their requirements make them vulnerable. The requirements of the terrorists are:

(1) Friends who will help them with food, with medical attention, and with information about the movements of security forces. Such friends may exist because there are genuine sympathisers or can be obtained by intimidation. The forces of law and order can, however, combat the terrorist through their friends by producing a better cause for the genuine and by providing a degree of protection which makes intimidation an unrewarding practice.

(2) Money or funds in some other form in order that the necessary recruits and replenishment of arms and ammunition can be obtained. Help from abroad may provide the latter (as it did in Greece); whereas large initial stocks of arms are an aftermath of the war which in Palestine and Malaya gave the terrorists resources which made them largely self-contained in arms and ammunition.

(3) A secure base area and line of retreat. In the type of country like Malaya and Kenya this is easily obtained because the depth of the country can give security even to the extent of allowing extensive food-growing areas. Where such conditions apply the terrorists have the advantage that they can restrict their activities to such incidents as give them almost a certain chance of success. Even when the security forces get the initiative this advantage remains, and the success of the security force is usually seen in the lessening of incidents rather than preventing them altogether. This is a great disadvantage to the cause of law and order, because in times of peace even minor terrorist successes, such as a small ambush resulting in the death of one or two officers or men, naturally attract publicity out of proportion to the success.

* * * * *

Dawn of the Nuclear Age

The First Atomic Shell. The mushroom cloud formation from the first atomic shell fired from the United States Army's new 280 mm. gun, in Nevada in May 1953

(Office of Public Information, U.S. Department of Defence)

(a) Boeing B.47 Stratojet Bomber
(b) Consolidated Vultee ("Convair") XC-99 military cargo–transport aircraft

Avro Vulcan 4-jet Bomber

SUBMARINE LAUNCH OF POLARIS A-3 MISSILE

U.S.S. *Enterprise* Displacement (full load) 85,350 tons. Length overall 1,102 feet. Maximum beam (flight deck) 252 feet. Aircraft carried—70–80. Nuclear powered. In company, nuclear powered cruiser *Long Beach*, 15,974 tons (full load)

PART VIII: 1957–1965
Strains Within N.A.T.O.: The Intercontinental Ballistic Missile

Introduction	235
48. 1957	
Defence Policy. A New Approach?	237
H. G. THURSFIELD	
49. 1957	
The Royal Air Force in a Time of Growth and Change	242
D. M. DESOUTTER	
50. 1958	
Those Against the H-Bomb	251
RICHARD GOOLD-ADAMS	
51. 1963	
West European Defence	258
AIR VICE-MARSHALL W. M. YOOL	
52. 1965	
NATO	263
53. 1965	
The Future of the Aircraft-Carrier	272
VICE-ADMIRAL SIR PETER GRETTON	

Introduction

1957 marked a turning point for Britain and her allies. The era of the I.C.B.M., ushered in by the successful launch of the Soviet Union's *Sputnik*, was to make the United States increasingly vulnerable to nuclear devastation and to raise doubts about the continuing credibility of her nuclear guarantee to Europe. This increased the determination of Britain and France to modernise their strategic nuclear delivery systems despite the considerable technological and financial difficulties involved. To this strain on the solidarity of N.A.T.O. were added those arising from the recognition that both American and European members were likely to have differing interests outside the N.A.T.O. area, a fact which had been painfully demonstrated at Suez in 1956. Already the withdrawal of France from N.A.T.O.'s formal military organisation was seen as a possibility.

In Britain, the Defence White Paper presented by the Conservative Defence Minister Duncan Sandys showed the weaknesses in her security posture. It assumed that the possession of an effective nuclear deterrent force would enable her to reduce her conventional forces and thus decrease the demands of defence expenditure on the national economy. National service was to be abolished and considerable reductions in manpower achieved, but without any concurrent cuts in commitments. The following years with their demands for conventional limited war capabilities outside Europe and the growing realisation that N.A.T.O.'s continental strategy was going to call for increasingly expensive land and air forces, showed how unrealistic Sandys's hopes had been. An additional complexity was the rise of the Campaign for Nuclear Disarmament (C.N.D.).

Furthermore Britain's failure to produce her own cost-effective missile system forced her to turn to the United States, first for the abortive *Skybolt* stand-off missile and finally for the *Polaris* submarine based system obtained under the Nassau agreement of 1962.

In these very uncertain conditions the Royal Air Force and the Navy were striving, increasingly in rivalry, to justify their claims on an overstretched defence budget, and the latter was beginning to face up to the possibility of losing the aircraft carrier force which it regarded as its most valuable contribution to the country's security. The fact that it would soon take over the responsibility for the strategic nuclear deterrent was little consolation to many naval men.

Recommended Reading

R. E. Hunter, *Security in Europe* (Revised Edition 1972).

48. 1957
Defence Policy. A New Approach?
H. G. Thursfield

"The time has come", says the Defence White Paper of this year, "to revise not merely the size but the whole character of the defence plan." On economic grounds, indeed, many good judges regard that revision as long overdue. It is true that the 1950 programme by which it was intended to spend £4,700,000,000 in a space of three years was quickly recognised as beyond the country's capacity, and it was accordingly both retarded and pruned year by year. As this year's White Paper very cogently remarks, "It is in the true interests of defence that the claims of military expenditure should be considered in conjunction with the need to maintain the country's financial and economic strength." The cynical commentator might well remark that this unexceptionable statement contains nothing new. The principle so defined was heard in Parliamentary debates even in the last century whenever, as in the 1880's, the latest war scare caused any substantial increase in the Service Estimates. It was operative in the years succeeding the first World War, causing severe pruning of the Estimates in the 1920's; and it was probably the chief influence in causing this country to enter into the Washington Treaty of Naval Limitation of 1921, and the later treaties that successively prolonged it. It is only in times of actual war that the economic factor is disregarded in deciding on defence expenditure, and even then it tends to reassert itself as time goes on. Since the last war, critics of successive Governments' Defence programmes have cited it time after time, as indeed have successive Defence White Papers. What it seems to convey in this latest pronouncement is that the time has come at which something more than lip-service must be paid to it, and at which the burden of defence expenditure *must* be reduced.

The fact is, of course, that all military thinking and planning, since the discovery of how to apply to the arts of destruction the immense energy hitherto locked up in the atom, has been in a state of uncertainty, almost amounting to confusion. When after World War II, from Russian maintenance in being of immense armies in Europe, and maintenance, through those armies, of rigorous Communist control over many states beyond what had been Russian frontiers, it became clear that the danger of aggression had not been eliminated by the destruction of Hitler, the problem of Defence became acute. The immense armed forces of the democratically governed Powers that had played their part in the defeat of Hitler's Germany, had been demobilised and disperse; and even if they could have been recreated they could not have matched the numbers of which Russia disposed. The danger of Russia's extending by force of arms over the whole of Europe the regime of Communist dictatorship that she had established in what became known later as the satellite States was clearly acute; and it might have become an actuality but for the existence of "the great deterrent".

America alone then possessed the atomic bomb, and if Russian armies had proceeded to overrun Europe—as they could have done, in the "steam roller" fashion visualised in 1914, for all the ground and air forces that were available to resist them—it was well understood that it would be used against Russia itself. That prospect was clearly sufficient to deter Russia under Stalin from a fresh adventure in European conquest through naked military aggression. Moreover, the situation was not materially altered when Russia, in the fullness of time—a fairly short time—working in the secret and remote spaces of Siberia, herself developed nuclear bombs. American bombs still constituted the great deterrent, for one atom bomb is no answer to another except as a counter-deterrent. In Russian eyes, it merely provided her too with a "great deterrent" against aggressive attack—of which, to be just, it must be conceded that the Russian leaders may well honestly have believed in the need. With nuclear weapons available on either side, another World War would almost certainly bring the end of the human race—or most of it. The deterrent should be just as effective still, even if it had become two-sided.

Faith in its efficacy as a "sure shield" against all possible attacks led some superficial thinkers to the optimistic conclusion that no other form of defence or armed forces would be necessary in the future. Merely let the free nations of the West unite, maintain a stock of atom bombs and the means of dropping them anywhere in the world, ran this comforting theory, and the free world would be safe from attack by Communist aggressors. Armies and navies were obsolete, and even air forces would before long give place to "push-button" warfare.

To express the theory thus is doubtless an over-simplification; and it must be said at once that the Defence expert advisers of the Governments of the Western world were much too clear-headed to be beguiled by so facile a generalisation. It is rather a favourite argument with armchair strategists who have no responsibility to their countrymen, when a new weapon makes its appearance on the scene, first to prophesy, with exaggeration, what its capabilities will be when it is fully developed, and then to advocate acting at once as if they had actually been already realised. But those on whom responsibility for defence rests cannot act on unproved theories, however economically attractive such action might be; and resistance to massive aggression is not the only function for which armed forces are maintained. Comparatively small disturbances, even the possibility of "limited wars", are not eliminated by the removal of the threat of such naked aggression as would inevitably lead to another World War. And moreover, the Communist dictators before very long demonstrated very clearly that the "European steam roller" technique was not the only means of aggression at their disposal.

War by proxy—as in Korea and Indo-China; the cold war, typified chiefly by the Berlin blockade; the fomenting of internal disturbances the suppression of which—as in Czecho-Slovakia and more recently in Hungary—left local satellite Communist regimes in control, or at least—as in Malaya and Kenya—involved the Western Powers in difficulties and kept their forces on the stretch; the covert support of rampant nationalism in many cases in the Middle or Far East; all these provided examples of emergencies calling for the use of armed forces in which the "great deterrent" was of no influence. The result of new developments, that is to say, was not, as the armchair theorists would have it, to render conventional armed forces obsolete. It was merely to add the obligation to provide the new and colossally expensive nuclear weapons, to that of maintaining the conventional armouries on which reliance was placed before their appearance.

Moreover, apart from the immense cost of developing and providing nuclear weapons, monetary inflation, and the increasing mechanical complexity of ordinary weapons, their carriers, and military equipment generally, necessitated a level of defence expenditure far higher than that to which Governments had been habituated before 1939. In the past, the close of a war was hailed with relief by economists as heralding a great reduction in defence expenditure. Not only should there be a reduction from war to peace establishments in the Forces, but the latter should be lower than before through the removal of the threat of war that had swollen them before war actually broke out. Since World War II, however, instead of the coming of peace bringing any relief of basic defence expenditure—underlying the cost of actual operations, which had, of course, ceased—it seems to have increased the burden. It is not surprising that military thinking and planning should have been in a state of uncertainty, almost amounting to confusion. How must the limited resources of States be rationed between the old weapons and the new? Are the new weapons going to affect the conduct of "conventional" warlike operations? If so, to what extent, and what provision should be made against those effects? To none of these urgent questions has any clear answer been so far available; and Defence Policy has been, it seems, based on providing as large a fraction of the material demands of the Services as successive Governments have adjudged that Parliament could be induced to authorise. Put simply, the fact is that we have been spending more on Defence than the country can afford.

It is hardly clear, however, how the policy defined in the Defence White Paper really amounts to "a new approach". It is now admitted that, for this country, there is no defence against the nuclear weapons of an enemy; and the White Paper states that that fact "makes it more than ever clear that the overriding consideration of all military planning must be to prevent war rather than to prepare for it." But again, there is nothing either very clear or very new about that statement. As to its clarity, surely "military" as distinguished from diplomatic planning, has no means of preventing war other than preparing for it—to the extent at least of providing the great deterrent. And as to its originality, that great exponent of military wisdom, Vegetius, wrote in the 4th century A.D. *"Qui desiderat pacem, preparet bellum"*—which may be freely rendered as "Who would prevent war must develop hydrogen bombs", though it is expressed with greater prolixity in paragraphs 13 and 14 of the White Paper.

What *is* new is not so much the approach to Defence Policy as frank recognition of the ineluctable fact that we have been spending on Defence more than we can afford, and must spend less. We must, in short, since there is no other way, reduce the Armed Forces; and the White Paper is concerned with setting forth the details of how that reduction is going to be effected. The reductions will arouse serious misgivings in many people's minds regarding the risks that must be taken in order to effect them. Some of those were recently, as these words are being written, expressed in Parliament regarding the absence of any mention in the White Paper of the hundreds of submarines being built by Russia, or of any adequate provision for dealing with them in the event of war. Other people have been disturbed by the reduction of British forces in Germany, others by the decision not to proceed with the development of a supersonic manned bomber aircraft. Such misgivings are inevitable whenever reductions in Defence forces have to be made, and it is not to be supposed that they are absent from the minds of those on whom the responsibility for framing Defence Policy lies. As so often happens in human affairs, the choice has to be made between two evils, and the choice cannot be postponed in the hope, perhaps, that Providence will make it easier soon.

There is, perhaps, one gleam of light in an otherwise gloomy prospect. This country is not the only one to feel the pinch of the inevitable increase in defence expenditure; for the scientific and material progress of this materialistic age has not simplified the business of destroying one's fellow men, or of preventing them from doing the destruction. If modern death-dealing weapons are more effective than yesterday's, by the same token they are more elaborate and expensive. Man's material ingenuity has increased the proportion of his resources that he must devote to defending himself against destruction by an equally ingenious enemy. Thus it is that the cost of defence continues to grow—this country's defence expenditure, the White Paper tells us, has absorbed 10 per cent of its gross product in the last ten years and the proportion in countries of potential aggressors cannot be greatly different. That liability may well come to be recognised as being a deterrent as effective as the hydrogen bomb. Then indeed will be the time for a New Approach to Defence Policy.

49. 1957

The Royal Air Force in a Time of Growth and Change

D. M. Desoutter

After a long period of enfeeblement the Royal Air Force is regaining its power. Paradoxically it was during 1957, when the newspapers and the public at large seemed to think that the R.A.F. was going into a sharp decline, that British air power really began to build up.

To appreciate how such misconceptions could gain currency it is necessary to look back some years to the end of the Second World War. At that time the country was impoverished and the government of the day was anxious to finance new schemes of social welfare. It was in such political circumstances that a military appreciation of "no major war for a decade" was presented to the cabinet. In consequence of that hope the re-equipment of the R.A.F. was postponed for a period of about ten years. There were certain exceptions, but in the main that policy held good and it was only in the early part of 1957 that the last Meteor, for example, was retired from operational duties in Fighter Command.

The most notable exception to the ten-year plan was the Canberra: without it the service would have been even more enfeebled than in fact it was. Even so, the major part of the R.A.F.'s equipment remained at the level of performance appropriate to the last war. In direct contrast, both the U.S.A. and the U.S.S.R. maintained a logical policy of steady introduction of aircraft of successively higher performance; comparatively therefore, the performance of the R.A.F.'s machines did not remain constant, it declined.

The Korean war came at about the mid-point of the ten-year plan, and Members of Parliament were shocked to find that the R.A.F.'s fighters were no match for either the Mig-15s of our enemy, nor for the Sabres of our ally. Their surprise did not appear to be lessened by the fact that such weakness was the consequence of a policy approved by so many of their number.

Emergency action was taken in an attempt to mend the damaging consequences of what was inherently a bad policy. Canberras, Venoms, Swifts, Hunters, Meteor Night Fighters and so forth were ordered in large numbers, and with mixed success. Only a few years were to pass before those orders were either cancelled or greatly reduced. By that time it was seen that the "crash" programme of expansion had been extremely costly: very much more costly than it would have been had enough fighters of intermediate performance been built to give practical experience in even a few squadrons. The broad plan of a complete re-equipment of the Air Force after a period of about ten years was seen to be militarily unsound because it left a dangerous gap from the fifth year onward. From the point of view of engineering development it was even worse. All engineering innovations and improvements have their teething troubles: they need cause no special difficulty if they occur in succession over a long period because then they can be dealt with as they arise. But the ten-year leap idea simply meant that ten years' worth of troubles could be expected simultaneously at the end of the period. An outstanding example was the case of the Hunter—a new aeroplane, with a new engine and new guns. The combination presented formidable problems which would almost certainly have been solved earlier had one or other of the three seen some squadron service sooner.

Those troubles, indeed the whole of the R.A.F.'s fighter deficiencies, were well publicised at the time. Less was heard about the lack of bombers. Although in the Canberra the Service had an excellent aeroplane, it was of limited range. While the two main contestants in the cold war faced each other with nuclear weapons and the means to deliver them, the R.A.F. was more or less powerless to strike into Russia. Although the first British atomic bomb was made several years ago, it was not until the V-bombers began to come into service that such weapons could have any real meaning in the British strategy.

Although it is clear that no fixed moment in time can be stated when the R.A.F. changed from being an ineffective attacking force to an effective one, one can say that 1957, as a year, marks the turning point. It also happened that 1957, was the year when the first revelations were made about British missiles and when a Defence White Paper put new ideas before the public. Those two events had the effect of diverting public attention from the tremendous change that was taking place in the stature of the R.A.F. Up till that time only two World Powers disposed of the power of nuclear attack. Since that time forward there has been a third power with that capability. That is the measure of Bomber Command's contribution to British authority. One can only hope that statesmanship will measure up to the new situation.

Defence Policy

From the foregoing it will be seen that as far as the R.A.F. is concerned, the new defence policy of the 1957 White Paper coincided with the fulfilment of a policy laid down many years earlier, and at a time when the increase was still well short of its planned peak. Thus the White Paper showed not some drastic reversal of policy, but a sequence by which the new-found air power could first be prolonged and extended and, second, be tapered out in harmony with a gradual tapering in of new weapons. It is necessary to keep in mind a clear appreciation of the time scale if the process is to be understood. Details of the exact timing cannot be known, since the date of introduction of later marks of aircraft and of new missiles are held secret. But it is not necessary to know exact dates; one needs simply to be able to take a view whose perspective extends over a total period of between 15 and 20 years, beginning as we have already said, at the end of the Second World War. The implications can be followed by outlining the evolution of the R.A.F.'s strength from now henceforward Command by Command.

Before turning to a more detailed consideration of the prospects for each Command of the R.A.F. there is one general point to be made. Partly because of cost, partly because of the tremendous power of modern explosives, and partly because any expected major war would probably be of very short duration, the air force must henceforward be a small air force. If it is small in numbers it must be high in quality—an object to which small size is in fact an aid.

Bomber Command

It is commonplace to say that we find ourselves in a period when the attack is stronger than the defence. Only a very small proportion of the attacking missiles or aircraft need penetrate the defences in order to achieve damage far exceeding anything that could be done by even a thousand bombers of the Second World War carrying high explosive. Thus, in the Government's words, "the overriding consideration in all military planning must be to prevent war rather than to prepare for it." The preventive depends "primarily upon capacity for nuclear retaliation."

In that respect the White Paper of 1957 was in complete accord with the defence policy of the preceding years. The king-pin of policy is still to maintain a retaliatory force so powerful that aggression will have no semblance of being profitable. That force is found in the V-bombers of the R.A.F., armed with nuclear weapons. Happily a similar, and even more powerful force has been at hand for many years past in the form of the Strategic Air Command of the United States Air Force. Without such a force, British or American, the countries of the west would be defenceless.

It is worth remarking at this stage that from the financial point of view this central deterrent force is quite cheap. The V-bombers with the nuclear bombs they carry account for only about one-tenth of the whole British defence budget. Much more costly are the provisions that have to be made against local wars and subversion in widely separated parts of the globe.

Now that Britain has gained this power it is necessary to preserve it and to prolong it. The writer is in possession of no special knowledge of the dates planned for the various stages, but it is possible to sketch the prospects against a reasonable time scale.

One can begin with 1956. Before that year Britain was powerless to stage an offensive, but in that year the Valiants were coming into the squadrons in numbers and the radius of action of the R.A.F.'s nuclear bombs became for the first time a threat to the Russian interior.

By 1957 all the Valiant Squadrons had formed and the first Vulcan Squadrons were being trained and equipped, giving a further extension of range for Bomber Command. As far as the bombs themselves are concerned, the government was able to say "British atomic bombs are already in steady production and the Royal Air Force holds a substantial number of them." When that statement was made the plans for the testing of the first British hydrogen bomb had been announced, with consequences which have since been widely publicised.

Another step which enhanced the range of Bomber Command about that time was the introduction of Valiants equipped to receive fuel while in flight. These machines are fuelled by other Valiants which have been fitted with the fuelling pack—a complete attachment which can be housed in the bomb-bay to make a quick conversion from bomber to tanker.

Now to 1958, with increasing numbers of Vulcans coming into the service and the formation of the first Victor Squadrons. Evidently we have a process of increasing strength, with crews becoming more and more familiar with their new aeroplanes, able to exercise over long ranges and so to gain a better appreciation throughout the Command of the way in which the force can best be used.

Deliveries of aircraft will still be continuing in 1959 no doubt, and by that time the extra benefits of "stretching" may begin to be felt. "Stretching" is something that can be done when one has new aeroplanes. It cannot be done when one has very old aeroplanes, nor when one has no aeroplanes at all, which was Bomber Command's condition not so long ago. It is in fact a process of steady improvement, of development: the same process that gave the Merlin engine its outstanding life history: the same that has enabled the Douglas 4 to grow through the Douglas 6 to the Douglas 7 over a period of nearly twenty years of service.

Both the Victor and the Vulcan are susceptible to stretching. The maximum altitude they can achieve over the target can be raised by aerodynamic improvements such as higher aspect ratio and increased wing area. With those changes will go greater engine power and improved fuel consumption. In all, both these kinds of aeroplane can be given greater height, greater speed and greater range during the next five years. Already they have many times been described by both statesmen and senior R.A.F. officers as being the equal of any other bombers in the world in height and speed. (Though not in range where bigger American and Russian bombers have the greater performance and the greater need of it.)

Perhaps 1959, too, will see the introduction of the powered and guided bomb. The date is pure hypothesis; the important point is that this weapon will be a continuation of the process of adding to the power of the V-force. The guided bomb is in fact a flying bomb—a small rocket-driven aeroplane, filled with explosive and its proper means of control and guidance. Great emphasis has been laid upon its importance by R.A.F. officers for several years.

The guided bomb must be about the size of a small fighter, rather comparable to a Folland Gnat. In its first form it will be released from a V-bomber at a point between 100 and 200 miles away from the target. Driven by its rocket motor it will then accelerate to two or three times the speed of sound and climb from the release height of about 55,000 feet to perhaps 70,000 feet. Only when it is directly over the target will it pitch forward and drop vertically at even higher speed. Even in its earliest form this weapon gives the bombers a direct increase in striking range and an even more important increase in operational flexibility. With the conventional bomb, the point of release lies somewhere on the circumference of a circle centred upon the target and with a fixed radius for any given height and speed. With the powered bomb there is an infinite number of such circles with radii extending inward from the range of the bomb itself to within a few miles of the target. In fact the bomber would probably keep as far away from the target as possible—hence the use of the title "stand-off bomb" for this weapon.

Like the aeroplanes themselves, the guided bomb is susceptible to stretching. In particular its range performance can be, and no doubt will be, increased. As later versions are introduced there is no reason why the stand-off distance should not rise to 300, 400 or 500 miles. Experience gained with tests of the early version will make it easier to perfect the guidance mechanism and the power plant for such a distance.

One sees in this stand-off bomb a step toward the long-range missile. It is in fact a rocket-driven missile of the aerodynamic, as opposed to the ballistic kind, and with the special operational requirement of being launched from an aeroplane. Here is a parallel to the ballistic missile launched from shipboard, be it from a surface ship or from a submarine. While this is not the place to go deeply into the possibilities that are opened by the missile-carrying aeroplane, it is evident that if an aircraft instead of a ship can be used as a missile launcher, then a great increase in mobility will result. That such a thing is ultimately possible is not in doubt—it is a matter of the time scale.

Mention must be made of the very important problems of navigation and guidance in the delivery of nuclear weapons. Suffice it to say that the bomber itself must have the highest possible degree of navigational accuracy, both to take the bomb to the best launching position and also to supply the bomb's guidance mechanism with an accurate "knowledge" of the location of the launching point, so that it can correctly lay a course for the target. The means are available in the Doppler system of measuring speed over the surface of the earth, and in the inertial guidance system which is completely self-contained and needs no outside source of reference.

What has been said should show that Bomber Command will still be evolving to higher levels of effectiveness beyond 1960. Somewhere about 1960 the British intercontinental missile of about 2,000 miles range may be nearing the end of its trials, and perhaps 1961 or 1962 the squadrons of bomber command will be able to begin training with it. In any event, it is clear that there will be a period of some years during which the ballistic missile and the missile-equipped bomber will be in service together. Whether, and how quickly the manned bomber is then tapered out of service remains to be seen. Much will depend upon the tactical applications that are seen for manned aircraft carrying new kinds of weapons. In other words, it will depend upon the relative merits of fixed and mobile missile launching bases.

Fighter Command

"It must frankly be recognised that there is at present no means of providing adequate protection for the people of this country against the consequences of an attack with nuclear weapons." So said the Government in its 1957 White Paper. Although the statement shocked some, it was no more than some students of military affairs had already been saying. No fighter or missile defence able to protect the country as a whole is within sight of practical achievement; that is the conclusion of the Minister of Defence. What must be done is to protect the V-bomber force against surprise attack. It must be accepted that the United Kingdom will not be the initiator of nuclear warfare. Thus if ever the V-bombers do have to set forth with their nuclear weapons it will be in response to an attack upon the United Kingdom or her interests. No potential aggressor must be given any opportunity for thinking that he could dismiss those bombers from the field in the first surprise onslaught. Their bases must be protected against air attack so that the Valiants, the Victors and the Vulcans have time to get into the air. To protect them is the duty of Fighter Command, whose forces for the purpose will evolve through three distinct stages.

Stage one is the stage which had already been reached in 1957, when Mark 6 Hunters and Javelin two-seaters, each armed with four of the 30-millimetre Aden cannon, had become the standard equipment of all the home fighter squadrons. No less important than the aeroplanes themselves, and their armament, the warning and control system had also by 1957 been overhauled and brought to a new high level of effectiveness. What an efficient warning and control system means in air defence was shown in the Battle of Britain and during the subsequent German attacks on the British Isles. The number of countries with direct experience of modern aerial warfare are few indeed—one might justifiably limit them to Britain and Germany alone. The modernised British radar system has the benefit of that tactical experience allied to considerable technical advances, which simplify the organisation and speed its work.

In stage two comes the introduction of the supersonic English Electric P-1 fighter. Although a single seater, this aeroplane is not excluded from the all-weather fighter role. Advances in aircraft interception radar permit the single seater to intercept unseen targets. An auxiliary rocket motor will give the power necessary for fighter agility at the high speeds and great altitudes appropriate to bomber operations in 1958 and thereafter. It is reasonable to suppose that the P-1 will be "stretched" just as the bombers will be stretched, both in itself and in the power of the weapons that will be made available for it to carry. One should think not of one kind of P-1, but of a succession of improving P-1 versions, just as the Spitfire was not one kind of aeroplane, but a series of versions with successively improved performance.

With the introduction of the P-1s comes the first improvement in the armament not only of that fighter but also of the Hunter and the Javelin as well: the airborne anti-aircraft missile. The first missile of this kind is the rocket-driven de Havilland Firestreak (the smaller Fairey

Fireflash is not intended for operational use, but only for training and for tactical research).

Firestreak shows two principal advantages over the guns of earlier fighters: it has greater range; it carries a very much bigger explosive charge. Whereas cannon shell can be fired in large numbers to raise the chance of a hit, only two missiles of the size of Firestreak can be carried by a fighter. It follows that the certainty of each missile scoring a kill must be assured by guiding it positively to its target. In the case of this particular missile the guidance depends upon the detection of infra-red radiations from parts of the target aircraft which are at a high temperature—or in other words, from the gas turbine exhausts. Needless to say, this principle of homing upon heat radiations within certain selected wavelengths opens the way to an attacker to use decoys, possibly on the lines of strongly burning flares fired out from the aircraft. There are fruitful fields for scientific counter-measures in the future, among which one of the most important weapons must be some form of missile which can home on to the enemy's radar transmissions and so destroy his chief source of tactical intelligence. Of such devices we have as yet heard little, and it may be that we can expect no better in view of the need to achieve surprise.

But to hold to Fighter Command: almost simultaneously with the introduction of the air-launched anti-aircraft missile in the form of Firestreak will come the first ground-launched missile. This is the Bristol Bloodhound, with which is the associated control system developed by Ferranti. In this case the missile is driven by a pair of ram-jet engines whose fuel consumption is more economical than that of a rocket, allowing a relatively greater range. Although few facts have been revealed concerning these new missiles, it seems likely that Bloodhound can be used on targets at more than 50 miles slant range—perhaps up to 100 miles.

Following the Bloodhound system is another, rocket-driven, which has been designed and made by the English Electric Company. At the time of writing this has not been officially named, but a version for use by the army to give cover in the field has been named Thunderbird.

Both the English Electric and the Bristol missiles are of the "semi-active" homing kind. That is to say they direct themselves on to the target by means of reflected radiations—the source of the radiation being a radar transmitter on the ground. Both missiles in fact share a common tactical radar on the ground. By contrast, the fully active missile carries its own transmitter, while the passive homer responds to some intrinsic radiation of the target, as Firestreak responds to heat radiations from the engines.

The details are somewhat shrouded in secrecy, but there is no doubt that the separate components—P-1, missiles, radar—of British air defence are all planned to fit together into a single pattern. As time goes by it may be possible to see more clearly the way in which all the parts are being co-ordinated. The whole plan marks quite a distinct departure, and in broad one sees that 1957 has initiated a fresh phase in which steady development will be possible. As is the case with Bomber Command, the new equipment will present plenty of opportunities for evolution for at least the next ten years.

* * * * *

50. 1958

Those Against the H-Bomb
Richard Goold-Adams

No single weapon in military history, not even poison gas, has aroused such universal doubt and debate as the hydrogen bomb. To the military expert this degree of discussion may seem remarkable, since even the H-bomb is in a sense just another weapon. Admittedly its power is so tremendous that its effect on the nature of war, or at any rate of global war, is fundamental. Not for nothing has it been called the absolute weapon. As seen by the general public, however, the cataclysmic character and the devastating effects of the bomb make this vast debate and hesitation inevitable. This, it is said, is not a weapon of war. It is a weapon of world suicide. Where does the truth lie? It may be that as yet we do not have sufficient evidence to know. But in so far as the old dictum remains true, that war is no more than the continuation of policy by other means, no serious student of military affairs can afford to ignore the political relationship between the H-bomb and public opinion.

The first broad group to consider, therefore, of those against the H-bomb is that which talks not in military terms but on moral grounds, and sometimes with a mixture of pacifist and political motives. These are the main people behind the Campaign for Nuclear Disarmament, which was launched in Britain early in 1958, but which has not so far made the headway that at one time seemed likely. The reason is partly that both the major political parties, Conservative and Labour, have so far adhered to the policy of supporting Britain's retention of the H-bomb; the original decision to go ahead with British nuclear weapons, it must be remembered, was taken by a Labour government, and, although the pacifist element which is the driving force behind the Campaign for Nuclear Disarmament has attracted as allies a good many of those who, for quite other reasons, doubt the wisdom of Britain possessing the H-bomb, the instinct of the British people has so far remained fairly solidly behind the government's present policy. If we had had the H-bomb in the nineteen-thirties, this would probably not have been so. But even if politicians do not always learn the lessons of history, the public sometimes does. In this generation at least there is great wariness about being caught a second time disarmed in the face of a ruthless dictatorship.

The biggest figure in the anti-nuclear campaign has undoubtedly been Bertrand Russell—Earl Russell. From the start he threw the weight of his own great prestige into the scales, and this has given a sense of direction to the wide variety of differing views and opinions. He himself has strongly rejected the whole concept of poising the world on the niceties of what can only be a series of calculations about the balance of terror caused by the nuclear deterrent. Some years ago he seemed to believe in the use of the old-fashioned atom bomb, but after the growth of Soviet nuclear parity with the Western powers following the development of the H-bomb he appeared to change his mind.

Those who have been more or less loosely associated with the general point of view which Lord Russell represents may be subdivided into roughly four kinds. First of all, of course, there are the pacifists, whose basic rejection of war as a method of policy has naturally been greatly stimulated by the horrors that nuclear warfare would entail. A second kind might be regarded as the extreme left wing of the Labour Party, which views these matters in terms of three broad political objectives, namely; to reach an understanding with Russia; to break the Anglo-American alliance; and to stop the rearmament of Germany.

In more extreme association with them are the Communists, who have put a great deal of effort into the anti-nuclear campaign for their own ends. They were prominent in organising the demonstration march to the Atomic Weapons Establishment at Aldermaston last spring, and they have since fostered rallies both in Trafalgar Square and to lobby M.P.s at the House of Commons. It is, incidentally, interesting how the press has tended to play down these anti-nuclear meetings. Quite apart from the strong Communist element behind them, a cross-section of ordinary worried citizens has often attended a considerable number of local meetings up and down the country and at least one major national women's rally. Yet almost all have gone virtually unreported. This seems due in part—though perhaps only in part—to the low calibre of speakers, who have failed to do justice to the calibre of their audiences.

The other two trends of opinion in this first general group are those who take a more dispassionate view of the situation in Europe but who draw from it political conclusions rather different from those of the general trend of opinion behind the government. The main protagonist of one such trend may be said to be Commander Sir Stephen King-Hall, whose latest book "Defence in the Nuclear Age" carries very much further than any previous work his general argument that Russia and Communist strategy can never be dealt with by purely military means.

Hence, he says, the possession of the H-bomb and the threat to use it appear to be not only disastrous but also useless.

Finally, there is certainly a trend against the H-bomb policy in the growth of the general idea that there can be disengagement in Europe. Although after due consideration the Western powers rejected the Rapacki Plan for a nuclear-free zone in Central Europe in which the manufacture, storage and use of nuclear weapons would be banned by international agreement, we have probably not heard the last of these Polish proposals or of others like them. Admittedly, the idea of disengagement is not intrinsically the same as a nuclear-free zone. But the two conceptions clearly march together, and if during the next few years negotiation for the partial withdrawal of military forces of all kinds from Central Europe is going to find more favour—as seems likely—then this will almost certainly encourage certain sections of public opinion to question even more thoroughly than hitherto the basic outlines of the H-bomb strategy.

The initial object of the movement against the H-bomb has been to secure the abandonment of any further nuclear tests. In this respect, of course, Russia's lead in making a unilateral declaration last spring that no more tests would be made in the Soviet Union—after the elaborate and extensive series which had just been concluded—made very good propaganda, quite apart from the question of whether it would be indefinitely observed or not. In Britain one of the most significant moves so far has been the letter sent to the Prime Minister last April over the signatures of 618 prominent scientists. This was forwarded with a covering letter written by Lord Russell as president of the Campaign for Nuclear Disarmament. There was, however, a significant divergence between the text of what the scientists had signed and Lord Russell's covering letter. The scientists asked for an international agreement to stop the testing of nuclear weapons, while Lord Russell said something quite different in favour apparently of suspending British nuclear tests unilaterally. Mr. Macmillan made the practical reply that "we cannot secure peace simply by wanting peace". Instead our task must be to "use the United Kingdom's bargaining position, not to throw it away".

* * * * *

As distinct from the mixed views of the pacifists, left-wing politicians and those who adopt the doubts of the scientists, criticism of the H-bomb in this country on military and economic grounds, and on the basis of what might be termed political realism, is concentrated mainly on Britain's own possession of the bomb rather than on its existence as such in the hands of the two super powers, America and Russia. One of the best ways of expressing the fundamental criticism of the Conservative government's present defence policy, as expounded by Mr. Duncan Sandys, is to quote the confusing discrepancy which the critics assert exists between the policy laid down in the Defence White Paper of 1957 and that of 1958. In the first White Paper the government says that: "It must be frankly recognised that there is at present no means of providing adequate protection for the people of this country against the consequences of an attack with nuclear weapons." In the 1958 Defence White Paper, on the other hand, the government says that: "It must ... be understood that if Russia were to launch a major attack upon them [the democratic Western nations], even with conventional forces only, they would have to hit back with strategic nuclear weapons." What Mr. Sandys is in fact saying, the critics assert, is that if Russia were to launch a major attack against the West—whatever that may mean—we in this country should so act as to cause the destruction of our society and the death of most of our people.

This line of comment goes on to doubt whether the government and Mr. Sandys mean what they say. It is questioned whether Britain ever could afford to use its H-bomb at all. Indeed, it may well be that thermonuclear warfare would destroy even huge countries like Russia and America. But there can be no doubt whatever that these small islands are too crowded and too compressed ever to serve as launching sites for nuclear weapons. In that case Britain's possession of the H-bomb is nothing but a gigantic bluff. It cannot even therefore act as a deterrent. And so, the critics say, this is a waste of money and effort which we can ill afford.

This type of argument is used with convincing effect by a good many people, not necessarily members of the Labour Party, who genuinely believe that British defence policy is in a state of confusion from which there is at present no issue. The argument is used by people who are in no sense whatever pacifist, and indeed by many who do believe in the great deterrent as such. They are wholly agreed that, dangerous as it may be to conduct a foreign policy with the implicit threat of the H-bomb behind it, the Soviet Union has nevertheless been deterred from a far more aggressive policy by the West's possession of thermo-nuclear weapons. But they believe that, since America already possesses nuclear weapons and the means of delivering them in sufficient strength to destroy the Soviet Union several times over, it is nonsense from the West's point of view for Britain to possess this weapon too. The phrase used in both the 1957 and 1958 Defence White Papers that "Britain is making a contribution ... to the Western nuclear deterrent" is regarded as either naïve or dishonest or both. What Britain is doing, it is argued, is developing its own deterrent independently of the

United States simply for its own purposes. This may or may not be sensible, and here the critics disagree. But the line of thought which has been outlined asserts that the government is doing wrong if it deliberately misleads people into thinking that Britain's contribution to the Western deterrent as a whole is of any real significance at all.

On the question which this raises, whether Britain should in fact equip itself with its own H-bomb and V-bombers for its own political purposes, a remarkable development of the recent past has been that the Labour Party has officially gone on record against any British abandonment of the H-bomb. When the question was formally debated in March, at least a hundred Labour M.P.s were almost certainly in favour of such abandonment. But, because of Mr. Aneurin Bevan's decision at the 1957 Labour Party conference in Brighton to support a minimum British H-bomb policy, Mr. Hugh Gaitskell and his shadow Defence Minister, Mr. George Brown, were just able to carry the vote against any unilateral British nuclear disarmament. What the Parliamentary Labour Party endorsed was a statement in favour of immediate suspension of British nuclear tests; a general disarmament agreement, including a declaration banning the use of all nuclear weapons; no more H-bomb patrols by aircraft based in Britain; and no physical steps to set up missile bases in Britain before a fresh attempt had been made to negotiate with Russia. But there was no decision to urge that we should throw away our own H-bomb altogether. On the other hand, whether this will go on being so indefinitely must always remain in some doubt.

* * * * *

This raises, however, the question of costs, and here we have the second line of political realism. This type of economic criticism is to be found both in the Labour Party and among Conservatives. Indeed, Mr. Peter Thorneycroft, the former Chancellor of the Exchequer, came out publicly in a speech at Newport last spring with the views which he had been known to hold in private for some time. "We should ponder well," he said, "the merits or demerits of duplicating the American effort. To fling resources at an ever-increasing rate into competitive manufacture of nuclear missiles is a questionable policy. ... Our prestige will be rated not by the bombs we make nor by the money we spend, but by the contribution we can make" towards the solvency and economic strength of the West. This line of argument takes on new force with the growth of the view that the thermo-nuclear stalemate has now switched the emphasis in the cold war from military preparedness to long-term economic competition. These critics of our defence policy feel that, apart from anything else, we cannot in this day and age afford approximately £300 million a year on the luxury of an H-bomb merely in order to stand up to the Americans.

The third angle to this type of criticism also springs in part from calculations of economy but takes two forms: one to be found in both the main political parties, and the other almost exclusively among Conservatives. The criticism is in essence that Britain's concentration on the nuclear effort has made it impossible to retain conventional forces of adequate size and efficiency to do the type of job for which they are constantly being called on in various parts of the empire. During the past seven years, it is argued, there has in practice been a considerable rundown in the scale of our forces. At 1951 prices, to-day's forces, apart from what is spent on the nuclear deterrent, cost not much more than half what they did then. By the time we have met our number one commitment in Germany—though even that has been reduced in scale—the result is that our forces available for employment in Cyprus, Aden, Kenya, Malaya and wherever else in the Commonwealth trouble may occur have now been spread too thin to do their job. One consequence, as these critics see it, was the fiasco of Suez. Indeed, it may well be that this year's further emergency in the Middle East has already changed the trend of thinking in Whitehall. Although British forces were in fact available to fly to Jordan at short notice, the strain of even this comparatively minor effort does appear to have shaken faith in the wisdom of our present run-down.

* * * * *

Lastly, there stands ranged against the government's H-bomb policy the school of thought which believes that distinctions can and should be drawn between tactical atomic weapons and the type of major thermo-nuclear explosion which produces widespread fall-out effects. As is widely known, this school owes its inspiration to Rear-Admiral Sir Anthony Buzzard, who has argued with skill and logic that the whole H-bomb concept runs counter to the basic military doctrine of using the minimum effective force to achieve any given end; indeed, his thesis is that adequate conventional forces are essential to the carrying out of any policy of true graduated deterrence. There are now varying degrees of those who appreciate his views. In both Britain and America it would appear that much modern staff planning is based on at least some of the premises which he has expounded. On the other hand, only a relatively small number of people go all the way with his tactical atomic theories in their purest form. But, whether or not the idea is accepted that it may be possible to fight a war limited to tactical atomic weapons only, there is one strong and valid objection which this line of thinking brings to bear on Britain's present H-bomb policy. This is that, with only a limited amount of money to spend on nuclear weapons of all kinds, it is a great mistake not to devote more of what is available to developing the smaller type of atomic weapon and to equipping the forces more rapidly with these weapons as they become available.

In recent years, it is emphasised, experimental work in the military nuclear field has been increasingly directed towards two objectives. One is to reduce the amount of fall-out from major nuclear explosions, and the other is to develop smaller and smaller atomic weapons for limited use in the field. Progress is being made in both directions, but, for the West which is incapable of meeting Communist manpower on equal terms with conventional weapons, the development of the small atomic weapon is becoming more and more important. This is because the possibility of global war with the use of the H-bomb is diminishing, owing to the risk of suicide that it presents to both sides. Under the umbrella of this thermo-nuclear stalemate there may well be a temptation, and a growing one, for the Russians to risk a limited war on the Korean pattern. If so, it seems clear that the West would be bound to need small tactical atomic weapons in order to be able to hold its own. In these circumstances it is particular folly for Britain, so it is argued, to skimp money on a type of weapon which has such a crucial role to play.

Those who support and defend government policy at the present time may well feel that none of these criticisms really affect the issue. On the other hand, the above outline of trends against the H-bomb represents such a broad band of opinion and sentiment that it can hardly be ignored. Perhaps the greatest need is for the exponents of the existing policy to take more account of these various lines of criticism when public statements are made. Many people, some with the sincerest of motives, are genuinely puzzled and confused. The government obviously has a puzzling and confusing problem itself. But there is much to be said for the view that, having clearly taken certain decisions at the present time, it would invite less criticism if it explained them more often and in simpler terms. At present too much is either being taken for granted or else hidden under the deliberate but mistaken idea that the less the public is told the better. In an age when public opinion is more powerful than ever before and when the nature of the problem is such that the lives of perhaps everyone on these islands is at stake, a hush-hush nuclear policy may well boomerang on its authors. The danger is that the consequent irresolution could prevent Britain playing its full and effective role in the Western alliance in its position as the world's third nuclear power.

51. 1963

West European Defence
Air Vice-Marshal W. M. Yool

For some time the allied defence system in N.A.T.O. has been under considerable strain and has looked at times as though it was splitting apart at the seams. There have been two main causes of this unease, both having their origins on the other side of the Atlantic, and related initially to the policy for strategic nuclear forces and subsequently to the system of defence in Central Europe.

Under the first heading such questions as whether or not France and Great Britain should have their own nuclear forces, and the shape of the proposed N.A.T.O. nuclear force, have been exercising the minds of many people. Under the second heading there has been discussion on the American pressure for increased conventional forces resulting from the growing missile threat to the United States, differences of view as to the extent to which reliance should be placed on nuclear weapons to defend Central Europe and differences between France and the other allies.

Strategic Nuclear Forces

The first difference arose from the refusal of the United States to share its nuclear knowledge with its allies. This led Great Britain, following the war, to develop her own atomic weapons, a decision taken by Mr. Attlee's Government in 1946. Some years later the French Government decided that they also would develop their own weapons.

After President Kennedy came to power in 1961 there was a further hardening of the American attitude, and their reluctance to share their nuclear secrets developed into a determination to prevent, if possible, any other country having its own weapons. This attitude culminated in the speech by Mr. McNamara, the Secretary of Defence, on June 16, 1962, when he said that national nuclear deterrents operating independently are provocative and dangerous. This speech was widely interpreted as being directed at both Britain and France, although it was later denied in the United States that it was directed at Great Britain.

But in spite of these official denials it would not be surprising if Mr. McNamara's remarks had been aimed at Great Britain as well as France. Apart from their genuine fear that a nuclear war might be sparked off by their allies it would obviously make life very much simpler for the American Government if they were the sole arbiters of Western defence policy, which they would be if no other ally possessed nuclear weapons.

The reason why Britain, and subsequently France, decided to develop their own nuclear weapons was simply that the United States could not be relied upon to come to their assistance in all circumstances if they, or Western Europe as a whole, were threatened with attack by Russia. There is a growing feeling in the United States that America might be able to contract out of a nuclear war by the use of conventional forces or possibly reduce the effects of a nuclear war by the so-called "counter-force strategy" (attacking military targets only). And there is no doubt that whatever the views of the American Government might be at the time, if Russia at the outbreak of war said that she was attacking Europe only and that the United States would not be attacked if she did not intervene, there would be tremendous public pressure in the States to prevent the American Government declaring war. At the same time there is a growing feeling in Europe that the United States may eventually pull out of Europe altogether.

There are also British interests in other parts of the world which are not covered by the North Atlantic or other treaties, where it is also by no means certain that the United States would intervene. Whatever the rights and wrongs of the Suez operation, when Britain was threatened by Russia with nuclear attack, it is at least doubtful whether the United States would have taken any action if Russia had fulfilled her threat.

One of the arguments of the opponents of Britain's independent deterrent is that the force is so small that it could not influence events: but a force that is capable, as the V-bomber force is, of causing millions of casualties and devastating large areas of Russia, can hardly be ignored. The possession of nuclear weapons by Great Britain and France would make the Russians consider seriously whether the damage such forces could inflict would not greatly weaken her *vis-à-vis* the United States, should the latter decide eventually to intervene, and might well prove the deciding factor in the preservation of peace. The British point of view was put by the Prime Minister during the defence debate following the Nassau agreement. On January 20, 1963, Mr. Macmillan said:

> "Second—I do not believe that our western alliance could really stand permanently if in this vital field the United States were given for all time the sole authority. We are allies. We must remain allies but we must not become satellites. I can understand why

the French Government, which is a world power as well as continental power, wishes to develop its own nuclear force, but I must frankly say that I hope they accept that such a force has obligations as well as rights.

"I believe it would not be a happy position for the United States themselves, in relation to their allies, if the other countries not so great as they but still great, were to hand over for ever the complete control of this unique weapon to the American administration of the day...."

"Fourth—and this is the most vital argument of all—there may be conditions, there may be some areas, in which the interests of some countries may seem to them more vital than they seem to others. It is right that a British Government, whatever the conditions and the principles under dispute, should be able to make its own decision without fear of nuclear blackmail. It is wrong and dangerous to particularise further, but members will realise what I have in mind."

This is the British case for the nuclear deterrent and, though more tactfully expressed than the views of General de Gaulle, it is also the French case.

Another reason why the United States is opposed to the possession of nuclear weapons by its allies is the danger that America might become involved in a nuclear war against her will through Great Britain or France taking independent action in defence of their own interests. But Cuba has shown that this is a two-way danger and that a nuclear war might be started by the United States acting independently of her allies on an issue of no direct concern to them. At least in such circumstances Europe might be able to preserve her own neutrality, which she would be unlikely to do if there were no European nuclear forces.

Up to the time of the Nassau meeting, discussions in Europe on the possession of nuclear weapons by the alliance as a whole, as distinct from the nuclear forces of Britain and France, had revolved mainly round the provision of nuclear weapons for tactical use. General Norstad, for some time before his resignation, had been pressing for medium-range missiles to supplement the nuclear weapons with which his tactical air forces were equipped. This proposal was resisted by both Great Britain and the United States, on the ground that such missiles were strategic rather than tactical owing to their range and that to place them under the command of the Supreme Commander would duplicate the control of allied strategic nuclear forces.

But shortly before the Nassau meeting new proposals had begun to gain ground in Europe for the creation of a European strategic nuclear force, which would be under the control of the European members of N.A.T.O. only. Assuming that Great Britain and France contributed to this force, this proposal would have meant that there would be two allies strategic forces, the American and the European.

The fact that such a possibility was even being discussed alarmed the Americans and led, during the latter part of 1962, to a rapid change in their attitude towards independent nuclear deterrents. At long last they recognised that nothing they could do would prevent Great Britain and France from continuing with their nuclear policy, and that it was therefore better for the United States to accept this as a fact of life and to learn to live with it. They therefore decided to evolve a nuclear system which would ensure them some measure of control over the French and British nuclear deterrents, whilst giving the European allies in return some measure of control over part of the U.S. strategic forces.

This was the background to the Nassau meeting. The main decisions reached at that meeting were that the manufacture of Skybolt was to be abandoned and that it would therefore no longer be available in the latter half of the sixties for the British V-bomber force. In its place Polaris missiles (with Britain providing the nuclear warheads as would have been the case with Skybolt) would be supplied for fitting in British-built submarines.

At first the abandonment of Skybolt appeared to be a political decision, designed to force Great Britain to abandon her nuclear deterrent, but as the decision also foreshadowed the end of the strategic bomber in the U.S. Air Force it seems that the decision was taken on military rather than on political grounds. The fact that Polaris missiles are to be supplied in place of Skybolt reinforces the view that the decision was a military one.

There were strings attached, however, to the Nassau agreement. To forestall the possibility of the creation of a European nuclear force outside American control, it was agreed in para. 6 that a N.A.T.O. strategic nuclear force should be created which would "include allocations from U.S. strategic forces, from U.K. Bomber Command and from tactical nuclear forces held in Europe. Such forces would be assigned as part of the N.A.T.O. nuclear force and targeted in accordance with N.A.T.O. plans." Later the Prime Minister offered to place the whole of Bomber Command under N.A.T.O. instead of only part of it, subject to Britain retaining the right to withdraw the force if necessary should it be required in defence of national interests outside N.A.T.O.

It was also agreed in para. 7, "that the purpose of their two Governments with respect to the provision of Polaris missiles must be the development of a multilateral nuclear force in the closest consultation with other N.A.T.O. allies."

Para. 8 went on to say, with reference to the Polaris missiles being supplied to Great Britain, that:

"These forces, and at least equal United States forces, would be made available for inclusion in a N.A.T.O. multilateral nuclear force. The Prime Minister made it clear that, except where Her Majesty's Government may decide that supreme national interests are at stake, these British forces will be used for the purposes of international defence of the western alliance in all circumstances."

It can be argued that under the Nassau agreement Britain has lost some measure of control over her nuclear forces in that, whilst hitherto her strategic nuclear force, the V-bombers, has been completely under her own control, it is now to be allocated to the joint control of N.A.T.O., and that when the V-bombers have faded out and are replaced by the Polaris submarines, the whole of Britain's strategic nuclear forces will be under N.A.T.O., although admittedly with the proviso that they can be withdrawn when supreme national interests are at stake. Nevertheless the agreement does involve some loss of British sovereignty, as it will not be easy to withdraw our forces once they are fully committed to N.A.T.O. At the same time the United States have gained some measure of control over one allied nuclear force, the British, and to that extent have moved a step towards their declared policy of concentrating the control of allied nuclear power in their own hands.

* * * * *

52. 1965
NATO
Alastair Buchan

For the past year or more hardly a week has gone by without the appearance in one of the world's leading newspapers of an article about "the end of N.A.T.O.," or "the dissolution of the Atlantic alliance." Much of this comment inevitably over-dramatises the tensions that have developed between the allies and minimises the continuing core of common interest which it expresses. But there can be no doubt that we are moving either towards some climacteric in which the major allies will have to hammer out a new kind of agreement or else split apart, or at any rate towards a quite new phase of the European-American debate.

Before examining which of these alternatives is the most probable dissolution or reform, I think it is worth noting that the choice is likely to be with us sooner rather than later. The year 1969, when the optional clause governing adherence to the North Atlantic Treaty becomes operative, when member nations can withdraw at one year's notice, would, it is true, provide a natural moment for a fundamental review. But, in my view, events are moving too fast, and the necessity of resolving fundamental doubts about the future relationship of both the United States and France to their allies is too urgent to make it conceivable that the alliance can persist in its present disorganised fashion until an artificial date nearly four years hence. The real political forces at work do not permit such a leisurely approach.

At the end of 1964 a cycle of American-European argument which had opened some seven years earlier came to a close when President Johnson decided to abandon American pressure for an immediate resolution of the negotiations regarding a multilateral nuclear force. Since then the common assumption has been that there is to be a year's lull, until after the German elections in September, before the next phase of the dialogue on the future scope and nature of the Atlantic Alliance is resumed, even though any successful outcome to it may have to wait until the attitudes and policy of post-de Gaulle France are clear.

The era of European-American argument which closed last December had opened in 1957-8 with the occurrence within a short space of time of five largely unrelated developments. The first was the perfection of the I.C.B.M., which raised questions as to the viability of the American guarantee to Western Europe in the light of the increasing vulnerability of American civilisation. The second was the signature of the Treaty of Rome, which offered the promise of a political as well as an economic role for a unified Europe to replace the hegemonic position of the United States and to share its strategic burdens. The third was the conclusion, in 1958, of an agreement on the sharing of information on nuclear weapons between the United States and Britain. This gave a new impetus to the old wartime special relationship between the two countries, while the return of Charles de Gaulle to the Elysée in the same year made certain that France would not rest content with the leadership of the Anglo Saxon powers. Finally, there were the years of the consolidation of Khrushchev's leadership within the Soviet system—the beginning of a period of greater sensitivity to the risks of war with the capitalist world, and consequently a desire to reach some sort of détente in Europe.

The fact that it was these events which set the argument in train tended to create the framework of discussion. The main characteristic of this phase of the debate has been, first, concentration on the role of nuclear weapons and the control of nuclear strategy; second, concern with nuclear proliferation primarily as a European phenomenon; third, a tendency to regard the Communist threat in its military aspect as principally a European one and the European confrontation as potentially the most dangerous source of war through escalation or miscalculation; fourth, an assumption that the European-American strategic problem was largely *sui generis* and unrelated to other alliances of relationships. Finally, success or failure tended, especially in the later period, to be judged by reference to a theoretical model of partnership between the United States and a unified Europe, a relationship of units of roughly equal size and economic strength.

The agenda of debate was dominated by certain questions about the American nuclear guarantee to Europe, the connection between the virtual American hegemony in nuclear strategic power and its right to lay down and alter the strategy of the alliance, and the relations between this commanding position and its responsibility for the pursuit of a special American dialogue with the Soviet Union, even on questions that might affect the interests of the European powers. These points implied further questions about the rights of the European allies, the freedom of action which the larger powers in Europe may legitimately exercise without damaging the solidarity of the alliance: what control the most exposed non-nuclear power in Europe, Germany, can exercise over its own security and prospects of reunification; what influence a Europe that is politically still a plural group of middle and small sovereign states can exercise over the evolution of political and strategic plans whose implementation remains primarily an American responsibility.

Unfortunately, during the latter stages of this argument, the opportunities for tackling these formidable questions in any broad context became gradually foreclosed by President de Gaulle's extremely narrow definition of the French national interest and by his preference for exploiting such nominal freedom of action as the strategic stalemate between the great powers permitted him, by British confusion about its relationship to Continental Europe, by German provincialism; and by a certain dogmatism in Washington about the solutions to be pursued. In consequence, the choices in terms of actual policy by 1964 had become narrowed down to one proposal, a multilateral force to give the non-nuclear powers in Europe some physical sense of participation in the handling of strategic weapons and the planning for their control. And when this became blocked by French pressure on Germany, by the tabling of an alternative British proposal and by an American decision to withdraw temporarily from the argument, it was clear that the time had come to reformulate the whole agenda, though not all the problems had by any means been solved.

It was largely for this reason that the meeting of the Ministerial Council in Paris last December was so unproductive. Having decided to abandon pressure for an M.L.F. the Americans had no concrete alternative to propose, although Mr. McNamara displayed a clear desire to lay the groundwork for an alternative approach. The British alternative of an Atlantic Nuclear Force had only just been formulated. German policy had been thrown into confusion by the sudden American abandonment of the M.L.F., and by French pressure, exercised particularly in the E.E.C., to desist from entering into a new strategic relationship with the Americans. At the same time, France herself was not yet ready to force the issue about offering proposals that would convert N.A.T.O. from an integrated military system to a classic coalition of co-ordinated national forces.

Before trying to discern what are the new questions which will dominate the agenda when full scale debate on the future of N.A.T.O. is resumed, it is first worth examining what unfinished major business remains from the last cycle of debate.

One question is obviously the control of strategic planning and decisions. The demise of the M.L.F. in the form mooted by the Americans since 1960 and officially proposed in 1963, has not settled the question of how the non-nuclear allies in N.A.T.O. are to develop a more satisfactory relationship with nuclear policy than they have had hitherto. The British proposal for an Atlantic Nuclear Force, which envisages the commitment of R.A.F. Bomber Command to N.A.T.O. (less those units required east of Suez) and the 4 *Polaris* submarines when they are built, plus an equivalent number of American *Polarises*, plus an unspecified mixed-manned element, did not commend itself as a readily desirable alternative to continental governments. Mr. Wilson made a serious tactical error in withholding part of the British strategic force rather than offering to commit it all to N.A.T.O. with provision for withdrawal in the event of an Asian emergency. Mr. Healey aroused continental suspicions by suggesting that the A.N.F. should be outside the control of S.A.C.E.U.R., though he altered his position at the meeting of Defence Ministers in May, 1965. As it stands, the A.N.F. provides nothing for the Germans which the M.L.F. did not, while the British proposal for a permanent American power of veto over the use of the force has alienated the enthusiasts for European political union. And of course the opposition of France remains as strong to the idea of an A.N.F. as to an M.L.F. even though the A.N.F. makes specific provisions for the later adherence of France.

A second unresolved question is the doctrine for the defence of Europe. The conflict between the American desire to introduce the maximum flexibility into the posture of the N.A.T.O. forces in Europe and the desire of Germany, backed up by France, to ensure the deterrence of any Soviet move by a more or less automatic doctrine of nuclear retaliation in the event of any aggression, has not been pursued in the past year with quite the same vigour as in the days of the Kennedy Administration: but it is still unresolved. And it affects not only doctrine but levels of forces maintained in Europe with all the economic and political ramifications of such questions, especially for Britain.

A third question that recent events have made more urgent without leading to any solution is the relationship of N.A.T.O., as a political structure, to developments outside Europe or the N.A.T.O. area. Ever since the Suez debacle, and the report of the Pearson-Lange-Martino committee of the Council which followed it, there has been an explicit understanding that N.A.T.O. is to be considered by all its members as more than just a regional security pact, in fact as the master alliance of the West. The powers with extra-European interests accepted a specific responsibility to use the N.A.T.O. Council as a clearing house for their views, policies and, if necessary, action. Yet the past two years have witnessed one instance, the early phase of the Cyprus debate, in which it was impossible to get the European members of N.A.T.O. to agree to take co-ordinated or joint action concerning a problem on their immediate flank. These two years have also seen the development of a French foreign policy in relation to the Far East, and towards

Eastern Europe which has been discussed with no other N.A.T.O. partner, let alone agreed. And since the Johnson Administration was confirmed by the American electorate in November 1964, American policy in the Far East has been made without even the formality of an attempt to develop a consensus between the United States and her N.A.T.O. allies.

The failure to resolve these questions has, of course, been intimately connected with the attitude of President de Gaulle whose hostility to N.A.T.O. as an organisation, that is to say with an integrated military command system, certain integrated forms of military planning, and a political authority, the N.A.T.O. Council, has been growing steadily. De Gaulle appears to have little quarrel with the N.A.T.O. Treaty as a form of collective insurance in case of a threat to any member country: he explicitly accepts the need for an American "nuclear umbrella" over Europe, and much less has been heard of late of substituting French for American nuclear force as a guarantee of the security of Western Europe. What he detests are all forms of international authority which appear to denigrate the historic role of the nation state, particularly in the field of military planning. There has been a growing fear that France might opt out of the web of multilateral arrangements out of which the military and economic fabric of N.A.T.O. has been woven over the past fifteen years, leaving, by reason of her central geographic position, a gaping hole in it. This could be patched but not mended by a diversion of American supply lines to Germany through Bremen and Hamburg: it would drastically reduce the military options open to S.A.C.E.U.R. in a crisis: and it would weaken the whole air defence of Western Europe: all this quite apart from the political damage that would be caused at a time when Soviet policy is concentrating on increasing the divisions between the European countries.

Whether the General will carry out his implied threat to withdraw France from all participation in N.A.T.O. is unclear—perhaps even to the General himself. His actions in abrogating various forms of naval co-operation, and refusing, in May 1964, continued French participation in the semi-annual Fallex exercise, point in that direction. On the other hand, he has to reckon that, while he may discount in advance American and British anger, a final French withdrawal from military participation in N.A.T.O. would arouse intense fear and hostility among France's partners in the E.E.C. and perhaps seriously and finally jeopardise its existence at a time when France's material interests are more deeply committed to its success than ever.

But whatever the real probability of a French withdrawal from N.A.T.O., the possibility has greatly strengthened the General's hand and has made the other N.A.T.O. governments unwilling to insist on final answers to the unfinished business of the European-American debate, even if they were all of one mind. Moreover, it has gradually come to be recognised that many of the positions adopted by the French government are unlikely to be wholly reversed when President de Gaulle leaves the Elysée, and that the French conception of N.A.T.O. as a multinational coalition is a permanent factor to be reckoned with, and that it need not conflict diametrically with the conceptions of the other major partners in N.A.T.O.

The first evidence that the United States and France might wish to build a bridge towards each other rather than force a showdown came at the Ministerial meeting in December 1964, when M. Messmer agreed with Mr. McNamara that the French nuclear force should co-ordinate its targetting with S.A.C. when it becomes operational. This was only a modest gesture and in the succeeding six months the general American line—at any rate from the State Department—was that the United States would make no concession to France, would probably revive the concept of a mixed manned nuclear force after the German elections, and would be prepared to operate a "two-tier" system, co-operating closely with those who wished it (Britain, Germany and Italy, for instance) and ignoring the others. But the crucial relationship to the defence of Europe is more widely appreciated in the Pentagon. And at the important meeting of the N.A.T.O. Defence Ministers in Paris in May 1964, Mr. McNamara explicitly rejected the "two tier" approach, and proposed that nuclear policy should be studied and presumably made in a new Executive Committee of the N.A.T.O. Council consisting of the United States, Britain, France, Germany and one other country (either Italy or the smaller powers in rotation). This suggestion met initially with a considered French decision to study the idea although in July the French Government decided that it would not take part in further negotiation on the proposal. It has also met with a good reception in Germany, although for the record Chancellor Erhard still says that it does not preclude the need for a multilateral force. But if the French government should finally accept the proposal, and if the present or future German government decides that the attempt to revive the idea of a multilateral force is not worth the internal political opposition it would generate, then this more political approach to the question of allied war strategy may successfully replace the preoccupation with mechanical means such as mixed-manned forces. But the United States will have to clarify its proposal before it will command general assent.

On the question of the level of forces that should be maintained in Europe, a great deal of useful discussion has been taking place in an official working group in Paris, the Force Planning Exercise, which, for the past two years, has been examining and challenging some of the assumptions about European defence that have been made by the military authorities over the past fifteen years. But the main impetus for a revision of N.A.T.O. doctrine has been provided by the rapidly rising cost of equipment and manpower, and the strain on the balance of payments of the external powers, the United States, Britian and Canada, which maintaining troops in Germany involves.

On this subject, Mr. Denis Healey, whose own Ministry is deep in a fresh review of what levels of commitment Britian can afford in Europe and the Indian Ocean without overstraining the economy, has made a valuable contribution. He pointed out at the May meeting of Defence Ministers that N.A.T.O.'s military planning is still primarily geared to an increasingly implausible contingency, that of a deliberate full-scale Soviet attack on Western Europe. He has suggested that the chances of confining such an attack, if it ever came, to central Europe or to ground action alone are negligible, and that it would rapidly escalate to strategic nuclear war. But the contingency which is more likely is some confused situation on the border arising most probably from miscalculation, which N.A.T.O. forces at present lack the mobility and conventional air power to deal with rapidly. The old force levels of 30 divisions with another 30 in reserve are as unrealistic as they are unattainable, and in Mr. Healey's view N.A.T.O.'s ground and air forces should be reorganised in the light of what is the most probable danger and to scales which the N.A.T.O. governments themselves can afford. Though there is a certain suspicion in N.A.T.O. that the British only advance their arguments as a rationalisation for their own desire to reduce B.A.O.R. still further, there is a growing acceptance in N.A.T.O. governments of the logic of the British argument—at least while there is no change in the overall strategic stalemate between the two super powers.

The groundwork has been laid, therefore, for a more constructive argument among the principal N.A.T.O. powers, on questions of European and Atlantic strategy, than has been possible in the past three or four years. It is true that there are several unresolved questions. One of them is how a fair division of labour on defence production, as between the United States and the European countries, is to be achieved. There are acute fears, and no little resentment, at what appears to be increasing American dominance of the advanced industries in Europe, aircraft, electronics, computers, and what is considered to be unfair American official pressure behind the sales of American equipment to European governments. This particularly affects Britain which, like the United States, seeks to offset the cost of its continental military commitments by the sale of tanks, artillery and aircraft to Germany and other countries. Mr. McNamara has spoken of the need for a "N.A.T.O. Common Market" in equipment, but the only practical signs of a new policy have been hints of a new American willingness to shift some of its naval procurement to Europe, which hardly meets the problems of the European aircraft and ordnance industries. There may have to be a considerable improvement in the co-ordination of military research and development among the European members of N.A.T.O. before a satisfactory bargain can be struck with the United States. It may meet the case to create a European Research and Development Authority of which Britain would be a founder member, independent of when or whether she eventually joins the European Economic Community.

But the most important change that is developing in the whole conception of N.A.T.O. concerns its relationship to threats to international security that develop outside Europe. For what has happened is that changes within the Communist bloc and the general shift of Soviet policy to an indirect strategy—in line with China's—of gaining the support of the underdeveloped countries, has invalidated the assumption that European security problems can be handled by a separate Western system from, say, Asian or Latin American ones. A serious extension of the war in Vietnam, for instance, is almost certain to have direct repercussions in Europe, and a fiasco like Santo Domingo affects the standing of the European powers in the uncommitted world almost as much as the United States. It is no longer possible for a global power like the United States to operate its different alliances on a more or less local basis. This will be particularly evident when China becomes an operational nuclear power, and Japan or Australia, for instance, begin to be confronted with much the same problems as Germany or Britain.

In this situation, the logical way for the United States to retain the confidence of its European allies in its Asian strategy and of its Pacific allies in European strategy, would be to capitalise on the central position of Washington, and to develop more effective long-range intra-allied planning machinery in its own capital rather than in Paris for N.A.T.O., or Bangkok for S.E.A.T.O. So far there is no sign that President Johnson is interested in the views of the European allies on Far Eastern or other non-European questions. Indeed American policy has been based on the supposition that only those that are actively involved in a particular area have a right to a voice (which is perhaps one of the main reasons why the present British government is so anxious to retain a presence east of Suez). But this policy could well lead to a neutralist attitude on the part of the European allies to American problems in the Far East. Perhaps when this becomes clear in Washington we may expect to see the United States at last display a willingness to use its central position to draw all its major allies, European and non-European, into a more confidential relationship with its own planning and policy-making than it has been hitherto considered necessary or practical.

Unless it does, the shift of the real focus of international tensions from Europe to elsewhere, from strategic threats to more diverse forms of pressure, risks at best the degradation of N.A.T.O. from the master Western alliance to a local defence system. And at worst its division into a European and a North American half, and the collapse of the concept of an Atlantic community of interest on which the whole of its political and military system has hitherto been founded.

53. 1965

The Future of the Aircraft-Carrier
Vice-Admiral Sir Peter Gretton

This is a particularly difficult time to write about the future of aircraft-carriers owing to the Government Defence Review which is now considering defence commitments and which will decide the fate of the new carrier. The designing of the ship was approved by the last administration, and "long lead" items have been ordered, but no tenders have yet been called for and no contract placed.

The changes in the aircraft production programme have also affected the situation. The death of the TSR-2 may have increased reliance on carrier-borne aircraft; the end of the P-1154 concept has diminished the practicability of an early use of carriers as mobile airfields for R.A.F. as well as naval aircraft.

Politically, the strains in South Arabia have made the long-term retention of Aden as a British base more improbable; on the other hand the possible joint development with the Americans of island bases in the Indian Ocean has increased the chances of keeping shore-based aircraft in that part of the world.

Strategy is highly fluid; and it is clear that commitments must be cut but where the cuts can come is not yet decided. I will try therefore to avoid specific current problems and to set out clearly the fundamentals of the argument for including an aircraft-carrier force in the Royal Navy during the next twenty years.

The Role of Carriers

The role of carriers has changed over the last forty years. In the 1920s, it was to provide the eyes of the fleet, and to help protect all ships from air attack; the strike role was looked upon as unimportant, and secondary to the gun. During the course of the Second World War, the carrier gradually replaced the battleship as the capital ship of the fleet. No naval force within range of the enemy was safe from air attack without the protection of its fighters, and its torpedo planes and bombers had taken over the task of destroying the enemy main fleet whether this was composed of battleships, carriers or both. The carrier had also assumed great importance in the attack on enemy shore targets.

With the coming of nuclear weapons, the carrier—at least the American carriers—became one of the wielders of the strategic nuclear deterrent, in addition to its contribution to fighting a nuclear or conventional war. Recently the Polaris submarines have assumed the deterrent role which is no longer one which it is sensible for surface ships to perform.

What then remains? I believe firmly that in a general nuclear war, it is most unlikely that the carrier will have a part to play. Improved reconnaissance, the increase in the number of nuclear weapons and the more accurate means of delivery achieved, have made the operation of surface ships under all-out nuclear conditions extremely hazardous, if not impossible. Anyone who has studied the delicate electronic gear, the radio and radar antennae and the carrier-borne aircraft themselves must doubt whether a carrier can operate effectively after even a nuclear near miss. Ballistic missiles have made the protection of carriers more difficult, reconnaissance satellites will make their concealment impossible. I believe therefore that the main task of any surface warships which survived a nuclear exchange would be the relief and rescue of any survivors. But the nuclear stalemate has made the probability of general nuclear war extremely low and while rational human beings continue to direct the destinies of the super powers, the danger of disaster is small.

What the world faces now and what the world will continue to face for many years to come—indeed until a system of world government with power to enforce its decisions is set up—is a permanent state of political instability. An instability due mainly to the policy of world communism which, having rejected major war as an effective instrument, depends increasingly on wars of liberation, on subversion, on internal revolt and on blackmail to achieve its ends. An instability which is increased by the exploitation of the envies and quarrels of so many of the new sovereign states with their nationalistic outlooks. Thus it is with limited wars and subversion which the West must be prepared to deal and the importance of preventing small conflicts growing needs no emphasis. There are two forms of limited war in which carriers might be needed. The first covers the peacekeeping operation, the war of intervention and the "brush-fire" war which has been a regular feature of the international scene since 1945 and which continues today in South Arabia, in Malaysia and in other areas such as Viet Nam in which we have no direct interest.

It is becoming increasingly widely accepted that the mobile joint service task force, reinforced when possible by the strategic reserve forces flown from the United Kingdom, is the most effective and practical way of bringing force to bear overseas, and the political difficulty of securing free overflying rights in crises has reinforced the need to have strong forces in or near the operational area. Similarly the reduction in our military bases overseas has emphasised the importance of the joint task force—composed, it must be firmly said, of all arms and not of naval forces alone.

The carrier must form the core of the task force. It has many tasks. Firstly, it must defend the task force from enemy attack from the surface, the air and under water, both at sea on passage and in an operational area during a landing. For this, aircraft and helicopters of several types are needed. Next, it must provide, if necessary, the long-range air power needed to neutralise potentially dangerous forces which might interfere with a landing. Then, during the landing of troops, the carrier must be ready to supply close air support until shore airfields have been secured. It must also supply reconnaissance planes for the command.

It is sometimes tempting to say that this is using a steam hammer to crack a nut and that air power is superfluous, that intervention will be by invitation. But whether invited or not, the inevitable conditions found ashore will be those of chaos and there will always be elements hostile to intervention. To attempt any such operation without strong air power, remaining in the background if necessary, would be to invite humiliating disaster. Many countries have been provided with modern weapons and aircraft with which to defend themselves and it would be folly to assume that they would not be used.

A second illusion is to believe that because Britain is unlikely ever to undertake a major intervention alone, American air power, operated from American carriers, will always be available to support us. Can we be certain of American support, possibly at short notice? Can the time for consultation always be afforded? In intervention and peace-keeping operations, speed is the essence of success. A battalion early is worth a brigade a week later.

Many small nations can produce a modern defence and it would be rash in the extreme to undertake any operation without the guarantee of strong air power being available. For these reasons, and also because any force offered to the U.N. should be self-contained, operationally as well as logistically, the aircraft-carrier should form a permanent part of the British defence effort overseas.

Limited War against Sea Communications

There has been a debate raging in N.A.T.O. circles for some years about the possibility of a conflict in Europe limited to conventional weapons. The "trip-wire" replaced the "plate glass window" and now the need seems generally agreed for conventional forces to counter incursions and to withstand pressures on land in Europe which would not be considered sufficiently serious to warrant nuclear retaliation and all which this implies.

The possibility of "limited" actions at sea by the Communists, of pressure, of blackmail or blockade has also received some unofficial attention. But in the debate on the Navy Department estimates in March 1965. Mr. Mayhew, the Secretary of State for the Navy, put forward, for the first time, an official view.

He described the vast merchant fleet of the West—twenty thousand ships in all—as a hostage to fortune. He gave examples of the type of action which might be undertaken by an enemy in the certainty that the risk of escalation to a nuclear level would be negligible. He pointed out the need for conventional forces, ships and aircraft, to counter Communist harassment at sea and he made clear that he intended to seek allied agreement to this new interpretation of the role of N.A.T.O. naval forces.

Now that the German army has been built up to its present strength, there seems to be general agreement that the dangers of major Communist incursions on land in Europe have almost disappeared and that the lightning coup which might present a *fait accompli* is most unlikely. Faced with this unpromising position on land, the Communist leaders may well turn to other areas where they can probe western weakness and thus continue to move towards their long-term aims. As Mr. Mayhew said, it may well be that the sea will be thought a profitable environment and the West must be ready to counter any such plans, not only in the N.A.T.O. area but also east of Suez, and indeed everywhere in the world.

To protect shipping, whether it be cargo ships, tankers, fishing vessels, or oil drilling barges, and to maintain freedom of access on the oceans, warships and aircraft are needed. Convoy has been proved to be the most effective method of defending shipping, and an aircraft carrier is the most important part of a convoy defence, whether against attack from the air, the surface or from under the sea.

Role of the Carrier

I believe therefore that the carrier will have two roles in the future—the first is to provide the main share of both the defence and the power of the joint service task force, and the second to help protect merchant shipping which is threatened or attacked, or whose movements have been restricted in any way.

Fortunately, these roles are very similar and require the same types of aircraft, so that no additional technical complication is introduced.

To implement these two roles, the carrier must embark aircraft of the following types:

(a) An aircraft capable of providing early warning against low-flyers and against surface ships armed with surface-to-surface missiles.

(b) A supersonic fighter to reinforce the missile defence for targets outside missile range or outside the missile radar "envelope."

(c) An aircraft capable of attacking enemy surface ships and especially ships fitted with ballistic rockets capable of out-ranging the gun defence of the force. This same aircraft must also be able to attack targets ashore.

(d) Helicopters for the defence of the force or convoy against submarine attack (anti-submarine fixed-wing aircraft such as the American S2F are also desirable but the size and number of British carriers does not allow them to be embarked and reliance must be placed on long-range patrol aircraft based ashore).

(e) An aircraft for close support of troops ashore, and for providing reconnaissance for the Command.

Fortunately the types of aircraft available are versatile in operation and can cover several roles each. Thus the Sea Vixen is a competent all-weather fighter which can also be used for the close support of troops and with the aid of in-flight refuelling can carry out long-range missions. Its successor, the Phantom, is even more versatile. The Buccaneer's primary task is to attack enemy ships and also enemy targets ashore, but it too can be used for direct support of forces operating ashore and also for reconnaissance.

The future complement of carriers will therefore consist of Phantoms, Buccaneers, Wessex helicopters, and the Gannet A.E.W. aircraft. Recent announcements have indicated that the new aircraft-carrier design will include arrangements for troop-carrying in emergency, but it is unlikely that this will affect the aircraft complement by adding a new type. It is more probable that a temporary transition from general purpose carrier to commando ship could be made without losing all operational effectiveness in the usual role.

* * * * *

Vulnerability

No discussion of carriers is complete without mention of vulnerability, and many of the critics concentrate on this aspect of the problem. All ships are vulnerable, and in a nuclear attack, I have no doubt that the carrier is the most vulnerable of all. But we are not considering the use of these ships in a nuclear war. The need for carriers is in limited wars or peacekeeping operations when conventional explosives only need be expected.

With modern methods of controlling damage to ships, the bigger the hull, the more difficult it is to sink it. And with the very strong defence which it is possible to provide in a carrier task force, there should be a very high probability not only of survival but of continued effective operation after attack by high-explosive weapons fired from aircraft, surface ships or submarines.

* * * * *

Aerial Flexibility

Piasecki PV-3 8-seat Helicopter of U.S. Navy

HAWKER P.1127. FIRST-EVER VERTICAL LANDING BY JET AIRCRAFT ON CARRIER. H.M.S. ARK ROYAL

PART IX: 1966–1981

Defence Reviews: N.A.T.O. and the Increasing Soviet Threat

Introduction	281
54. 1966 The 1966 Defence White Paper and Debate MAJOR-GENERAL J. L. MOULTON	283
55. 1966 Review of the Military Situation in Europe MAJOR-GENERAL E. K. G. SIXSMITH	287
56. 1967 People's Revolutionary War SIR ROBERT THOMPSON	293
57. 1969 NATO's Northern Flank CLAUS G. M. KOREN	299
58. 1970 American Defence Policy and Strategy for the 1970s COLONEL R. D. HEINL JR	306
59. 1971 NATO from a SACLANT Viewpoint ADMIRAL E. P. HOLMES USN (ret)	312
60. 1972 The Transformation of Strategy MICHAEL HOWARD	318

61. 1972
    The Army in Norther Ireland   326
    MICHAEL BANKS

62. 1974
    The Defence of France and the Defence of Europe   333
    PIERRE DABEZIES

63. 1976–1977
    The Soviet Military Effort in the 1970s: Perspectives and Priorities   338
    JOHN ERICKSON

64. 1978–1979
    Perspectives of NATO Defence   342
    J. M. A. H. LUNS

65. 1980
    Intelligence—The Handmaiden of Policy   347
    LIEUTENANT-GENERAL DAVID WILLISON

66. 1981
    Civil Defence—A View for 1981   354
    C. N. DONNELLY

Introduction

Since 1951 Britain's defence weaknesses had become increasingly apparent. She had a significant excess of commitments over the armed forces and financial resources available. Denis Healey, who became Secretary of State for Defence in the Labour Government of 1964, struggled to restore the balance in a series of Defence White Papers. These involved considerable reductions in the manpower of the Services, for the navy the loss of its fixed wing air capability, and for the air force the cancellation of some of its most sophisticated aircraft programmes. Commitments in the Middle and Far East were planned to be ended gradually, but had to be hastened in view of a rapidly worsening economic situation. All these measures were strongly opposed by the Conservative Opposition, but when they returned to power in 1970 the hard facts of life forced the acceptance of the majority of them. Britain retreated from "East of Suez" and became a predominantly European power with a residuary capability for operations in the rest of the world.

On the wider NATO front there were more complex problems. The inevitably different points of view of the United States and the European partners continued and were exacerbated by De Gaulles's withdrawal from the military command structure and by European uneasiness about America's involvement in Vietnam. This last revived British interest in the countering of armed insurgency, especially as her own Army became heavily committed to the sectarian and ideological struggle in Northern Ireland. In the United States popular reaction to the tragedy of Vietnam led to an outburst of anti-militarism and a more practical demand for a substantial reduction in the number of forces overseas, including those in Europe. In an effort to prevent these tendencies turning into isolationism, some American defence thinkers began to talk in terms of an oceanic strategy based on greatly increased naval and air forces with capabilities for large scale global intervention.

This line of thinking was strengthened by changing appreciations of the Soviet threat. Not only did the conventional superiority of the Warsaw Pact continue in Europe, not only was there evidence of the

Soviet Union's success in achieving strategic nuclear parity, if not superiority, but she was now also taking to the sea. The transformation of the Soviet Navy from a mainly coastal force to one capable of worldwide operations at all levels from nuclear strike to presence in support of foreign policy, could not be ignored. In practical terms it meant that no longer could the United States and her allies count on being able to intervene outside the N.A.T.O. area without having to take into account Soviet counter-measures. The great increase in her maritime power, accompanied as it was by similar expansion of air capabilities, had given Soviety strategy added flexibility and longer reach. In N.A.T.O. too the threat was diversified. The ability of the Soviet Northern Fleet to inhibit, if not interrupt, Atlantic re-supply in times of tension as well as of war, was an added hazard.

The result of all this was growing concern about N.A.T.O.'s maritime strength, including its merchant shipping, and increased discussion on the thorny problems of extending the Alliance's boundaries beyond the Tropic of Cancer in order to protect its supplies of energy and raw materials.

Recommended Reading

G. Williams and B. Reed, *Denis Healey and the Policies of Power* (1971).
Bryan Ranft and Geoffrey Till, *The Sea in Soviet Strategy* (1983).

54. 1966

The 1966 Defence White Paper and Debate
Major-General J. L. Moulton

Some Progress towards Objectivity

After months of leaks, guesses and speculation, the 1966 Defence White Paper at last broke the news that Britain was to discontinue plans to build a new generation of aircraft carriers, and was to go ahead with the purchase of fifty F-111A to replace the scrapped TSR-2 project. The Minister of Defence (Navy) and the First Sea Lord resigned. The White Paper, also announced that, although Britain was to "retain a major military capability outside Europe," British forces would be withdrawn from the Aden base when South Arabia becomes independent in 1968. Some South Arabians and some British traditionalists were dismayed, but oil shares remained steady. All this provided plentiful ammunition for the Defence Debate of March 7th and 8th.

Then came the announcement of the General Election. The Conservative Party Manifesto said that a Conservative government would build CVA-01 and "fulfil our treaty obligations in the Middle and Far East," which implied staying in Aden. Apart, however, from some heckling about Mr. Wilson's advocacy in the 1964 campaign of a strong navy, neither the White Paper announcements nor any other aspect of defence excited much interest in the election campaign. British defence debate seems doomed to remain in its usual mood of grudging apathy periodically rent by flashes of partisan rancour.

Yet it will be a pity if the 1966 White Paper remains in memory as just a pre-election announcement of the cancellation of CVA-01 and of impending departure from Aden. Whatever opinions one may hold on these issues, the White Paper contains far more than CVA-01 and Aden. As Mr. Healey's earlier White Paper showed, a Minister of Defence is at last making a serious attempt to define to the public the problems of British defence, to gauge the resources the nation is willing to set aside as insurance against attack on its interests, and to analyse the functions of defence and the allocation of expenditure to them. The information in Mr. Healey's second White Paper is neither as complete as it might be nor really adequate to the need, but the intention apparent in it deserves the gratitude of those who believe that responsible public discussion of defence is vital to its effective prosecution.

The full title of the White Paper is, of course, "Statement on the Defence Estimates 1966." This year it appears in two parts: Part I, "The Defence Review"; and Part II, "Defence Estimates 1966–7." Part I is, as the name implies, a brief review of the main points of policy and strategy. Part II deals with the activities of the past year and with the current annual estimates.

Within its limits Part II tries to be functional. However, the single item "General Purpose Combat Forces" accounts for rather more than a quarter of the total expenditure, and at second stage each function disintegrates into single Service elements. Nor can one isolate in it the costs of such urgently relevant things, as for example the Aden base and the war in the Radfan.

The difficulty is real enough—naval and air forces especially are transferable from one theatre to another. Total national forces have some meaning. Yet it is hardly creditable that, for example, if the whole air effort now centred in Aden were to cease there would not be some considerable saving. Everything in Germany—ground staffs, installations, civilians and all—is not really transferable to Aden or Singapore. Yet this is how the figures make it appear, with little or nothing for the navy and air force under theatres, and almost everything under "General Purpose Combat Forces." From these the analysis continues into tactical functions: air defence, light bomber reconnaissance, etc. It looks rather as if someone had ordered the Air Department to be functional and that the department had no intention of doing anything of the sort. So for a rough idea of what our east of Suez policy is costing, one must turn to Mr. Mayhew's presumably informed opinion, stated in his resignation speech of February 22nd, that without it we might get away with a bill of £1,800 m. a year instead of £2,000 m. What would the Stock Exchange think about a major industrial concern which treated its cost accountancy so cavalierly?

Costing Future Policy

In the 1965 White Paper, Mr. Healey described one of the most important tasks of a Minister of Defence: "The fundamental problem here is how to reconcile strategic requirements with financial resources and the forces that can be provided from them; . . ." The policy and strategy with which Part I of his 1966 White Paper deals is increasingly and properly decided on long-term costings—ten years has become the accepted term. It is here that a hiatus appears, for the figures in Part II, although interesting enough in themselves, are not only inadequately analysed by strategic function, but, dealing as they do with 1966–7 estimates, are not really relevant to the strategy and policy announced in Part I.

The Ministerial speeches in the Defence Debate of March 7th and 8th produced some of the missing information, and the columns of "Hansard" are really more relevant to Part I of the White Paper than its own Part II. It is natural for Ministers to hold back information which the Opposition in time-honoured fashion will try to turn against them, but doing so depresses the value of debate. The acrimonious opening and closing, which was the least valuable part of the 1966 debate, was caused by the Opposition's attempt to bring home charges of misrepresentation of costs to refute the Government's charges of past Conservative mismanagement. Alarming though it is that it should be so, even £100 m. one way or the other in a ten-year programme is today not very important, and debating points scored on arithmetic generate more heat than light. Their usefulness this year was in twisting the Minister's arm to extract more information from him. Dare one suggest that this rather undignified business could be avoided, and the debate made more constructive, by producing the facts in advance without the arm twisting? Mr. Healey is the sort of man to make the change.

In his press conference of August 4th, 1965, Mr. Healey expanded on the theme of reconciling strategy and finance. He said that British defence resources were overstrained in the three vital areas of manpower, foreign exchange and gross expenditure, and proposed to reduce the strain first by rationalisation, then if necessary by reduction of commitments. He set himself the target of reducing gross cost by 1969–70 to £2,000 m. at 1964 prices or about 6 per cent of the Gross National Product. Part I of the White Paper opens with a restatement of this policy, and notes that this year's estimates fall within the £2,000 m. limit.

There are those who say that, because survival is essential to all other activities, defence should have the first claim on national resources. Advocacy of security regardless of cost, always a little unrealistic, loses its force when a considerable part of the cost stems, not from defending our own territory and national life, not even from supporting allies or international justice, but from the pursuit of self-imposed responsibilities in distant areas. Idealistic though such policies may be, they are open to challenge. In playing the self-appointed policeman, it is possible that we foster irresponsibility and resentment and so in the end do more harm than good. Be that as it may, physically remote responsibilities cannot have the urgency of immediate survival.

Therefore it is sensible and logical to say openly what the financial limits are, and to argue from there what one can reasonably expect to do. One must welcome Mr. Healey's definition of this problem in its real terms. A little more of this sort of thing from his predecessors might have saved us from the many abandoned projects which today litter the field. Far better that the Cabinet should say plainly what the limits are and will be and tailor its foreign policy to them, than that it should allow the Foreign and Commonwealth Relations Offices to play the super-power and the Defence Ministry to embark on programmes in the vague hope that money will somehow be found to complete them.

This is admirable. If the new limit is a little below what had become the customary one of about 7 per cent of the Gross National Product, well economics is an abstruse subject, and whoever's the fault, national finance are strained. Two questions remain worth arguing. Are Mr. Healey's solutions the most effective to achieve his declared aim of reducing the overstrain in the three areas of manpower, foreign exchange and gross cost? And, has he persuaded the Cabinet to honour his pledge that, having done everything possible to spend the available money wisely, he would then reduce political commitments to what it could be expected to achieve?

* * * * *

55. 1966

Review of the Military Situation in Europe
Major-general E. K. G. Sixsmith

The decision made by General de Gaulle on France's participation in N.A.T.O. has overshadowed all other events in the year. In some respects it has brought to light the determination of the other partners that the alliance should continue, but it has also afforded countries on both sides of the Iron Curtain and both sides of the Atlantic the opportunity to review the meaning of the Cold War today and the future of deterrence. It would be an over-simplification to ask ourselves what sort of an alliance we should make if we started anew today; politics and history do not start anew. A cool and critical look at N.A.T.O. shows that despite the changes in the international situation since 1949 it is sound in principle and in design. Its strength is that it is in no way aggressive; no nation with peaceful intentions has any cause to fear its will—it has not ever imposed its authority on its own members in dispute. In defence against aggression its purpose is clear, an attack against one is an attack against all. Its organisation may be cumbersome, an alliance based on unanimous will can hardly be streamlined—but in essentials it is efficient. The integration of command and staff and the assignment of forces give the strength and flexibility which previous peacetime alliances have lacked. Against these very real assets it has two weaknesses both of which arise from the stark fact that, even in these enlightened days, political power depends upon military strength. The first is that a defensive alliance is much more valuable in keeping things as they are, than in developing new policies. For this reason some critics say that the Atlantic alliance has been unable to take advantage of the easing of tension with the Soviet, and above all, that no progress has been made towards arms control or any degree of disarmament—progress on which long-term hopes of peace depend. The weakness is that the preponderant nuclear strength of the United States and the very nature of nuclear strength, upsets the balance of the alliance and undermines the belief that authority in the alliance is shared. There can be no doubt that General de Galle believes sincerely that hopes for the future depend upon listening to the experienced voice of Western Europe, and that this means the leadership of France. It does seem, however, that he is wrong to try to improve the alliance by discarding the integrated organisation which is its chief source of strength.

The facts of General de Gaulle's decisions are that all French forces assigned to N.A.T.O., and all French officers and men in integrated staffs, were withdrawn on July 1st, 1966, and that allied military headquarters and installations located in France must be moved by April 1st, 1967. Permission for allied military aircraft to fly over France will only be given on a monthly basis. The main contribution of France to N.A.T.O. is in the orders of two divisions serving in Germany. Under the allied arrangement they were completely under national command and control—just as is the British Army of the Rhine—except when put under N.A.T.O. Command for an exercise or in a crisis. In the latter circumstances they formed part of the Central Army Group, under American Command, but the American Commander was under the French General Crepin who commanded the Allied Forces in Central Command. Incidentally the withdrawal of General Crepin from his allied appointment will be one of the most serious results of General de Gaulle's decision; he has shown himself to be one of the outstanding allied commanders, his thoughtful and imaginative ideas on training have been a tremendous asset in Europe. It is difficult to know what the French really want done about their forces in Germany. The Germans and N.A.T.O. as a whole want the forces to remain in Germany with a definite commitment that, if necessary, they will co-operate with other allied forces. France is playing a very non-committal line, seeming to show she does not mind whether they stay or not. France and the other fourteen countries have agreed in principle that the French forces can stay in Germany and will co-operate with other N.A.T.O. forces, but France seems in no hurry to state the limits to that co-operation and to show any reality for the change to take place on July 1st. The move of the headquarters has been already decided S.H.A.P.E. moves from Paris to Wavre near Brussels, not far from where Wellington and Blücher concentrated to crush Napoleon at Waterloo. The headquarters of the Central Command. AFCENT, goes to Trier, in West Germany between Luxembourg and the Rhine. The new locations of the headquarters do not in themselves offer any real disadvantage over their present locations, but the move will obviously be a large and unnecessary expense, and involve a dislocation and renewal of communications. No move of the political headquarters of N.A.T.O. is contemplated at present. This is all to the good; curiously enough the first rumours of de Gaulle's impending action were that he intended to demand the removal of the N.A.T.O. organisation from their new building in the Dauphine; that has happily proved false and the other

N.A.T.O. powers have been wise to build on the fact that General de Gaulle has stated that he wished France to remain a member of the alliance.

The actual moves of headquarters and even the possible loss of forces are not therefore insurmountable obstacles to the continuance of N.A.T.O. in strength. It would however, be madness to pretend that the alliance can be strong without the participation of France. The facts of geography are plain and France is an essential participant in Western European strategy. In the question of air space and of communications her position is cardinal. The question of air defence obviously depends on integration of early warning and of fighting forces, but there are other needs too for freedom of the air over France. Imagine the necessity of a reinforcement or even an exercise flight for the southern and south-eastern fronts in Italy, Greece and Turkey. In the absence of liberty to fly over France and remembering the neutrality of Switzerland and Austria, there is no access from Britain, the Benelux Countries and West Germany except via the Atlantic and Gibraltar or by agreement with Spain. Possibly France has no intention of withholding its monthly authority to overfly but this example illustrates the importance to the alliance of France's geographical position.

Flexible Response

The organisation of N.A.T.O. and its military structure have not been materially changed since the inception of the alliance. It is fairly obvious that what was right in 1949 is not necessarily good now. The real reason changes have not been made is that satisfactory and radical changes are extremely difficult to work out. General de Gaulle, the strongest critic of the organisation, has so far made only one positive suggestion, that put forward in 1958 for a political standing group. No greater tribute to the present organisation can be paid than a sincere and single-minded attempt to work out a better.

Perhaps the best way to begin to see the improvements that might be made is to examine the reasons behind General de Gaulle's dislike of the present arrangements. Apart from his inherent belief in the destiny of France, this probably stems from the conviction that integration means subjection to the will of the United States. This belief in its turn arises from the imposition of a policy of flexible response on N.A.T.O. This strategy was not the result of any decision by the N.A.T.O. Council, or even by the Standing Group or Supreme Commanders. It was an American conception worked out by Mr. McNamara and his advisers and accepted by President Kennedy. The French do not and have never believed in flexible response; they believe that it is policy that could lead to the sacrifice of Europe to the whim of the giants, the

United States and the Soviet. This may be an exaggerated and unjust view, but it is not one which is confined to General de Gaulle or which has been imposed on France by him. General Ailleret, the present Chief of Staff of the French armed forces is a known opponent of the concept of flexible response. In 1964 he gave a lecture to the N.A.T.O. Defence College in which he explained his objections. His thesis was that a conventional force was required only to gauge the intentions of the enemy. The Soviet could then only attempt aggression in the knowledge that it would be answered by nuclear weapons. He considered flexible response a reasonable policy outside Europe, but inside Europe it must be a case of no war at all, or it would be a nuclear war. There is really little between the French view and the American view of the theory of deterrence, but the French seem to be suspicious of the way the theory will be applied. They fear that because of the danger of attack on each other's cities the United States and the Soviet could, in the last resort, do a deal which would restrict a war to conventional and tactical nuclear weapons. Such a war could only lead to the destruction of Europe and might be avoided altogether by the certainty that all the weapons and strength at the disposal of the United States would be used if necessary. There is not a shred of evidence to support this cynical French view of United States intentions, and it is contradicted by the wholehearted way in which the United States has remained involved in Europe for the twenty years since the end of the war. If the United States did have any leanings in this direction surely the way to increase them would be to persuade her to remove her forces from Western Europe; yet General de Gaulle's policy seemed designed to make it as difficult as possible for them to remain in Europe. On the other hand Germany, who has as much to lose from war in Europe as has France, realises that her security and the possibility of peace depends on the presence of American forces in Germany.

One of the main ways in which dislike of flexible response has shown itself is the French objection to the Mobile Force. This eminently sensible idea provides for a small allied force to be available to fly to a trouble-spot, say in the south-eastern or northern region, and to ensure that if possible a minor incident can be dealt with at once. The French have consistently refused to take any steps to provide a share of this force.

One other fear which General de Gaulle seems to have used to persuade the French of the dangers of American policy is the fear that United States involvement in Viet Nam and the Far East generally might drag Europe into unnecessary war. Without going into the pros and cons of United States policy in Viet Nam it may be said that the dangers of the situation are minimised by the nature of the N.A.T.O.

alliance. Alliances only work under the cement of a common interest, and much as we sympathise with the United States in the difficulties in which she finds herself no good could come from the involvement of N.A.T.O.

A Solution

The most understandable reaction to General de Gaulle's policy has come from General Norstad, who showed during his twelve years at N.A.T.O., half of it as Supreme Commander, that he always looked at a problem as a N.A.T.O. man rather than only as an American. He sees that the fears of France stem only from the United States exclusive control of nuclear weapons. Not only France but all European nations see that this subjects the whole N.A.T.O. alliance to the decision of the United States President. If the Supreme Commander in Europe, working as he does under the Control of the N.A.T.O. Council, had at his disposal intermediate-range nuclear weapons, there could be no doubt that the decision of N.A.T.O. strategy and policy could really come from the N.A.T.O. Council. Here, for the uninitiated, it must be stated that the members of the N.A.T.O. Council are not the ministers or permanent officials who attend routine meetings, but the nations themselves. The level of representation at Council meetings depends on the subject being discussed, and can be from Prime Minister downwards. General Norstad's idea is for a small sub-committee of the N.A.T.O. Council to be given authority to take majority views on nuclear questions. Such a solution would obviate the necessity of forming special forces, such as the Multi-lateral Force or the Atlantic Nuclear Force. Since membership of the Committee would presumably be from the United States, United Kingdom, France, Germany and one to represent the remaining countries, it might be said also to answer General de Gaulle's requirement for a small controlling body for N.A.T.O., giving Europe as much weight as the United States. Certainly such a committee would give to non-nuclear powers some say in the strategy, of the alliance and the power to decide against what targets and when nuclear weapons could be used. An essential feature of the suggestion is that it introduces for the first time, majority decisions into the N.A.T.O. machinery. This is an effective way of ensuring that America does not have undue influence but it may have one grave disadvantage. The Soviet is learning to live with the nuclear balance between herself and the United States; the one thing she fears more than anything else is that Germany should obtain control of nuclear weapons. Membership of this nuclear committee hardly lends itself to irresponsible action, but it is possible that the Soviet would regard it as going outside the limits of nuclear sharing which she considers compatible with an arms control agreement. The risk of this misinterpretation of our intentions is one which is worth taking if France and Germany do accept this as an effective solution of the problem of nuclear control in N.A.T.O.

* * * * *

56. 1967
People's Revolutionary War
Sir Robert Thompson

Since the Second World War there has been a formidable list of People's Revolutionary Wars. Some have been successful, as in China, Indo-China and Cuba, and some have failed, as in Greece, the Philippines, Kenya and Malaya. All have been protracted and many still simmer, but, of them all, the present war in Vietnam is the classic example and will undoubtedly prove to be the test case. It could well determine the future history of the world as decisively as the two world wars of this century.

In the study of People's Revolutionary War attention has rightly been paid to its prophets, such as Mao Tse-tung, Vo Nguyen Giap and Che Guevara, but there has been a tendency to concentrate on the guerrilla or military aspect and to ignore the strategic concept and the political base of such wars. Most people are romantically fascinated by the tactics and operations of guerrilla warfare, and Mao has added to this attraction with the cloak of such descriptive and compelling phrases as: "The strategy of guerrilla warfare is to pit one man against ten but the tactics are to pit ten men against one"; or, "The people are like water and the army is like fish." People's Revolutionary War is not really as simple as all that.

In order to give some idea of the scope of the problem it is worth stating, from the examples of Malaya and Vietnam, the initial size of the forces on both sides and showing how they developed. In Malaya, with a population of about seven millions, the initial strength of the guerrilla forces when the Emergency broke out in 1948 was estimated at 4,500 with a base of possibly 50,000 positive active supporters within that population. The initial strength of the government forces was about 21,000—an unfavourable ratio for the insurgents of $1:4\frac{1}{2}$. The guerrilla forces were only able to increase their strength to 10,000 within three years and thereafter declined, but it took twelve long years before the Emergency could be declared at an end. Even then about 500 hardcore guerrillas remained in the jungles on the Malayan/ Thai border to await a future favourable opportunity.

In South Vietnam, with a population of about fourteen millions, the initial strength of the Viet Cong guerrilla units in 1960 was 5,000 with a base within the population which at that time probably did not exceed 100,000 of positive active supporters. The strength of the government forces was well over 200,000, an unfavourable ratio of 1:40. Within seven years, at the beginning of 1967, the armed strength of the Viet Cong had been increased to 280,000, of which perhaps one-fifth represented infiltration from North Vietnam. These forces now control a population base within South Vietnam of about five million people, i.e. one-third of the population of the country. Opposed to them is an American and Allied strength of 400,000 combat troops and South Vietnamese forces of about 500,000. Well over one million men are therefore engaged in combat and there is no conclusion in sight.

The most striking feature of these figures is that the initial strength of the insurgent moment in both countries represented less than 1% of the total population. While this hardly qualifies an insurgency for the title of a People's Revolutionary War, it does raise the fundamental questions: How with such inferior strength does an insurgency ever get off the ground? How does it survive and develop? And finally, how can it possibly win against all the odds?

Before considering the three well-known phases of a People's Revolutionary War—the building-up, the guerrilla war period and the war of movement—it is essential to understand five points about a People's Revolutionary War which I will call its organisation, the time dimension, the space dimension, the use of subversion and selective terror and the exploitation of "contradictions."

* * * * *

The Three Phases

Great prominence is given in the study of People's Revolutionary War to its three main phases: the subversive build-up phase, the guerrilla war phase, and finally a war of movement, as if these were all an essential part of the progress to victory. In fact these phases are a natural evolutionary process when victory can only be won by a long, protracted and arduous struggle; in other words, when the insurgent cause and support are weak at the outset and have to be developed through the course of the insurgency. If the insurgent cause was well founded and popularly supported and the government cause was correspondingly weak, then victory might be achieved during the subversive phase, assisted by selective terror. The need to promote an "armed struggle" and to advance into the second phase of guerrilla warfare is in itself an admission that subversion by itself was not enough to overthrow the Government. The move to the second phase is of course also an indication that the subversion has been sufficient to get a guerrilla war going successfully within a country. The length of time which the build-up takes, in spite of the advantage which can be taken of favourable events during that period (*e.g.* the Japanese War), is a measure of the weak base on which the insurgent cause was originally founded.

In the second phase of guerrilla warfare the political aim of the insurgents is to gain control over the population, starting in the remoter rural areas, and gradually to destroy the Government's prestige and authority throughout the country. The military aim is not to defeat the Government's armed forces (because this would be beyond the insurgent's capability) but to neutralise them and render them powerless to save the State. Here again, given a modicum of support within the population, an insurgent victory could be achieved without the third phase. This was in fact the point almost reached in South Vietnam early in 1965 before the arrival of American combat troops.

In small under-developed countries, where space does not permit and where government resources are limited, it is unlikely that People's Revolutionary War need go beyond the second phase to achieve victory though this may mean that the second phase could be very prolonged and even hover on the brink of a war of movement. Certainly there is likely to be a period of large-scale guerrilla actions with at least regimental forces engaged. Because of the space element a war of movement was the essential concluding phase in China where the Nationalist Armies had to be defeated piecemeal in the vast areas which they occupied. On a very modest scale this might have been necessary in Vietnam during the course of 1965 to conclude the war and achieve victory if American combat troops had not been committed to redress the balance. The insurgent appreciates that foreign troops, however massively inserted at this stage, cannot win the war by themselves. They can only hold the ring while the local government recovers and set the stage for the local government to achieve victory. It must be the insurgent's aim, by making use of time, space and manpower, to thwart this purpose without necessarily having to defeat such foreign troops in battle.

Throughout a People's Revolutionary War the insurgent will be able to profit from the threatened Government's mistakes and tactical errors. He may be able to compel them. For example, in an area under government control, the insurgent will create a high rate of activity and incidents in order to induce the Government to take strong measures which may alienate the population, whereas in areas under insurgent control, which the Government is trying to regain, there may be no incidents in order to lull the Government into a false sense of security.

There is also every prospect that the Government, and its allies, may be diverted from the main issue by a number of alluring distractions. The most likely of these, because it has some significance, will be infiltration and this in turn will give rise to the bombing of sanctuaries and a call for negotiations, all of which are of secondary importance and in some ways are almost irrelevant to the main issue.

An insurgent movement does not depend on infiltration. It is the other way round; infiltration depends on the success of an insurgent movement within a country, that is, on its capacity to absorb infiltration. This absorptive capacity is probably a constant percentage of the strength of an insurgent movement at any given time, perhaps between 10 and 20 per cent, though the quality both of men and materials infiltrated may make this more significant than the quantity. If an insurgent movement expands, so will the infiltration increase. This has been remarkably demonstrated in Vietnam, where the infiltration at the beginning of the insurgency was a mere trickle, but where it has now increased to about 8,000 men a month, including whole North Vietnamese regiments, in spite of the bombing of infiltration routes in North Vietnam and southern Laos. It is interesting to compare this situation with that in Malaya and Borneo during the Indonesian confrontation with Malaysia, where infiltration completely failed because there was no insurgent movement within the country to absorb it. Infiltration is therefore a booster to the insurgent machine and can accelerate the place of the insurgency; it can never provide its main driving force.

* * * * *

Decision

In this situation the insurgent cannot be defeated by foreign troops alone. He can only be defeated if these troops set the stage for the local government to win its own war. In respect of South Vietnam this means that the country must be restored as politically and administratively stable and economically expanding. Not only must the local government clear and hold lost territory permanently but it must reduce the insurgent underground organisation within the population so that it can no longer support guerrilla units and ceases to be a threat. At the same time the political structure of the government must be re-established, its administrative machine rebuilt, its armed forces remodelled, its police force reformed and its economy revived—all this in the midst of war.

If the insurgent can maintain the tempo and thwart this constructive programme, he will retain the strategic initiative with two alternative routes to victory. Either, through international and domestic pressures, the foreign supporter of the local government will lose heart and accept a solution which in effect would be defeat, or the local government itself will collapse thereby destroying the political foundation of continued resistance. The insurgent strategy is more likely to be directed at the second alternative because it is the one which he can most directly influence by exploiting all the side effects of war— inflation, refugees, corruption, incompetence, war-weariness, disillusion and loss of hope. By so doing, he also keeps the first alternative wide open.

If, however, the insurgent military units are contained and the right constructive measures are systematically taken, then the insurgent must slowly lose space and sell time. But he is not yet beaten. An insurgency which has taken twenty years to build up its popular base and perhaps five to ten years to achieve a high level of guerrilla activity, can take almost as long to unwind. By taking "one step backward" and accepting a lower scale of activity and a gradual loss of space the insurgent can spin out time in the hope that events may still swing in his favour and enable lost ground to be retrieved. The local government is still faced with a colossal task requiring a clear view of objectives, unity of effort and great resolution. The insurgent will use every trick to "spoil" the Government advance.

His final gambit, when defeat appears inevitable, may be to withdraw or sacrifice his military units in order to save his political underground organisation within the country. Either he can simply disperse them and fade away, thereby compelling the eventual departure of foreign troops and a reduction in government troops, or he can offer to negotiate their withdrawal and disbandment. In both cases he will hope to retain his political underground organisation as intact as possible. This represents, in its build-up, the greatest element of time. There will be every prospect, in the aftermath of war with much of the pressure off, that political or economic chaos will ensue and a fresh opportunity occur to seize power, perhaps without a shot being fired.

In the case of Malaya, when defeat became inevitable, the insurgent offered to negotiate. Tempting though the prospect of peace was, the Malayan Prime Minister refused terms which would have allowed the Malayan Communist Party future freedom of political action. While an offer to negotiate in Vietnam might happen (it is still a long way off), it is less likely than simply fading away. Hanoi will not want to run the risk of the International Control Commission being given more teeth and more effective guarantees being internationally agreed for the future security of South Vietnam. By fading away this risk could be avoided and, unless the stability of South Vietnam is maintained and the Viet Cong underground organisation checked, at some point in the future an opportunity might occur to start all over again.

If People's Revolutionary War, as an instrument of Communist Policy, is to be foiled, the war in Vietnam must be won. It is a contest of will and a battle of wits. It will require skill, patience, determination and nerve. It is not just the future of the people of South Vietnam which is at stake but the future of all peoples in small, under-developed countries—the "countryside" of the world—who wish to determine their own political future and choose their own way of life. The "cities" of the world in Europe and America can only ignore this at their peril.

57. 1969
NATO's Northern Flank
Claus G. M. Koren

'The withdrawal of the British forces from east of Suez and the availability of more British forces for the defence of Allied Command Europe as a whole has given an entirely new complexion to the defence of Northern Europe.' These words were spoken in May 1969 by General Sir Kenneth Darling then Commander-in-Chief Allied Forces Northern Europe. N.A.T.O.'s Northern European Command consists of Norway. Denmark and that part of the Federal Republic of Germany which lies north of the river Elbe, plus adjacent sea areas. Headquarters Allied Forces Northern Europe is situated under a mountain top some 12 miles outside Oslo, the Norwegian capital. Here officers from five N.A.T.O. nations sit together to plan how to handle military contingencies, which might occur anywhere from the small mining town of Kirkenes on the Soviet border in the extremeties of north Norway to the metropolis of Hamburg, some 1,200 miles further south.

The five N.A.T.O. nations directly involved in the defence of northern Europe are the three "host" nations and the two most important Western maritime powers. Great Britain and the U.S.A. To these latter nations, the territory of the Northern European Command constitutes a first line defence, which, with the growing strength of the U.S.S.R., particularly at sea, is of increasing importance.

The defence problems of the Northern Command are dictated by geography and by political conditions. If one looks at the map of Europe through the eyes of a strategist, it becomes apparent that somewhere in northern Germany the military problems acquire another dimension when compared to the rest of Western Europe. North of this area, the defence problems of the West are closely linked to naval strategy and to the somewhat special political conditions in Scandinavia, whereas the defence problems south of this area are those of the traditional land/air war. A dividing line here groups the defence problems on each side around one main dimension of strategy. The commanders-in-chief responsible on each side of this dividing line are therefore able to concentrate on their main strategical problem without being distracted by problems which are secondary to the solution of their main task.

From a military point of view, therefore, the boundary line between Allied Command Central Europe and Allied Command Northern Europe is drawn at its most logical place, namely along the river Elbe. This boundary decides the military situation in the whole of the northern region. This we shall discuss later.

The Special Characteristics of Northern Europe

A normal map can give a false picture of the physical features of the Northern European Command. If, however, we take station somewhere near the Faroes and look at the Command with Hamburg on the right and North Cape on the left, we get a very different view.

We see the considerable width of the Command, its lack of depth, how it is split in half by the Baltic and the approaches thereto, and the fact that nowhere in the Command are we very far from the sea. Most of the land communications, particularly in Norway, are very close to the sea; they frequently have to cross the sea by ferries and bridges. Most airfields are within easy reach of the sea; some are virtually in it. Climatic conditions vary enormously, and they produce some very special problems in the north. In general, it can be said that the command stands out on a limb from central Europe, with the sea at its back.

The second main factor which bears on the defence of northern Europe is political. The history of the Scandinavian countries as well as the close kinship of their peoples draw these countries together. Only reluctantly would they cross each other's interests. Yet, history shows that their foreign political interests differ. This is why unions of the Scandinavian countries never could last, and this is why Norway and Denmark joined N.A.T.O. in 1949 while Sweden remained neutral. Norway felt close ties with the West, Denmark's security was inevitably tied to the fate of her southern neighbour, but Sweden felt that her security and that of Finland were best safeguarded by Swedish neutrality.

Had Sweden joined N.A.T.O. in 1949, she would have risked that Finland would have lost what independence she is now allowed to enjoy. Similarly, it was considered that the prospect of closer military ties between Sweden and the West would influence Russian moves vis-à-vis Finland.

In this way, a balance is established in the Scandinavian relationship with Soviet Russia. Sweden, being the strongest of the Scandinavian countries, has her back covered by N.A.T.O. and is able to direct her considerable military potential towards the east. The possibility of these forces aligning themselves with N.A.T.O., should Finland be threatened, improves the latter's position as an independent nation. On the the other hand, Finnish independence is extremely important to the security of north Norway. The Scandinavian governments are therefore most sensitive to anything which might upset this balance.

In addition to this, which virtually amounts to a common Scandinavian security policy, there is a feeling that the Soviets have legitimate security interests in areas adjacent to Scandinavia. For this reason, the Scandinavian partners have led the Alliance into accepting a somewhat restrictive military policy in northern Europe to avoid provocation. At the same time, however, determination to defend this area is clearly demonstrated—often in the face of loud Soviet protests.

A consequence of this Scandinavian thinking is that Norway and Denmark have no foreign forces stationed on their territories in peacetime, except for exercise purposes. Furthermore, Norway and Denmark, "under the present circumstances," as it is termed, have declined to establish stocks of atomic stockpiles. The military value of atomic stockpiles is not contested, but it is, in the opinion of the governments of the two countries, outweighed by political considerations. Political consideration have, however, not prevented Norway from taking active steps to strengthen her defence in the extreme north, as a result of recently demonstrated developments in the Soviet military capabilities.

Nowhere is the imbalance of forces between the Warsaw Pact and N.A.T.O. greater than in northern Europe. The combat-ready Soviet land forces facing Finmark, the northernmost province of Norway, run to some two divisions, and further south, within the same military district, there are half a dozen more to back them up. A most significant point is the tremendous growth in the maritime strength of Russian in the last 20 years. Much of this sea power is based at the very doorstep of Norway in the Murmansk area. In the air, also, Soviet forces greatly outnumber the air forces of the Northern Command.

In the Baltic approaches area a similar situation exists. Here there are some six to eight divisions facing Schleswig-Holstein and Denmark. The Soviet Baltic navy based on Leningrad is similarly powerful, and the air superiority of the Soviets here also is a fact of life.

Denmark and Norway, being small countries, naturally can only maintain comparatively small standing military forces in peace. These have, however, over the years been built up on a sound foundation and are, within the limits of what is practical, equipped with modern weapons. It is worth noting that both Denmark and Norway have considerable mobilization forces, and in each country the mobilization system is highly developed.

Although only a small part of Germany is included in the Northern

Command, she makes a sizeable contribution of military forces to it. Thus, assigned to AFNORTH are an armoured infantry division, the whole of the German Navy, including its naval air arm, and air squadrons of the German Air Force.

If the situation in Norway and the Baltic approaches were looked upon in isolation, it would appear that the forces in these areas were totally inadequate for the purpose of dealing with the massive threat against them. However, by the presence of the Allied command organisation, and through arrangements made, planned and practised in peace, the total joint strength of N.A.T.O. can be brought to bear in the exposed areas of northern Europe. Political considerations are built in even in the command structure, but sound principles of command and control have nevertheless been established throughout the area.

* * * * *

Reinforcement

It will be evident that one of the main problems in northern Europe is to make provision for speedy reinforcements to arrive.

The availability of such forces for deployment to northern Europe does not in itself solve the problem. Adequate provision must be made in peace to receive such forces. In this respect, the N.A.T.O. infrastructure programme has made a great contribution. Under this programme great sums have been invested for the construction or improvement of airfields and for the installation of electronic equipment for communications and control of airspace. The N.A.T.O. Defence Ground Environment (N.A.D.G.E.) system alone, now being installed in the Northern Command, amounts to some £117,500,000. Naval bases and depots have also been built. In addition national funds have been devoted to further improvements.

The command organisation, satisfying both national and Allied requirements, which has been established over the years, makes it possible to absorb forces coming from outside and to direct them in a co-ordinated effort with forces organic to the command.

If it is to work in war, a defence posture which relies heavily on outside assistance calls for almost incessant practice. The techniques of strategic mobility must be practised to give the command staffs and all others concerned the necessary co-ordinated experience to ensure maximum efficiency. SACEUR has at his disposal an international mobile force which he can project quickly to his flanks to protect these against limited incursions, should the host nation require such assistance. This force has been exercising regularly in north Norway and was for the first time deployed to Denmark in September 1969.

With the withdrawal from east of Suez of the British forces, the military situation in northern Europe is being improved. More navy, more air forces and more strategic reserves are now available for possible employment in the Northern Command area. Vivid examples of this are such exercises as "Bold Adventure," which was conducted under N.A.T.O. auspices in Denmark and Schleswig-Holstein in early 1969, and an increasing number of bi-national exercises of a unit training nature, in Denmark and Norway, involving British forces.

The availability of British forces not previously committed to N.A.T.O. gives a better credibility to the defence of the Northern Command, increasing its deterrent value. In spite of the fact that the forces organic to the Northern Command would seem insufficient, the situation does present major problems to a would-be aggressor.

Strategic Assessment

In the south, an aggressor's aim might be to gain control of the exits from the Baltic and at the same time to secure the flank of his forces operating in western continental Europe.

With a proper command organisation in this area, such as the one which is now firmly established, it is possible to exploit to the full the defensive capability of both the German and the Danish forces available. They are mutually dependent on each other, and proper co-ordination has been made possible through the joint Allied command.

From a military point of view, it does not seem likely that an attack against any part of this area could be carried out without involving the forces of the Central Command. The aggressor would have to regard the forces of N.A.T.O. as an entity. If he attacked the forces in Denmark and/or Schleswig-Holstein, he would have to reckon with these forces being supported by powerful forces of the Central Command and must plan accordingly. The defence of this area cannot, therefore, be considered in isolation, and it would not be correct to measure the isolated strength of the Northern Command forces in this area against that of a possible aggressor's without taking the whole of N.A.T.O.'s strength into account.

As long as Schleswig-Holstein and Denmark are held, there is no immediate threat to south Norway. For this reason, the bulk of the standing Norwegian forces can be deployed to the north, where they are able to fulfil their essential role in a defence of northern Norway based on strategic mobility. Concentrated in north Norway, the standing Norwegian forces make a N.A.T.O. defence of the north not only possible but also credible.

Because of terrain difficulties and the distance involved, an overland attack against north Norway would not be an easy operation. The most likely threat in this area is, therefore, a limited land advance combined with amphibious landings on the coast and complementary air and airborne operations. An aggressor would, however, realise that the Norwegian forces on the other side of the border are not the only forces he would have to deal with if he crossed that boundary.

The Norwegian forces are deployed and organised in such a way as to secure the arrival of augmentation forces. Through the Allied command, reinforcements can be brought to bear. To the aggressor, therefore, the Norwegian forces represent the joint N.A.T.O. defence and not the limited forces of a small isolated nation. Therefore, the problem in the north is the same as it is in the south: first, to provide adequate facilities to enable reinforcements to arrive speedily in the area and, secondly, to be strong enough to hold out until they do arrive.

Between the two exposed areas in the north and in the south, south Norway constitutes a logistics and reinforcement area, particularly for north Norway. Mobilisation reserves, trained for north Norway, will be available here when alerted, and suitable aircraft are now being made available for their speedy transfer.

Contrary to a feeling some years ago, it seems that the strategic importance of the Northern European Command is growing rather than diminishing. With the growth of Soviet maritime power, the areas close to her bases assume new importance—for the Soviets and for the West. The approaches to her most important naval bases in the Leningrad and the Murmansk areas run either literally through the command area or in close proximity to the lightly defended Norwegian coast.

Because of the maritime orientation of the Northern Command, the development of Soviet sea power is looked upon with growing concern in northern Europe. It was not least the new Soviet amphibious capabilities backed by modern surface vessels which caused Norway to take a new look at her defences in the extreme north. The appearance on the scene of the N.A.T.O. Standing Naval Force in the Atlantic (STANAVFORLANT) is indeed welcomed, but considered by many inadequate to balance the Soviet development, which, it is felt, will impair even further the strategic picture in northern Europe.

Political Developments

Some Norwegian political scientists call for analysis of this development in order that long-term policies may be arrived at. It is felt that the Soviet build-up of naval forces is part of an effort to maintain the strategic balance with the U.S.A. of mobile missile forces, and that it therefore has an element of legitimacy about it. The situation could, therefore, in the long run be extremely difficult to handle from a

Norwegian political, as well as from a military point of view, unless all possible developments were analysed in advance and a sensible long-term policy, backed by Norway's allies, is adopted. Others draw attention to the increasing danger which they feel that the membership of N.A.T.O. embodies.

In this context it is interesting to note that the 'Out-of-N.A.T.O.' campaigns—which started bearing their drums in both Denmark and Norway as the 20th anniversary of N.A.T.O. was coming up, and possibilities of withdrawal presented themselves—have collapsed. Opinion polls show that public support of the N.A.T.O. policy of the two countries is as strong as ever.

In Norway, where more detailed analysis of the N.A.T.O. polls is possible, the amazing fact is that young people between the ages of 21 and 30 are the strongest backers of N.A.T.O. of all groups. This shows not only that the danger of a revolt by youth against N.A.T.O. is more remote than anticipated but it is also an indication that young people, who have done their national service, generally are of the opinion that their military service makes sense. This again indicates that the armed forces do a good job in training the national servicemen. This, of course, is not only of the greatest importance for an efficient defence but it is vital for the defensive will of the people.

A similar analysis of the attitude of Danish youth is not possible because relevant information is not available at this time. It is a fact, however, that the general trends in opinion surveys in both Norway and Denmark show a striking resemblance, and there is no reason to believe that Danish youth differ greatly from their Norwegian cousins in this respect.

* * * * *

58. 1970

American Defence Policy and Strategy for the 1970s
Colonel R. D. Heinl, Jr.

With apologies to John Reed, "ten years that shook the Pentagon" might be a fair description of the 1960s. Eight years under Robert S. McNamara, the Cuban missile crisis, the Vietnam war, Russia's rise to nuclear parity, and something approaching a national revolt against established priorities in defence and foreign policies—these are but a few of the momentous events which have shaken, deflected or overtaken the defence establishment of the United States in the past decade, and whose effects promise to dominate American strategic and defence decisions in the decade ahead.

Yet it is the past five years which, by several orders of magnitude more than the preceding five, have trained enormous, quite unforeseen pressures against the US defence establishment, its policies and politics, its strategies and, in fact, the underlying rationale for American national defence and national security.

As a result, the defence establishment and those who make and support its policies—including important elements of industry, labour, Congress, and even the academic community—are literally at bay in American society as the 1970s commence.

The Mood of America

Amid the prolifery of diverse, intense forces being brought to bear today against American military policies and stance, and on the responsible people behind them, three overriding pressures predominate.

(1) First and foremost is the tidal wave of generalized anti-military hostility which is sweeping the United States. This anti-militarism is far more profound (and, like the times, infinitely more violent) than the anti-war revulsion of the 1920s and 1930s. It draws strength and bitterness from deep domestic divisions of the country as a whole—

from inconclusive prolongation of the Vietnam war, from student and black militants in revolt, from continuing clashes over national priorities, and from the exploitation of these divisions by an apocalyptic, anarchic, radical-fascist New Left.

But there are other, more specific causes that have produced America's mood of anti-militarism. This mood results (it cannot be repeated too often) from having to fight a distant, unpopular and seemingly endless war; from using conscripts drafted by an unfair, perverse and maladministered system presided over until recently by a crusty septuagenarian general; and from the almost intolerable sense of national fatigue arising over the fourth major war America has had to fight in 53 years.

This generalized anti-militarism makes itself plain in widespread vituperation of "the military-industrial complex" (which in better times we spoke of as "the arsenal of democracy"); in the academic revolt against defence research and against college officer-training programmes—in fact, the brains of the country ostentatiously turning their backs on the common defence; and in the evangelical anti-military "New Left" faction in the Senate, which seems bent on dismantling the defence establishment of the United States.

(2) Closely related to, and to an extent productive of America's anti-military disenchantment (though intellectually more rational and more honest), is a general demand throughout the country for a lower American silhouette around the world—a profound distaste for the Kennedy–Johnson interventionist role of the world policeman, coupled with a closely related feeling that the rest of mankind ought to do more about its own problems and take on considerably more responsibility for its own defence and security. Some applaud, and some decry, this mood as one of US neo-isolationism.

(3) Finally, the dawn of the 1970s sees rising—and for the most part, wholly legitimate and reasonable—demands for more economy in national defence, for heightened efficiency, for improved defence procurement and for more effective organisation for national security.

These great pressures—anti-militarism, neo-isolationism and a no-nonsense attitude toward the Pentagon and its men—are no accident.

* * * * *

Toward an Oceanic Strategy

The result of diminished public confidence in the defence establishment and of hostile pressures exerted against U.S. defence policies and their underlying assumptions is a wide-spread demand for fundamental re-examination and fundamental modifications, both in national strategy and, to a lesser degree, in the defence establishment itself. These demands (which politicians in and out of Congress have been quick to take up) are already being projected in proposed new directions and developments in post-Vietnam deployment and strategy.

Both the military usefulness and the political appropriateness of large American garrisons in Europe, Japan, Thailand, Taiwan, the Philippines, Okinawa—and of course Korea and Vietnam—are coming into serious question. The once worldwide network of American land air bases is shrunken and impotent. The loss of Wheelus Field in Libya—at the height of the 1969 Lebanese crisis—almost the same day as Turkey announced that the U.S. Air Force base at Adana could not be used in connection with that crisis—is only the latest example.

MIGs are flying daily from the $1·4 billion US-built air base complex in Morocco. Dhahran on the Persian Gulf has been retroceded to Saudi Arabia. General Franco has declared the three American air bases in Spain off-limits to nuclear weapons—the original and only purpose for their construction. In France, Charles de Gaulle deprived the United States (and NATO as well) of nine major air bases when he took over the US-constructed Evreux complex.

It has not escaped the American public that the U.S. had over 900 air force bases overseas in 1945, 120 in 1954, and only 30 in 1970. Even so, the United States still maintains over 400 major and nearly 3,000 minor overseas installations in 30 countries throughout the world. Many of these are obsolete or being used for strategically obsolete purposes.

Leaving out South-east Asia, as well as American overseas dependencies, the United States armed forces still have 653,000 officers and men serving in shore bases and garrisons abroad.

Proceeding from the foregoing, three future developments seem logical if not desirable to most Americans.

Despite the Nixon administration's so far contrary position, pressure is strongly mounting for a sharp reduction in American troop deployments to NATO. If Vietnamisation is logical in South-east Asia, many say that Europeanisation is equally logical—at least to the point at which European nations are spending (as the United States spends) 8 per cent of gross national product (GNP) on defence. There is a widespread feeling that Western Europe in general (and France most particularly) is getting a free ride in the U.S. defence budget. Those who hold this opinion (including 51 out of a hundred U.S. Senators) feel that two U.S. divisions in West Germany—and many thousand fewer

American dependents—can provide as much of a trip-wire against the Russians as the present six divisions.

One way of stating this feeling is the often-heard remark that if the United States is providing the nuclear shield for NATO, then it seems unreasonable for Europe to ask that the U.S. provide the major manpower shield, the cannon fodder, as well. Putting this point, former Secretary of State Dean Rusk used to advert privately to the inequity of having to conscript Kansas farmers to serve 5,000 miles from home in NATO as long as some NATO members, a lot nearer the front lines, did not feel it necessary to conscript their own farmers.

Or, to quote *The Times* (of London), at present the 250 million people of Europe rely on 200 million Americans to defend them from 200 million Russians, although the Russians have 700 million angry Chinamen at their backs.

To do him credit, Mr. Denis Healey, at least on the NATO issue (if not on aircraft carriers and some of his other hard decisions), has earned American applause for what many consider timely and realistic warnings that Europe had best be thinking toward the day when the U.S.—possibly taking a leaf from Canada—firmly proposes Europeanising the conventional forces of NATO.

The resounding failure of European allies to help (or even sympathise very much with) the United States in its Vietnam agony—indeed, the degree of vocal anti-Americanism emanating over Vietnam from Western European peoples and politicians—has unquestionably reinforced American doubts if not disillusion as to NATO.

Looking to the other side of the world. Americans widely feel, the time is approaching (if not now is)—short of joining the nuclear club— for Japan to undertake to defend herself. Increasing numbers of people (including if we are to rely on campaign pronouncements, President Nixon himself) consider it inappropriate that the nation with the world's third largest GNP today spends about 1 per cent of the GNP on security while American taxpayers do the rest.

In Vietnam, especially since the Cambodian incursions, we are facing intense public pressure to limit operations, withdraw forces and accelerate disengagement. The situation seems almost analogous, as this is written, to the internal pressures against Mendès-France to withdraw in his day, from Indochina, regardless of cost.

Similar, though so far not nearly as pressing, demands are now being made in some political circles for a major troop cutback, if not outright withdrawal, in our two-division force in Korea, which is still there 17 years after the armistice.

On any reasonable basis, it would seem essential that the U.S. maintain substantial military presence both in Korea and South-east Asia for many years to come. But these are not reasonable times, and

American public opinion at the moment is in no very reasonable frame of mind.

The corollary, perhaps one should say the resultant of all these factors is emerging as a new American defence strategy. This strategy might be put into an equation.

General overseas roll-backs + sharply reduced foreign commitments + declining usefulness of overseas bases + "Nixon (or Guam) Doctrine" (Asians to fight their own conventional wars) = return to an oceanic strategy.

For the first time since the 1930s, it now seems increasingly clear, American defence is again going to be based on an oceanic strategy.

* * * * *

What an American oceanic strategy for the 1970s implies can therefore be simply stated.

Such a strategy means that the U.S. defends itself by maintaining absolute control over the oceanic and aerial approaches to the Western Hemisphere; that the U.S. defends vital interests beyond the hemisphere (as in the Mediterranean or Asia) by fleets rather than garrisons, by aircraft carriers more than by land bases in other people's countries, by floating expeditionary forces of Marines rather than by airborne brigades tied to oversea airfields.

In terms of strategic (ie. nuclear) warfare, an oceanic strategy implies serious re-examination, now actually in progress, as to the practicability of putting the entire nuclear-warfare forces of the U.S. at sea. Such a strategy would ultimately entail abandonment of our fixed, vulnerable and not wholly reliable Minuteman ICBM system in favour of Polaris and Poseidon submarines (and their ultimate successor, the ultra-long-range missile submarines (ULMS) to be expected in the mid-1970s.

* * * * *

For the American defence establishment, the prospect of the 1970s ranges from obscure to bleak. Challenges, retrenchment and changes, mostly unsought and many unwelcome, lie ahead.

One element that cannot change, however, is geography. And geography has ringed the United States with oceans.

The Western Hemisphere, North America, primarily, is the last great island. Today, North America stands in much the same relationship to the Eurasian land mass as the British Isles stood to Europe in Napoleon's time. Then, 22 miles of Channel separated Britain from

Europe. Today, 22 minutes' flight by ICBM separate North America from Russia.

But the moats are still salt-water and the walls if no longer wooden, still float.

Not to retreat into neo-isolation, but to look, again to the oceans and the skies over them, for national security, and to do so without scuttling truly essential commitments and interests beyond the seas—this is the central challenge for American national security policy in the decade ahead.

59. 1971

NATO from a SACLANT Viewpoint
Admiral E. P. Holmes, USN (ret)

NATO Today

The headquarters of the Atlantic Command is located on the east coast of the United States at Norfolk, Virginia. The view of NATO from there is different from that at other locations but it is an excellent view nevertheless. It is wide-angled and has depth. It gives the viewer a perspective across the whole area of the Alliance from the azimuth of the Channel ports and Central Europe, sweeping on around to the Gibraltar approaches and terminating, now questionably, along the Tropic of Cancer. Within these limits a vast area is encompassed having features of special strategic importance to all who would understand NATO thoroughly.

Viewing this area from Norfolk is quite like watching a sector scan on a radar scope. It is never twice the same. There are areas in it of high luminosity such as those of special strategic importance, e.g. the various islands, maritime "choke points," etc. The principal axis of the sector scan, of course, runs in the direction of Central Europe through the area of responsibility of the Commander-in-Chief of the Channel Command (CINCHAN) on to the front maintained by the Supreme Allied Commander of the European Command (SACEUR). There is a general luminescence, however, over the whole area. This is representative of the concern that exists in varying degrees for what transpires in the whole area, realising that areas of heightened interest may glow with increasing intensity from time to time.

* * * * *

Author's note: Recent completion of more than three years' service as Supreme Allied Commander, Atlantic (SACLANT), has left the author with certain reflections regarding NATO in general and the role of the Atlantic Command (ACLANT) therein in particular. Foremost is a conviction of the transcendent importance of the North Atlantic Alliance to the security of its member nations. The Alliance has been a great success for more than twenty years. A periodic re-examination and re-evaluation of it is justified wherein an appraisal of the role and functions of ACLANT within the Alliance is pertinent. The latter is the purpose of this article wherein the thoughts expressed are solely those of the author.—E.P.H., 16th March 1971.

A stabilized situation being the case in Central Europe the Russians in the mid-sixties shifted or enlarged their strategy so as to continue to press outward by going around the centre. They have "gone to sea" and in a metaphorical sense are flowing around the midstream rock of Central Europe with its military, economic, and political strength. That they are going around end is an assertion, not a supposition, because the evidence supports it; numerous statements of authorities in Moscow verify it. It is but logical, moreover, from the standpoint of events in the early sixties (the forced turn-back from Cuba and some failures elsewhere, for example) which demonstrated forcefully to them that ability to operate effectively in the maritime realm is essential to attainment of their ultimate objective. More recent events in the Eastern Mediterranean and again in Cuba serve to reinforce the conviction that a maritime strategy is a very large part of their total strategy. As has been said, the evidence is clear and there for all to see.

The situation which has evolved comes down, then, to:

(a) Security of NATO in Europe is being maintained by forces in Europe. It is as essential as ever that this role is continued. This ingredient alone, however, is not enough to assure that the full purpose of the Alliance is being achieved.

(b) Security of NATO outside Europe has become more necessary than ever and calls for an ability to cope with Soviet adventurism and opportunism on and across the sea, or through and around the flanks of NATO. This is the other ingredient essential to meet the objectives of the Alliance—in the military sense.

The reason that the latter ingredient must now be added lies in the fact that the interests of the member countries of NATO have widened from simply being secure within their own territories to being able to protect vital functions outside their boundaries. Lifelines across the seas and the ability to compete for influence offshore must be secure as well. Nowadays strategic and economic materials must come from overseas in greater quantity than ever. Dependence on "externals" is now greater and more imperative. This is true for each nation of the Alliance if it is to remain viable in freedom.

How ACLANT Fits

Looking more closely into what this evolution means from a SACLANT viewpoint, one finds that the requirements on the Allied Command, Atlantic, have grown in parallel and are unremitting.

To take the first ingredient. Implicit in SACEUR's task of preserving the stability in Central Europe is SACLANT's ability to insure that the reinforcement necessary in case that stability is threatened, either by

Soviet aggression of limited scope or in the event of a major war, may reach Europe in timely fashion and adequate measure. The role of the Channel Command in concert with the Atlantic Command is one of high importance in this regard. A further requirement upon SACLANT and CINCHAN is to insure the inward flow of materials essential to the life and economy of the European members. What is said here is merely to restate the classical function of naval forces to protect the movement of men and materials across the seas to where they are needed—but to state it with greater emphasis.

As to the second ingredient, something more is required. In one sentence, it is an ability for NATO's naval forces successfully to engage the Russians at sea and to remove their forces from it, if hostilities should erupt.

Just that; now why.

Given that the Russians have developed and are enlarging a competent, modern navy, the options available to them to extend their influence or to pursue an aggressive course have increased substantially. They have the option in the first instance to attack and harass the shipping by which reinforcements will be carried to SACEUR and by which the life and economy of Europe will be sustained. One immediately thinks of submarines in this connection but their surface forces could have considerable effect as well. (The submarine problem will be expanded a little later.)

Another option realized from their surface and amphibious forces is to assail the flanks of NATO, either north or south. These same forces could also attempt the seizure of strategic island locations in the Atlantic as a natural extension of this option.

A third option, and one which fits the original defensive role of the Russian navy, is to attempt to drive off or at least neutralize the direct combat support which can be rendered to SACEUR by the striking forces of ACLANT. It is a good assumption that the heavy Russian naval emphasis on surface-to-surface missilery is designed toward this end, just as are their frequent demonstrations of air activity deep into the middle Atlantic.

A last option is to exert influence by basing naval units in or near to neutral or otherwise uncommitted areas if for no other reason than to inhibit these same areas from Allied use.

It is to be noted that each of these options tend to circumscribe the erstwhile freedom of action that the Western navies have characteristically enjoyed at sea *vis-à-vis* the Russians, and more account must be taken of this fact rather than being inclined to assure that somehow or other it will take care of itself.

To nullify these options; to preserve the security of the Alliance and all of its multitudinous off-shore interests; to retain for the Alliance the ability to blend its military endeavours together and support the area where an ultimate result might be decided, i.e. in Europe, is why an ability to thwart the Russians at sea is a *sine qua non* if NATO is to maintain a relevant posture and, if need be, to prevail in any conflict.

That these operations may seem to pertain directly to an environment in which hostilities are in progress does not diminish the validity of the requirement to maintain forces in time of peace to be successful in war. Naval power is not built overnight nor may we expect by some good turn of fortune to have the time to build it after the need has arisen. It is needed in fact day-by-day because of the pertinence which forces in being provide to the authorities of the Alliance in their dealings in the international political field, particularly when other parties are dealing from a position of strength. As has been said many times, only if our arms are strong enough may we be sure that they may not have to be used. This truism loses none of its logic when applied to naval arms.

The Russian Submarines

Now to return to the submarine problem, one shared of course by CINCSOUTH in the Mediterranean and by CINCHAN (who also, it should be remembered, is in the Atlantic Command as CINCEASTLANT under his other NATO hat) but nevertheless of fullest and greatest moment to SACLANT. Here is where a SACLANT view is unique, because the problem predominates in the Atlantic.

Russian submarines represent the outstanding challenge to NATO at sea. Whether it be in opposition to the shipping sailed for Europe and SACEUR, or in opposition to the striking forces bolstering the flanks, or in the strategic warfare role whereby the industries and populations of Western Europe and North America can be held under forfeit, these units are the ones which most explicitly run around the ends in Europe and pervade the ocean domain linking the major segments of NATO. They make up that element of the Soviet navy which will be the toughest to overcome, and the prospects of NATO's prevailing in a contest will become increasingly bleak as time passes and such suppression has not been accomplished or substantially progressed.

Even if the situation has not come to one of actual hostilities, the presence of these submarines, be they in the Atlantic, in the Mediterranean or even in the Indian Ocean, greatly influences the thinking of NATO, as perforce it must. Of particular interest to SACLANT is the fact that some 40-odd per cent of *all* Russian submarines are based in the Northern Fleet areas and hence have open access to the Atlantic around North Cape. They do not have to sortie through the narrow, controllable waters of the Danish Straits or the Straits of Gibraltar to get there. From their northern sortie points, moreover, they have relatively shorter distances to traverse to reach one of the more critical areas where the trans-Atlantic shipping lanes converge to and from the Channel ports of the British Isles, Belgium, and the Netherlands where much of the support for SACEUR must be expected to be off-loaded.

It is principally from the northern Russian ports, too, that we expect the long-range, nuclear-powered boats to emerge to reach into the South Atlantic. The situation in the South Atlantic is new since 1967 and needs to be widely understood. With the Suez Canal closed, a large part of the oil for Western Europe coming from the Middle East oil fields now passes around the Cape of Good Hope and through the South Atlantic. If the time comes when the Suez is again open to traffic, the situation will not change greatly because the oil shippers have shifted to the giant tankers which would not be able to use the Suez Canal in any case and have proven to be more economical than if sailed by the shorter route with the Egyptian tolls added. Routes through the South Atlantic will continue to be heavily used whether the Suez Canal is reopened or not. We may be assured that this point is not lost on the Russians.

The Tropic of Cancer

This circumstance brings back the matter of the Tropic of Cancer and brings it into clearer focus. This Tropic was established as the line at sea marking the southern boundary of NATO's cognisance. The reason why this arbitrary limit was set in the first place is not clear but such a limit no longer makes sense. It would in fact make much better sense to remove it entirely, for three reasons:

(a) Lines on the open ocean are meaningless from the standpoint of naval operations because there is no practical distinction—sovereignty, geography, manoeuvring space, or whatever—between what is on one side of a line and that which is on the other;

(b) NATO interests patently include the protection of shipping at sea no matter where it may be; and

(c) the existence of such a boundary tends to limit an awareness of the important sea areas south of the Tropic and thus inhibit the planning which should be in hand because NATO maritime operations are required in the South Atlantic willy-nilly.

It is too much to expect in event of a war that Russian submarines, or surface forces for that matter, will wait until NATO ships have passed to the north of the Tropic of Cancer before attacking them. This boundary presently tends to impose a limit on NATO but none on the Russians and so the area to its south could constitute a sanctuary which, like most sanctuaries, would be militarily intolerable. NATO should, and hopefully will, reappraise this boundary and do away with the Tropic of Cancer as a southern limit at sea.

* * * * *

60. 1972

The Transformation of Strategy
Michael Howard

For the purposes of this article I take "strategy" in its original sense, the activity of the *strategos*, the general, as it has been extended in nineteenth- and twentieth-century usage; the employment of force— functional and purposive violence—to achieve the objectives of policy. In particular I shall apply it to the employment of force by states in their dealing with one another to achieve the objectives of national policy; the attainment or retention of national independence or of national unity, the deterrence of other states from or their compulsion to certain courses of action.

The object of strategy, as Schelling and others have reminded us, is to make one's adversary conform to one's will by reducing the other options available to him to the single alternative: subjection to an intolerable level of destruction to his property, and of deprivation, pain, mutilation and death to his population. In the socially inchoate and technically limited societies of which mankind was largely composed before the seventeenth century, the capacity to prevent the infliction of such damage was restricted. Little could be done to prevent raids, the burning of crops, the sacking of cities, the seizing of hostages. Such destruction might be, within its limits, total. Carthage was not the only society to disappear as the result of military defeat. The only real defence against such raids was retaliation. But it became increasingly difficult, as communications, fortifications and weapons improved, to inflict such damage without first defeating the organised armed forces of the enemy. And once his armed forces were defeated, his surrender usually made further destruction politically unneccessary.

During the eighteenth and nineteenth centuries, warfare thus came to consist almost entirely in the conflict of armed forces, and the object of strategy became confined essentially to the defeat of the opposing army. Thus crystallized, war could take quite restrained and even elegant forms; though the problem might arise, as it did for the French in 1870, that the defeat of its highly specialised army was not accepted as decision by the society on whose behalf it fought.

This period, however, was brief. By the end of the nineteenth century, social organisation and communications had destroyed all real distinction between the armed forces and the society they defended, and the former could not be defeated until the composite will-power of the latter had been exhausted. The two World Wars saw the full development of the situation, which had first become apparent in Grant's campaigns in the American Civil War: the object of strategy had become not so much to defeat the opposing army in battle as to impose on it a rate of attrition which would sooner or later exhaust the will of the society supporting it to continue the war. This attrition was increased where feasible by naval blockade and, eventually, by air attack as well. But command of the sea, which made blockade possible, and command of the air, which made effective bombing possible, had both to be fought for. The enemy's armed forces might be the last active ingredient in an exhausted and starving society, but it was only when those forces were defeated that one had achieved the object of one's strategy; to be able credibly to threaten annihilation, to have one's adversary *at one's mercy*, to kill or to spare.

The Threat of Annihilation Reappears

The second half of the twentieth century introduced a new era. Thermonuclear weapons made it possible credibly to make such threats without having to *fight* at all, and for both adversaries to make them simultaneously. The whole process of war could now be, as Clausewitz had believed it could never be, concentrated into a single act. Yet this technological development did not simplify war. The total annihilation of another society, however hostile and dangerous it might appear, has seldom been a state-objective in international politics. Such annihilation has, indeed, always been possible after military victory, and in antiquity, as we have seen, it was not unusual; but self-interest as much as common humanity has prevented more recent conquerors from carrying it out. They might eliminate the intelligentsia, the ruling class, and any inconvenient racial minorities in the conquered country, but normally they have found the resources and the manpower of their defeated adversaries too useful to be destroyed. The process of achieving this position of total dominance, in fact, was politically more important than dominance itself; for this, once achieved, might be anticlimactic, embarrassing and often surprisingly expensive. For that process, the fighting required to attain this position was in itself a form of bargaining. It was possible at any moment during its course for one party or the other to call a halt when victory became too expensive to buy, or when it seemed wise to surrender while one still had something to bargain with—even if it was only the capacity to maintain order within the defeated state and to pay reparations to the victor. The longer the process of fighting, the greater were the opportunities for such negotiations. The shorter the process, the less need there was for them at all. Under exceptionally favourable circumstances it might be possible, as it was for the Germans in 1940, to capture the resources of the adversary and to destroy almost overnight both the capacity and the inclination of the opposing government to resist one's demands.

Thus thermonuclear weapons, unless they were supplemented by conventional weapons, did not necessarily increase the potential power of the state. They were like gold bars, beyond the reach of most of the actors in the international system, and not negotiable currency for the powers which did possess them. So for most of mankind since 1945 force of a traditional kind has, unfortunately, continued to be a perfectly viable instrument of state policy, and traditional strategic concepts have provided reliable guidance to military action between armed forces. A military textbook based on campaigns from the Korean War in 1950 to the Indo-Pakistan conflict in 1971 would contain little to surprise a survivor from the Second World War. All the lessons learned during that and earlier conflicts—the need to maintain command of the air over the battlefield, the value of speed, surprise and deception, the desirability of severing the enemy's communications while maintaining one's own intact, the effectiveness of superior strength at the vital point—all have been borne out. In all these conflicts force was being used for specific and limited political ends by professional armed forces operating primarily against one another. The principal difference from the Second World War is that in many of these cases—and the Korean and Vietnamese Wars are significant exceptions—both belligerents were comparatively poor countries with few military resources other than those committed to the campaign; and those, anyhow, were usually provided from abroad. Their strategic objective was not to impose an attrition which would bear just as hardly on themselves as on their victims, but to secure a complete military victory; or failing that, to be left at the end of the fighting in possession of territory which they could either retain or use as a bargaining counter in peace negotiations. To that extent they have resembled the campaigns of Frederick the Great and Napoleon more closely than those of Grant or of Eisenhower.

Limitations of Super Power

The conflict between the United States and North Vietnam was certainly different. The American aim was simple; to compel the North

Vietnamese to desist from helping the Viet Cong in South Vietnam. American technical superiority enabled the United States, without defeating the North Vietnamese forces in battle, to impose on the enemy society a degree of destruction limited only by considerations of domestic and international politics. That degree proved insufficient to break the resistance of the North Vietnamese government and people, whose aid to the Viet Cong continued uninterrupted. It lay well within the capacity of the United States Air Force to impose the Carthaginian solution advocated by, for example, General Curtis Le May, but the military benefits of doing so were judged incommensurate with the political costs. The military power of the United States had therefore to be deliberately geared down to a level of conflict at which it might be of value in political negotiation. Air fleets which might have flattened entire cities confined their attention to remote villages and jungle tracks. The United States Armed Forces congratulated themselves on their restraint. Others were horrified at the collateral damage caused by these giant bulldozers attempting to adapt themselves to work which called rather for a dentist's drill. And the North Vietnamese, by sheer endurance, won a classic defensive battle; that is, they set on victory a price which, in the last resort, the United States was not prepared to pay. It was not just that the gold bars in America's nuclear Fort Knox were not convertible into political currency. Vietnam could have been bombed into the Stone Age with conventional weapons. But to do so would have involved the United States in dangers of conflict with the Soviet Union, of internal disruption, and of moral isolation in the world as a whole, which no American government was prepared to incur.

It is, however, far too soon to say that the United States have been defeated in Vietnam. The North Vietnamese have not been compelled to desist from helping the Viet Cong, but the power of the Viet Cong itself has clearly been significantly reduced as a result of the operations of the United States Armed Forces in South Vietnam. Those operations may have been wasteful and brutal, but they appear to have been effective both in wearing down the strength of the Viet Cong and in creating a situation in which it is difficult for the remainder to operate successfully. Whether the society which will emerge as the result of United States intervention in Vietnam will be a happy and prosperous one, and whether the United States will be the more secure because of its existence, are problems which lie beyond the scope of this article to discuss.

Conventional Forces the Fuse

Events in Vietnam do not therefore invalidate the continuing value of conventional weapons in the nuclear age; whether in conflicts between armed forces in which a decision by arms is accepted as being also a political decision, or as instruments for bringing pressure on a hostile society in those cases where states have large enough armouries to engage in such wars of attrition. At very low levels of conflict, such as that on the southern frontiers of Sudan, the demands of these latter on the resources of undeveloped states need not necessarily be very high. But in major international conflict between sophisticated states, prolonged conventional hostilities are improbable, if only because states capable of sustaining them will be capable of producing nuclear weapons. The British and French examples indicate that the attainment of nuclear capacity by any except super-powers will be used as an argument for the drastic reduction of conventional forces. As yet only the Soviet Union and the United States (and perhaps the Republic of China, though it is a little early to say) have shown themselves willing to sustain the burden of substantial conventional forces and of nuclear forces as well. It could be argued that the real danger of proliferation lies not so much in the possiblity of nuclear weapons getting into the hands of irresponsible powers as in the probability that powers will acquire them at the expense of those conventional weapons which do continue to provide at least an option for the limited and purposive use of force.

Yet the value of such conventional forces is not easy to gauge. Arguments against them were voiced on both sides of the Atlantic during the great debate about flexible response or graduated deterrence which raged during the late fifties and early sixties. The existence of such weapons, it was then suggested, would be a standing invitation to use them, if only to get value for the money spent. A conventional capability in Europe, it was widely believed in Germany, would reduce the effectiveness of nuclear deterrence. If Mr. MacNamara had not built up the conventional forces of the United States so successfully, there might have been no involvement in Vietnam. I myself cannot believe that it can ever be sound statesmanship deliberately to deprive oneself of a possible option for fear that it might one day be used unwisely. The real question is, what and how many options can one afford to buy? And once one has these conventional forces, what in fact can be done with them?

There can be little doubt that such forces have, and are likely to go on having, considerable *political* value. The deployment of conventional forces beyond one's own frontiers in peacetime, whether on the high seas or on the territory of allies, is still the clearest, the most convincing demonstration of state interest in that area; of determination that no fundamental change in its status should take place without one's approval, on pain of involvement in an open-ended conflict whose level of intensity cannot be foreseen. Their presence shows a readiness to engage in such conflict; and most people would argue that the larger the presence, the more effective it is likely to be—whether in deterring an adversary from involvement or, if involvement does occur, of holding the conflict at a level below the nuclear threshold. The function of conventional forces in such a situation—the function, for example, of NATO conventional forces in Western Europe—is to act as a kind of fuse. The greater the length of the fuse, the longer the duration of the conventional conflict, the longer will be the time available for political intercourse, for crisis management. The chances of such intercourse continuing beyond the level of the first commitment of nuclear forces are almost impossible to assess.

But what scale of conventional forces is desirable? The arguments are not all on the side of the long fuse. More time available to take the decision to use nuclear weapons might well make it less likely that the decision would be taken at all. It would provide more opportunity for pressures to the contrary to build up, for doubts to grow, for the native hue of resolution to be sicklied o'er with the pale cast of thought. The best strategy for a hostile Soviet Union invading Western Europe might indeed be to do so with a comparatively small number of conventional forces combined with a profession of its limited aims. The longer NATO conventional forces were able to continue to fight, the more reluctant their governments might be to resort to nuclear weapons at all, and the more ready, ultimately, to acquiesce in a Soviet victory at the conventional level without invoking the unimaginable destruction of a nuclear deterrent which had now failed to deter. The shorter the fuse, on the other hand, the less time for second thoughts. A belligerent with a low level of conventional and a high level of nuclear power might command the terrifying credibility of the hijacker who threatens to blow up the plane and himself with it. His very irrationality makes it prudent to believe him.

Yet the credibility of such a threat will depend on the circumstances in which it is made. In defence of one's own territory it might be highly credible; defending that of an ally, somewhat less so; protecting distant interests, very unconvincing indeed. The less vital, or the less apparently vital, the interest, the greater will be the value of conventional forces in protecting it; and the degree of importance is still, rightly or wrongly, considered to be related to geographical distance. The United States was able to make very much more credible threats of nuclear war over Cuba than could the Soviet Union. Neither could do so very convincingly over Vietnam, a peripheral interest for both of them. But if the Soviet Union had been in a position to support her declared interest in Cuba or in Vietnam by the commitment of conventional forces in the area, there can be little doubt that the United States would have acted a great deal more warily—as warily as she is acting in the Middle East today. And her wariness would not be the result of any doubts of the issue if US forces were to engage Soviet forces in a local conflict. It would be the result of wondering what would happen after that.

Crisis Management Replaces Strategy

For however large their size and however remote the objective over which they may come into conflict, the conventional forces of nuclear powers cannot fight one another as if the nuclear stockpiles did not exist. The mere existence of nuclear weapons adds a new and inescapable dimension to the operations of the conventional forces of the states which possess them, and the strategist has to take it continually into account. Because there is no certainty that a power whose conventional forces are in danger of defeat will not introduce nuclear weapons at some stage in the conflict, the defeat of those forces can no longer be the unequivocal object of strategy unless one is assured of an absolute dominance of the nuclear level—a dominance so absolute, indeed, as to guarantee one against virtually *any* level of nuclear punishment. And because the involvement of, and with, conventional forces involves lighting a fuse which it might then be impossible to extinguish, such units even at the very lowest levels need to treat one another with the very greatest caution.

For nuclear powers in their dealings with one another one is tempted to say with Mr. Robert MacNamara: "There is no strategy any more— only crisis management." The armed forces of such powers no longer engage one another in order to secure a position of dominance from which to threaten the annihilation of the enemy. Thanks to their nuclear weapons-systems, they are in that position already. The function of conventional forces now is to indicate not, as in former conflicts, the degree of *damage* states are prepared to sustain in order to gain or defend their political objective, but the degree of *risk* which they are prepared to run. In fulfilling this function, the size of the armed forces above a certain level is not necessarily of major importance. The number of naval units which the Russians, for example, can deploy in distant waters will certainly increase their political influence in those areas and impose caution on their adversaries; but in dealing with the United States it is not clear that greater numbers will bring greater strategic advantages. The more heavily a Soviet fleet engages an American fleet, the greater risk it will run of compelling the United States to consider nuclear escalation. Thus small naval forces showing the flag of a substantial nuclear power in troubled areas are likely to be as effective as large ones. They are the end of a fuse which even an adversary deploying an overwhelming local superiority of strength would be reluctant to light in a hurry.

Violence without Rules

However valid traditional strategic concepts may remain in conflicts between, or against, non-nuclear powers, one thus must be careful in applying them to the operations of nuclear powers in conflict with one another. For them the lessons of three hundred years of strategic thought—years during which war became crystallized as a distinct social activity with its own rules and objectives—are now of doubtful relevance. In one sense it might almost be said that nuclear weapons really have abolished "war" in the sense that our fathers understood it; that is, as a condition distinct from "peace." Even if the likelihood of nuclear holocaust is remote, the total result is not pure gain. The institution of war restricted violence within certain limits of time, space and, often intensity. The legitimization of certain activities in wartime made them clearly illegitimate in time of peace. But today we live in an age when it is possible, and widely considered legitimate, to commit in "peacetime" acts of violence ranging from the intimidatory wounding of individuals for political purposes to the dropping on a small Asian country of a tonnage of bombs exceeding that delivered on Germany in the Second World War. In abandoning a system under which violence was legitimized under clearly defined circumstances in the hands of clearly delineated groups of men operating under close supervision, we have accepted one in which it is endemic; a return to an earlier and grimmer period in the history of mankind, in which violence, if not actual, was imminent at virtually every level of social intercourse. In such an environment, strategy appears an obsolete art—almost, indeed, a humane one.

61. 1972

The Army in Northern Ireland
Michael Banks

In last year's Brassey's *Hugh Hanning gave an historical and political background to the troubles in Ulster and dealt with some of the problems confronting the Army. Michael Banks, who recently accompanied infantry patrols on the border and in Armagh, now assesses military operations during the period leading up to the imposition of direct rule in March 1972.*

There have been swift and radical changes, usually for the worse, in the situation in Northern Ireland during the last year and a half. The gunman and the bomber have emerged, automatic fire has been directed at the troops, the Catholic community has, most regrettably, polarised its attitude towards the British Army which it now identifies as the traditional oppressor and enemy. The Army itself was no longer required to act as a buffer between the Catholic and Protestant communities, or even as a para-military anti-riot police force (an unsuitable role it was forced into). Instead it has taken a number of initiatives and has found itself fighting an urban guerilla warfare campaign reminiscent of its colonial days.

In this article I shall attempt to assess the threat, posed at the moment almost entirely by the IRA, and then describe the way the Army is meeting it. I shall also take a look at some of the new equipment which has been specially developed to meet the specific needs of the conflict in Northern Ireland.

The IRA

The original IRA, whose memory is much revered by the Republic of Ireland's legitimate Army, were patriots who today would be regarded as freedom fighters. Their aim, an independent Ireland, was nationalistic rather than political. The memory and tradition of these men are now safely enshrined in the pantheon of Irish history. The old IRA bears little relationship to the Official IRA who inherited their name.

The Official IRA is a Marxist-orientated organisation whose proclaimed aim is the creation in all Ireland of a united socialist workers' republic. This political aim, obviously, poses almost as much a threat to the Republic of Mr. Lynch as it does to Northern Ireland. For this reason the Official IRA is proscribed in the Republic, although attempts to eradicate it have been less than half-hearted, principally on account of the large measure of public sympathy its aims, but not its methods, in Northern Ireland attract.

Although adversity makes for strange bedfellows, it was asking too much for the Catholic/nationalist extremists of the North to make common cause with the Marxists of the Official IRA based in the Republic. Inevitably perhaps a splinter group appeared—the Provisional IRA—which is an organisation peculiar to Northern Ireland. Whereas the Officials generally favoured political struggle for the attainment of their ends, the Provisionals chose for one of their slogans "Liberty grows from the barrel of a gun." They opted for outright violence as the means of unifying Ireland, a patently hollow philosophy and one which has left a trail of blood, misery and bitterness behind it.

When the rift between these two factions of the IRA became irreparable, internecine warfare broke out between them in which an unspecified number of men were killed. An uneasy truce now exists, with the Provisionals very much in the ascendency except in a few small pockets in Belfast.

I want to keep as far away from politics as I can (which is fiendishly difficult when writing about Ireland), but it is worth explaining that the IRA saw very clearly that if the Catholic Civil Rights movement attained its aims of social justice within the Province, which under pressure from London it could well have done, then the IRA would have had the steam taken out of its campaign for the unification of Ireland.

It must therefore be conceded that, using military methods, the IRA has not only taken over the leadership of the Catholics from the Civil Rights leaders but it has also achieved two of its main aims; keeping the question of the unification of Ireland very much a live issue; and the abolition of the Government of Northern Ireland at Stormont.

The Nature of the Threat

Until the middle of 1970, the main threat to the troops, who were perforce spread fairly thinly on the ground, was the ability of the Catholics to drum up an enormous crowd at extremely short notice which would then threaten to overwhelm a small or isolated detachment. In mid-1970 the gunman made his appearance and the IRA began to manipulate the crowds which would contain a good proportion of out-and-out troublemakers. The troops were pelted with stones intermingled with petrol or nail bombs, sniped at from behind the crowd, and occasionally subjected to automatic fire.

The Army kept its head and did not over-react against the crowd itself, no doubt to the disappointment of the IRA. Instead the Army took on the IRA in a struggle it had been trained for—a firefight. Although violence was escalated during this period. Army morale was raised. The soldiers at last found themselves doing a military job, fighting the emergent urban guerillas, and they found it a welcome change from standing in stolid rows being stoned and taunted by hooligans. It was also a time of increasing danger on the streets. One sniper in the Ardoyne killed five men of the Green Howards in skilfully set up ambushes. Usually he would arrange for a disturbance to be staged. On the arrival of an Army patrol he would fire one or two shots from a well-concealed position and then quickly disappear. His "trade mark" became recognised but he has apparently disappeared from the scene, and it is possible that he himself eventually became a casualty or, perhaps, a detainee.

Naturally the big crowds had little enthusiasm for being caught in the crossfire in this confrontation and disappeared from the streets. It is very probable that the Army came off best in this firefight, as one would expect them to, because the IRA did not choose to sustain the trial of strength.

The IRA then switched to its current campaign of selective terrorism, including the cowardly and cold-blooded abduction and murder of three unarmed young Scottish soldiers who were off-duty and having a drink in a hotel. Public figures and members of the locally enlisted Ulster Defence Regiment became priority IRA targets.

IRA bombing activities also increased in this period, probably because it was found to be a less hazardous pastime than shooting it out with the Army. We have all become familiar with press reports of armed terrorists planting bombs, often in crowded shopping areas, and giving bystanders a few minutes warning to move away. The ferocity of the bombing campaign should not be underestimated. The total casualties up to 10th March 1972 were 63 dead and the appalling figure of 1,264 injured. Selective targets, such as the houses of prominent Unionists, have also been either bombed or burnt. Perhaps the most dangerous form of bombing is the Claymore mine, dug into the side of a hedge or placed under a culvert in the road, which is then detonated remotely by electricity when an Army patrol is in the immediate vicinity.

In summary, the threat the Army is countering at the moment includes the hit-and-run sniper, bombs in public places which have to be rendered harmless, the concealed mine in an ambush position and the threat of abduction and assassination if any of the security forces were to put themselves in a vulnerable position. At the same time the Army must enforce the ban on marches, be they Protestant or Catholic. As events at Londonderry showed, the mixture of a large crowd and a gun-battle between the IRA and the Army can create a tragic situation.

The Army Reacts

The IRA mounted its terrorist campaign in the hope, possibly to be realised sooner than they had imagined, that such disruption would be caused that, in the short term, the U.K. government would be forced to abolish Stormont and impose direct rule. In the long term the IRA envisage public opinion in Britian becoming so antagonistic to a continuance of the bloodshed that the British government will be forced to pull out of Ulster and leave the road open to the unification of Ireland.

In consequence, the operational scene changed considerably in 1971. The Army intensified its intelligence operations in the summer, having first improved its own intelligence organisation and, particularly, the co-ordination between the Army and the Police Special Branch. Numerous raids, usually at dawn, were carried out against suspects. Valuable intelligence, often supported by collateral information, began to mount up. The controversial decision was then made to intern suspects under the Special Powers Act of the Stormont government.

During the early hours of 9th August some 300 suspects were rounded up, many of them on information obtained during the recent intelligence raids. About 70 of them were released shortly afterwards. These early arrests have been followed by many others making a total of about 700 persons who have, at some time or other, been detained.

Detention was followed by the even more controversial "interrogation in depth" of about 40 prime suspects. Modern methods of interrogation first made their appearance during the Korean war when Allied prisoners were broken down mentally or brain-washed. These methods have remained standard practice in Communist countries and it therefore became commonplace in Western armed forces to require troops, principally those most liable to capture, such as aircrew, to undergo Communist-type interrogation so that they would be mentally prepared and fortified against this modern barbarism should they ever become prisoners of war. It was therefore a profound shock to the British public to learn that these very same measures had been used in peacetime against mere suspects. Deep interrogation was stopped pending a public inquiry by Lord Parker into the methods of arrest and interrogation. The Parker Report has now been completed and very stringent regulations have very properly been applied to future methods of interrogation.

The free-for-all deep interrogation was a political blunder and a departure from the accepted standard of British ethics and justice and it is reassuring to know that it has been stopped. With internment itself, it was in contravention of the United Nations' concept of human rights.

From the purely military point of view both internment and interrogation proved most productive. As a result of interrogation, further arrests were made and it is safe to say that a considerable number of potential murderers are now behind wire. The flood of information the Army and the police extracted enabled them on many occasions to drive up to a specific address and arrest the individual they wanted. This accuracy in identifying IRA members must have sent shock waves through the entire membership. No one in the IRA could feel safe any longer. However, success was far from absolute as witness the press conference held in Belfast by Joe Cahill, leader of the Provisionals, a few days after internment had been authorised.

No one imagined that internment would put an end to the IRA, if only for the simple reason that there is an almost inexhaustible supply of youngsters who have been nourished on hatred of the British and who are always ready to offer themselves as recruits. What internment achieved was the disruption of the leadership of the IRA. Evidence for this is, for instance, to be found in the number of bombers who were killed by their own explosives immediately following the internment of known skilled bomb manufacturers. The number at the time of writing is probably at least 22 and this does not include the 15 people killed in the explosion in McGurk's bar in Belfast which was an accident. An examination of the corpses after this accident indicated that some of them were probably leaning over the bomb tinkering with it when it went off.

As a direct result of the effects of internment, the IRA has moved from attacking hard targets towards attacking soft. The armed attacks on military or police posts have fallen sharply, and IRA efforts have been concentrated on indiscriminate bombings, hit-and-run firing, the remote-controlled Claymore mine and selective assassination, including the attempts in Armagh on the life of the Stormont Home Affairs Junior Minister, Mr. Taylor. Violence spread to England in the notorious and inept bombing of the Airborne Brigade Officers' Mess at Aldershot which killed five women, a gardener and a Roman Catholic Army chaplain.

The Civil Rights movement and the much discussed but so far chimerical Protestant backlash are at the moment, threats in a lower key which the security forces must watch carefully. It is a valid viewpoint that the Catholic population would have achieved social justice more effectively through a vigorous Civil Rights movement of Gandhi-esque proportions. This would have presented Stormont with a political rather than a military problem. It is, after all, eventually in the arena of politics, albeit under military pressure, that the problem must ultimately be solved. As far as the Protestant backlash is concerned, the Army is uneasily aware of the thousands of weapons "under Protestant beds" which might be brought out if an unpalatable political solution were forced on them. It is a possibility full of menace. The Army has not forgotten that it first came under heavy fire from Protestant gunmen in the Shankill Road of Belfast in 1969. However, at the time of writing, in the weeks immediately following the imposition of direct rule, all is relatively quiet in the Protestant communities.

* * * * *

Conclusion

Nobody realises better than the Army itself that there can be no purely military solution to the communal problems of the Province. Although the eventual solution cannot be seen in military terms, it is nonetheless true that the military situation will have an important impact on the political climate. If the IRA were running rampant, civil war would be one step the nearer. If law and order, of a sort, prevails, the political options are wider and the prospects of communal peace that much the better.

The path of events in Northern Ireland can only increase the soldier's inbred distrust of the politician. After the troubles of 1969, the soldiers held the ring steadfastly and with courage. They gave the politicians that most precious of elements, time, particularly in the crucial year of 1970. The subsequent deterioration of the situation is testimony that the politicians, particularly those at Stormont, failed to appreciate the urgency of the crisis. They squandered the time the soldiers had won for them and the Army has carried the can ever since.

This is a war no one can win. The performance of the IRA has varied between the amateur and the mediocre, and they can never win against well-trained, professional troops. Neither can the Army ever win in the sense that it will be able to eliminate an urban guerilla movement broadly based on a large and sympathetic section of the community.

A fair insight into the current situation was given me by a corporal leading a patrol to whom I spoke one night in Armagh. "The kids who stoned us this afternoon came over and had a friendly chat this evening," he said. Other men told me how a number of Catholics, in private and away from the threat of intimidation, could be friendly towards the Army. In other words the situation is not without hope despite the polarised attitudes taken in public.

For nearly three years the Army has stood firm and kept its temper in the face of extreme provocation. After all, there can be few countries in the world where you can hurl rocks at an armed soldier and not get shot for your trouble. The Army has been a model of restraint for which they deserve the very highest credit.

Northern Ireland has not been an entirely negative experience for the British Army. It has had to adapt its tactics and modify its equipment to meet the sort of unrest which is becoming increasingly widespread in the world today. The glare of publicity which has been focused on the Army has contributed significantly in attracting recruits who are now enlisting in greater numbers than for many years. Finally, it has become very much a soldier's war, a fact reflected in the currently high morale. Patrols are usually of section strength which throws heavy responsibility on the corporal and the private. These men have risen splendidly to the occasion and, in the process, have gained invaluable operational experience.

* * * * *

62. 1974

The Defence of France and the Defence of Europe
Pierre Dabezies*

Professor Pierre Dabezies is Director of the Centre for Political and Defence Studies in the University of Paris.

Strategy is the daughter of politics, and politics the daughter of history. Europe today is not just a tangle of contradictions, it is a tangle of misconceptions. To attribute its current divisions to a conflict of selfish interests would be an over-simplification at odds with the facts, for Europe is also the product of the individualities of the states or nations of ten centuries or more standing that go to make it up. The individuality of the British Isles is clearly identifiable; that of France perhaps less so, for her present attitude, deprecatingly dismissed as "Gaullist," is in truth deeply rooted in history.

Lacking alike England's sea barrier and the coherence of Germany, a country that has been at once the heart and the outpost of Europe since the days of the Holy Roman Empire, France has been forced over the centuries to struggle forward inch by inch, in face of internal disruptive forces and, across her frontiers, against over-powerful neighbours seeking to subjugate her. A monarchic tradition and a revolutionary will. A "state" before it ever became a "nation," France has always suffered at the hands of its xenophiles, the *Bourgignons*, even though, from Richelieu to de Gaulle, they have always been the defenders of the very spirit of national unity and independence which has finally got the better of them.

True, times have changed; since the last war, France has become "European"; the rejection of the European Defence Community by the Fourth Republic and of the Atlantic Alliance by the Fifth were no more than digressions on this long journey.

But why does France adopt this wavering and to some extent obsolete stance? Does she hope to shut herself off for ever in splendid

*Translated by Richard Simpkin.

isolation? By no means, and all the less because she, unlike many foreigners, does not feel herself capable of dominating Europe. That is not the problem; it is rather perhaps that she sees herself as representing a tenuous synthesis of Northern Europe and the Mediterranean, vulnerable in face of a predominantly Saxon, Protestant and Nordic Europe which, in the last resort, threatens to deprive her not only of her culture and her way of life but of her soul.

Hence her repeated drawing back. European from the head rather than from the heart, and determined to preserve her identity, France considers that union must be founded on a steadily growing community of interest rather than on decisions taken under foreign pressure. Her historical tradition of independence is, she feels, too strong for her to allow herself to fuse into a group dominated by one of the superpowers; and she sees in an "Atlantic Europe" the danger of Nordic influence and civilisation becoming so preponderant as to swamp her. The very cultural, economic and strategic weight that gives her a significant place in a "free Europe" would tip the scales against her in an "American Europe." Further, history has taught her that the Slav world provides her with a counterweight and safeguard against Germanic thrustfulness. To hitch Western Europe once and for all to the American wagon would at once strengthen the blocs, crystallise Europe for ever into two halves set against one another, rule out all hope for reunification and topple the equilibrium that alone can preserve her individuality. Again, as General de Gaulle writes in the Memoirs of Hope*:

> The aim of my policy is to set up a concert of European states so that they may grow together as links of every kind between them develop. There is no harm in looking further ahead to the point at which, perhaps in face of a common threat, this evolution might culminate in a confederation.... But it is equally clear that, if the Western peoples of the Old World remain subordinate to the New, Europe will never be European and will never be able to reunite her two halves. ...

Thus, however little those who choose to read our country's fate in their own parochial crystal ball may like it, "Gaullist" policy is not the policy of an individual, a clique or a minority. Too inflexible and too crudely expressed it may have been; but after the escapist tendencies and readiness to resign of the governments of the Fourth Republic, after the vagaries of the ideologues and technocrats of the fifties, this policy is at root a return to the policy of France—a statement that can readily enough be justified. Drawn as he was towards flexibility and European *entente*, Pompidou not only maintained de Gaulle's defence policy but, along with Jobert, was forced to return to the former rigidity to prevent everything being swamped in the American floodtide. Everyone knew the French were well disposed towards the

---
*Vol. I, *The Renewal* (Plon, 1970), p. 184.

European idea; none the less in the presidential election campaign 63 per cent of them showed their approval of the existing foreign policy. In the camps of Giscard and Mitterand alike, everyone emulated "Gaullism" by affirming, without ever questioning the need for friendship and the appropriate ties with America, that any Europe other than a free one was inconceivable. "Supranationalism" was hanging fire; Atlanticists of the right such as Lecanuet kept quiet just as did the Communists on the left. "Yes to Europe, but a European Europe"—this was beyond dispute what most Frenchmen wanted.

Naturally this attitude may be called "sanctified egoism"—if it is egoistic, after so much trial and tribulation, to wish to safe-guard one's personality, and to do so when there are no signs that anyone in Europe or the world particularly wants to see it crushed. Again, the cry of nationalism may be raised—but only by those who forget that, over a decade ago, France proposed to Europe the Fouchet Plan. Who can tell what strides might have been taken towards unity if it had been accepted! Last but not least comes the accusation that France is guilty of self-contradiction and hypocrisy in claiming to want a united Europe while constantly slowing down the building of its political framework. But there remains the need to know what kind of a framework it will be, and this is the focal point at which politics and strategy coalesce.

It may well be that for years the major issue in Europe has been the dilemma between a *Europe des patries* and integration; but today we can see that it is in fact Atlanticism which lies at the root of the problem. It was for fear of Britain's championing this cause that de Gaulle set himself against her. After all Churchill had said in 1944: "You should know that, if one day we have to choose between Europe and the seas, we shall always choose the seas." Likewise the basis of the protracted United States support for European union may very possibly be that Europe's weakness gives America reason to think that she could go on holding the reins. If Europe became her competitor, the USA would soon become hostile to the very structure she now supports.

In this respect, defence has played a fundamental part, NATO had to be, and at the start American leadership was natural enough. But it has gradually become an established fact, a habit, a convenience. Paul Valéry wrote: "Europe seems intent on being run by an American commission." He did not know how right he was. The nuclear deterrent started by discouraging Europeans from building up their own defence and gradually went on to make them renounce an independent existence.

The result is a vicious circle. In the absence of a credible European defence, any move towards a political structure inevitably veers towards Atlanticism. And any proposal to set up a common defence falls foul of the objection that this is impossible without a political structure.

France opposed this trend. First she opposed the European Defence Community because, whatever its merits, it was tending in the last resort to create, under cover of a defence community, a Continental community subordinate to the USA. In the same vein, however desirable she felt NATO to be as an alliance, France took a stand against its claim to become a comprehensive authority institutionalising American control across the board.

It remains to be seen whether France in her isolation has cut her coat according to her cloth. In European terms her cultural, economic and strategic strength is neither negligible nor dominant. Can she continue indefinitely to stem the tide? While she does, her position over defence will remain set in the historical and political frame depicted above.

National Defence

The development of France's nuclear capability is not, as is all too commonly supposed, due to General de Gaulle. It was initiated in 1954 by Mendès France, and continued under the Fourth and Fifth Republics to culminate in February 1960 in the explosion at Reganne. From the start, of course, the programme had met with the General's approval, for experience had convinced him that there could be no kind of independence without a defence that was at once national and up-to-date. This conviction implied a nuclear capability, the more so at a time when divisions within NATO were casting doubts on the credibility of the American shield.

It is superfluous to remind the reader of these altercations, which have never entirely died down. Despite Washington's repeated assurances, it was inevitable that Europeans should come to expect that the USA, now vulnerable to Soviet strikes, would suddenly change her strategy and invite her Allies to build up powerful conventional forces in prospect of flexible response and limited war. It is true that, however anxious the Europeans may have been, there was nothing they could do about it. On the other hand it was natural that France, which was already beginning to fret against the constraints inherent in a defence system in foreign hands, should spell out for herself the logical conclusions. There being at that time no question of a purely European defence, the French nuclear force was born, having been conceived on one key principle—nuclear power cannot be shared out; and no country could be sure beyond all doubt that on the day an ally would risk nuclear apocalypse to defend it.

The debate was a political one; although the Herter Plan and subsequent proposals for multinational and multilateral forces attempted to confine argument to the technical level, they served only to deepen the differences within NATO. Despite the smokescreen of concessions of principle offered to the Europeans, there was never in fact any question whatever of the Americans foregoing their hegemony, even if this had been possible. In like vein with their unilateral adoption of the strategy of flexible response, it was unilaterally that, to head off opposition in Congress, they decided on 20th February 1963 to replace the submarines planned for the Multilateral Force (MLF) with surface ships. Despite the acceptance at Nassau of Britain's wish to use her Polaris force for her own ends should her supreme national interests be threatened, the British themselves were shocked at the highly restrictive interpretation placed on it. Germany alone remained firmly in favour of the MLF project, for she saw in it the possibility of gaining access to nuclear power through the back door. France could only reject the proposal; Washington, it seemed, only became even more determined to make her toe the line.*

* * * * *

*See Didier Truchet, *Le projet de force multilatérale* (PUF, Paris, 1972).

63. 1976-77

The Soviet Military Effort in the 1970s: Perspectives and Priorities

John Erickson

The author is Professor of Defence Studies at the University of Edinburgh.

* * * * *

The inevitable focus, however, must be the expansion and the diversification of Soviet military capability, what is generally labelled the Soviet military build-up. The gross figures speak for themselves.*
Over the past decade the Soviet ICBM arsenal has grown in staggering fashion from a mere 224 to some 1,600 strategic missiles whose throw-weight has increased dramatically with each new generation of weapon: submarine-launched ballistic missiles (SLBMs) once amounted to a paltry 29 and now number almost 800 together with a fleet of 58 nuclear-powered submarines whose latest SLBM, the SSN8, outranges all existing SLBMs by no less than 1,600 nautical miles: the current total of ICBM/SLBM warheads has jumped from a few hundred to over 3,400, a figure which could double or even virtually treble depending upon which options the Soviet command exercises with its MIRV programme for the latest Soviet missiles:† the Soviet anti-ballistic missile (ABM) system, with 64 launchers deployed, is undergoing a new phase of development and testing with improved radars and missiles: Soviet strategic air power can now count on its latest bomber, the Backfire "B" of which at least 50 are in service, an aircraft with the capability of attacking targets in the United States as well as covering

---

*See under Appendix A (US-Soviet Strategic Nuclear Force Levels: as of January 1976) in Robert L. Pfaltzgraff Jr. and Jacquelyn K. Davis, *SALT II: Promise or Precipice?* University of Miami, Center for Advanced International Studies, 1976, pp. 40–2.
†The figure of 8,364 MIRVs or 7,764 MIRVs would depend upon the configurations of the latest SS18s, SS19s and SS17s: while US MIRVs have yields in the kiloton range, Soviet MIRVs are in the megaton range.

the Eurasian land mass and providing the Soviet Naval Air Force with a potent anti-shipping strike weapon.

Wherever one looks, there is expansion, modernisation and improvement. The Soviet Navy has dominated the headlines, reaching out and into the world's oceans with a new class of aircraft carriers (the Kiev class, whose first 40,000-ton ship is undergoing final tests with two more building), additions to the very powerful Kara-class guided missile cruisers, completion of the large Kresta-class missile cruisers and the "handy, handsome and well-armed" Krivak-class missile armed destroyers, extensive modernisation of the smaller gun destroyers and escorts, further additions to the missile-armed fast attack craft, steady improvements in ships for afloat support and a new class of large landing ships (the Ropucha-class, recently on sea trials in the Baltic).* The Soviet submarine fleet,† already the largest in the world, is being improved with the introduction of the new Tango-class diesel-powered attack submarine to replace the tried and tested Foxtrot boats: the production rate of the nuclear-powered C class has recently been stepped up to two a month (rather than three a year) and here is a boat with the unique capability of launching anti-ship missiles while submerged, using its eight SSN7 anti-ship missiles with a range of some 30 miles. Meanwhile the Soviet Naval Air Force (*Morskaya aviatsiya*) is taking the Backfire bomber into its inventory and thus adds a very powerful anti-ship strike weapon to its overall strength of some 1,200 aircraft and about 50,000 personnel.‡

That other element which provides greater "reach"—Soviet air power—has been similarly expanded and refurbished, showing the same profile of improved capability combined with a switch from predominantly defensive commitments to offensive tasks. There is the same infusion of numerical strength, accompanied by major qualitative improvement: between 30–40 plants supply the Soviet air force with over 1,800 military aircraft each year (half of them advanced first-line aircraft, plus 700 helicopters), furnishing a park of some 4,500 tactical aircraft since 1970. The thrust is now concentrated on tactical aircraft with offensive capability as opposed to the previous over-reliance on defensive interceptors. This is not to say, however, that the Air Defence Command (*PVO Strany*) has in any way been neglected: its air defence system, constantly thickened, comprises no less than 10,000 launchers (with 12,000 surface-to-air missiles), 5,000 surveillance

---

*See Captain John E. Moore RN, *The Soviet Navy Today*, Macdonald and Jane's, 1975, under separate classes of Soviet naval combat units. The *Kiev* is also at sea.
†The Soviet SSBN/SLBM programme is discussed under strategic weapons.
‡See Norman Polmar, "Soviet Naval Aviation" in *Air Force* ("Soviet Aerospace Almanac 1976" Edn.), March 1976, pp. 69–72.

radars and 2,600 interceptor aircraft.* Soviet tactical aviation (*Frontovaya aviatsiya*) has undergone a major transformation, reducing its commitment to an air defence role—keeping NATO aircraft "off the backs of Soviet tank commanders"—in favour of interdiction and counter-air roles: its new equipment makes this all the more feasible.

* * * * *

The Soviet Ground Forces (Soviet Army) have also shared in this proliferation and intensification of effort. In the past six or seven years the Ground Forces have improved their tactical nuclear capabilities while expanding their capability to conduct conventional operations for some sustained period: the new main battle tank (the T72) has finally made its appearance, but in that lapse of time no less than five battlefield air defence systems, five artillery systems and two infantry fighting vehicles (the BMP with its triple anti-tank armament and the BMD for use with airborne troops) have appeared, together with improved bridging equipment, battlefield engineer vehicles and trucks with greater load-carrying capacity. It is no longer a case of crying that logistics are the enduring Soviet weakness.

The present development is in the direction of combined arms operations, reflected in the increase in artillery, anti-tank and air defence capabilities, while the tank forces have begun to embark on a new relationship with the motor-rifle troops, albeit a complicated process but one which demonstrates that the Soviet command has definitely absorbed some of the major lessons of the Middle East war of 1973.* Though developing its impressive nuclear war-fighting capability (and the equally impressive arsenal of chemical weapons), the Soviet command has latterly concentrated on efforts to sustain relatively prolonged conventional operations even in the initial phase of a major conflict in the European theatre, hence the intense interest in and commitment to breakthrough operations with conventional forces—and the equally consuming interest in a form of preemption which catch NATO's defences unprepared and undeployed.

* * * * *

Amidst the detail of this massive programme certain objectives appear to be assuming tangible form—a drive towards superiority in

---

*See "Air Power for the Pact", *Flight International*, 5th June 1976, p. 1507: also "The Soviet Air Forces", *Air International*, June 1976, p. 277: also *Air Forces on the Central Front,* Assembly of Western European Union, Doc. 690 (1st December 1975).

*See Phillip A. Karber, "The Soviet Anti-tank Debate". *Survival*, May/June 1976, pp. 105–11 on the Soviet recognition of recent military lessons and the Middle East War (1973).

strategic weapons, the development of a diversified offensive capability, improvement in the effectiveness of Soviet general purpose forces and the expansion of means to support Soviet initiatives or commitments at a distance from the Soviet periphery. What is patently obvious is the steady emergence of a support system once intended only for a short war concept but which now can increasingly sustain all forms of operations. Here is the true substance of the Soviet principle centred on waging and winning any campaign at any level of weapons.

* * * * *

Is the Soviet aim, therefore, outright and unequivocal superiority? Here it is necessary to return to the notion of the phases in the build-up and the powerful additive of sustained force modernisation. At least one factor here demands closer attention: Soviet military thought, earlier mocked as crude and unsophisticated, has shown itself to be consistent, cogent and remarkably topical. The principle of survivability is seared into this outlook, as is the concept of developing a warfighting capability for all environments, first seriously adumbrated under Marshal Zhukov's reforms in the mid-1950s. With Khrushchev's "cheap deterrent" abandoned, Soviet military and political leaders concentrated on closing the gap between Soviet objectives and Soviet capabilities, all the while paying the closest attention to the survivability of what was a very limited strategic potential. The result was (and still is) a careful balance between a war-avoidance strategy and the development of a war-fighting capability, with hints of a war-winning capability as military strength grows. Hardening and dispersal of strategic targets was an early priority, combined with experiments in mobile missiles and an ABM system, all to secure by the mid-1960s the elements of a protected second strike force. This achieved, the Soviet command loosed its drive for parity and then numerical advantage in strategic missiles, followed in turn by a build-up in general purpose forces capable of operating in a conventional as well as a nuclear environment. There is also the interesting dialectic, if it can be put that way, of the relationship between the Soviet pursuit of matching MIRV capability and investing in strategic defence, with the latter coming once more to the fore once the former had gained operational status.

* * * * *

64. 1978–1979

Perspectives of NATO Defence

J. M. A. H. Luns

His Excellency Dr. Luns is Secretary-General, North Atlantic Treaty Organisation

In the spring of 1978, the Heads of State and Government of the North Atlantic Treaty Organisation will have met for the second year running to discuss the defence affairs of the Alliance. This attention at the Summit to NATO's defence capabilities is a clear indication of the concern at the highest governmental levels at the present trend in the balance in conventional arms between NATO and the Warsaw Pact, which is moving strongly to NATO's disadvantage, and of the determination at those levels that NATO must take the necessary remedial steps.

I do not propose to reproduce details of the comparison between the military capabilities of NATO and the Warsaw Pact. These can be found in specialised publications such as *The Military Balance*, published by the International Institute for Strategic Studies; the United Kingdom Defence White Paper for 1978 also presented some telling comparisons. But I will mention some of the salient features of the Soviet threat. I say Soviet rather than the Warsaw Pact as it is the Soviet political direction and the massive might of over $4\frac{1}{2}$ million men and women in uniform which could threaten Western Security.

A major development has been the achievement by the Soviet Union in recent years of a rough equivalence with the United States in the nuclear field. In parallel, the Soviet Union has made great efforts to modernise its conventional forces and to give them the capability of attacking the West within a much shorter period of preparation than has hitherto been possible. The emphasis in the Soviet military build-up is being placed on offensive capabilities. Characteristic of this offensive posture are the powerful amphibious forces in the Kola Peninsula and in the Baltic, and, facing the Central Region of NATO, the 19,000 or more tanks and large numbers of tactical aircraft which have recently been given a deep penetration capability in both conventional and nuclear roles.

The massive expansion of surface and underwater fleets provides further evidence of the determination of the Soviet Union to acquire a capability, at sea as well, to threaten vital Western interests. The submarine fleet is designed partly for the purpose of attacking the United States ocean strike fleets, even more so for offensive attacks on the Atlantic lines of communication. The Soviet surface fleet, though still inferior to Western navies, is present in all the oceans of the world and has the capability of interfering, both in peace and war, with the essential supplies of raw materials and oil upon which Europe remains dependent for its existence. The activities of the Soviet merchant fleet are also a matter of concern. In 15 years, the fleet has expanded from $3\frac{1}{2}$ to about 20 million tons. Most of the fleet is engaged in some form of intelligence-gathering but, apart from that, considerable harm is being inflicted on the maritime interests of European powers, through the use of direct State subsidies to make inroads into mercantile trade thus endangering the future of many European shipping companies.

The Soviet Union does not publish information about its defence expenditure in a way that can be readily compared with that of the Eastern nations. According to well-founded estimates, however, the Soviet defence effort as a whole absorbs additional expenditure of about 5 per cent in real terms each year. Since the growth rate of the Soviet economy is probably less than 4 per cent, defence is thus taking an increasing share of Soviet national wealth each year, a share now estimated to be as much as 13 per cent of the gross national product. By contrast, European countries participating in NATO's integrated military structure increased their total defence expenditure in real terms by only 5 per cent over the last 10 years; in other words, by less than $\frac{1}{2}$ per cent a year on a cumulative basis, reducing the defence share of the Gross Domestic Product for NATO Europe as a whole to 3·6 per cent compared with 4·2 per cent 10 years earlier. Within the Soviet annual defence budget, only 15 per cent goes on personnel costs—for NATO countries the figure is nearer 50 per cent—leaving 85 per cent for equipment, training and maintenance.

It is clear, therefore, that the Soviet Union is conducting a relentless long-term build-up of its military capabilities to levels far beyond that required for the defence of its homeland, even if account is taken of the need to control its East European allies and of the challenge which the Soviet leaders perceive in the growing economic and military power of China.

NATO's Position

At the meeting of NATO Defence Ministers in December 1976, a clear warning was signalled of how much the balance of conventional forces was shifting to NATO's disadvantage. NATO's response, measured against the force goals established for each country earlier in 1976 and covering the period up to 1982, was judged in the light of the sum of their current national plans to be inadequate. The clear message from the highest NATO military authorities was, that while NATO's current ability to deter and defend was not in doubt, that ability would be progressively eroded if the trends were allowed to continue.

To be more explicit, for a number of years commitments accepted by NATO countries have fallen well short of the goals which the Alliance as a whole has agreed are the minimum necessary to enable the military commanders fully to carry out their missions. In particular, there is evidence that a number of defence programmes were being reduced or considerably delayed despite the fact that only for a few countries could this be fully justified by economic constraints. The unwillingness of countries to allocate sufficient resources to defence has of course a cumulative effect in that backlogs of defence requirements eventually can become so great that they are almost impossible to overcome. Over and above this, the qualitative superiority of NATO equipment is gradually being eroded and so can no longer be counted on as a significant factor in the military equation. Should these past trends be allowed to continue it would place at increasing risk the ability of NATO forces in Europe to maintain the effectiveness of the deterrent and make it more than ever difficult for the maritime forces to keep open the sea lanes for the reinforcement and resupply of the European theatre.

Ministers reached their sombre conclusions after taking full account of a wide range of on-going programmes to improve NATO's forces. But all this, Ministers agreed, was not enough if the adverse trends in the balance were to be reversed. They, therefore, undertook to review their national contributions to NATO's collective defences in order to identify further specific force improvements and commitments to be declared to NATO during 1977, and they recognise that the achievement of that objective would call for both real annual increases in defence expenditure by NATO governments and for a greater emphasis to be placed on allocations within defence budgets to major re-equipment and modernisation programmes.

Shortly after the Ministerial meetings in Brussels, the new United States Administration took office and 90 hours later, Vice-President Mondale came to NATO carrying to the Alliance President Carter's reaffirmation of the American commitment and his deep concern about the Alliance defences; the President, the Council was told on that occasion, was prepared to consider increased United States investment in NATO's defence and in turn America would look to its Allies to join with her in improving NATO's defence forces to the limit of individual abilities and to provide a defence adequate to NATO's needs. In the ensuing period of intense diplomatic activity and consultation, in Brussels, in capitals and in NATO military circles, the outline of a new long-term look at NATO's most urgent defence needs took shape.

* * * * *

Re-establishing the Balance

During 1978, therefore, NATO's determination to improve its defence capabilities as a means of countering the alarming trends in the military balance between NATO and the Warsaw Pact is being made manifest in three ways. As a result of measures agreed by the Defence Ministers in December 1977, a programme of short-term measures designed to achieve improvements in anti-armour, war reserve ammunition, and readiness and reinforcement, will in most cases come to fruition by the end of 1978. Also, in the spring of 1978 the NATO Defence Ministers will have endorsed a new set of force goals—or targets—addressed to each nation and covering the whole spectrum of NATO's military needs up to the mid-1980s. There are some 1,300 of these goals with most countries receiving a hundred or more. There is already welcome evidence of the intention of most of the NATO countries to respond to the call for additional real increases in defence expenditure in the region of 3 per cent, although there are signs that in some cases greater increases than this will be required if countries are to make significant improvements in their forces over and above current national plans.

The third strand is the Long-Term Defence Programme which has concentrated on a limited number of areas where collective action is urgently required. The Programme will address both the medium up to 1985, and the long term up to 1990 and even beyond and the emphasis will be placed on the establishment of co-ordinated programmes for action. The areas selected are readiness, reinforcement, reserve mobilisation, maritime posture, air defence, communications, command and control, electronic warfare, rationalisation and logistic support. In addition, NATO's Nuclear Planning Group, which has already given much study to the modernisation of theatre nuclear forces, is developing both medium- and long-term programmes to ensure that these forces continue to play their essential role in NATO's Triad of strategic nuclear, theatre nuclear and conventional forces.

One of the major objectives of the Long-Term Defence Programme has been to improve the effectiveness of the forces of the Alliance by enhanced co-operation between nations in the areas of equipment, logistics and infrastructure, the three aspects of what is known in NATO as Defence Support.

In the area of equipment, the Alliance has to take into account the continued advance of Soviet technology and the steady increase in the quantity and quality of the weapons now coming into service with the Warsaw Pact forces. The threat that this poses is exacerbated by the rising cost of modern defence equipment and the pressure in most member nations for budgetary restraint in the face of economic difficulties. This situation has caused the Alliance to seek more effective methods of co-operation than in the past. The fact that our forces are often handicapped operationally by differences in their equipment, has given increasing importance to the concepts of standardisation and interoperability.

* * * * *

As I write this article, the Long-Term Defence Programme has yet to come to fruition. The special Task Forces, which were set up to produce recommended programmes in each of the selected fields, have only just reported and their reports now have to be scrutinised from the political, economic as well as the overall military points of view and, even more importantly, countries will have to decide on their individual responses to the inputs requested from them. The reports will have been submitted first to NATO's Defence Planning Committee, which meets at Ambassadorial level, then to the NATO Defence Ministers when they met in Brussels in the middle of May 1978 and then finally to the Summit meeting in Washington at the end of the month.

It would be wrong to underestimate the complexity of the task and the challenges which still lie before NATO in bringing the Long-Term Defence Programme, together with the establishment of a well-founded set of medium-term force goals, to a successful outcome in the spring of 1978. This will require the application of a large fund of political resolve and a continuing determination by the NATO countries to ensure that allocations to defence attract the necessary priority in national economic planning. I am, however, confident that the foundations have already been laid for a successful Summit meeting, from which should flow a substantial programme of work, endorsed at the highest levels of governments, and designed to bring about much-needed improvements in NATO's deterrence and defensive capabilities.

In conclusion, I believe that, in retrospect, 1978 will be seen as a year when NATO took measure of Soviet military capabilities and set in train remedial action towards the re-establishment of the military balance essential for the security of the West.

65. 1980

Intelligence—The Handmaiden of Policy
Lieutenant-General Sir David Willison

Lieutenant-General Sir David Willison retired from the British Army after 40 years service. From 1971 to 1978 he was successively Director of Service Intelligence, Deputy Chief of Defence Staff (Intelligence) and Director General of Intelligence in the U.K. Ministry of Defence.

When asked what one does to earn one's pay, the reply that intelligence is the way of life is apt to bring a somewhat strained expression to the face of the interlocutor. Responses vary. Some remark "oh, very hush hush"; other scoff "you people in intelligence always get it wrong—what about the 1973 Arab–Israeli War?" Another line is to look vaguely uncomfortable and to say "I suppose your spies are watching me". To scotch this last red herring once and for all, military intelligence at any rate does not deal with protective security and counter espionage. Nor is the intelligence product, where it is served up to the consumer, particularly secret. The CIA statement released in Washington, or the so-called intelligence source in Brussels, are well known ways of purveying intelligence headlines. At least since the time of Queen Elizabeth I, our islands have had a reputation for producing intelligence assessments of high quality. Yet to a great number of people it is still customary to treat this subject with deep reticence. This attitude stems from a praiseworthy sense of personal duty of the citizen to maintain security over national secrets. This is a right and proper thought which is still at this moment backed by the Official Secrets Act. As ever, the real aim of protecting national secrets is to safeguard the sources used. A particular end product may or may not go into two much supporting detail, the degree of secrecy attached to a particular report is always a factor of how much of the sources used to frame it might be revealed should the contents fall into hostile hands. In this paper I shall try to dispel some of the myths that surround the subject of intelligence, with particular reference to the art of defence intelligence I shall also try to illustrate how the process of intelligence, itself in no way secret, can work to the advantage of policy makers, whether these be concerned with the day-to-day conduct of affairs or with longer term planning.

Of course, intelligence gets it wrong on occasion; but successes are less well publicised for obvious reasons. So what exactly is intelligence? The term covers a number of processes that call for a wide variety of talents. The first stage in the game is to collect the raw material of the trade; facts, opinions, forecasts of intentions, military capabilities, scientific developments, economic data, the list is endless. In Communist states in particular much of this raw material is jealously guarded by tight security measures. But even in such states, and far more so in the Western world, the need to carry the mass of the people along with current policies causes public statements to be made. These may be found in radio or TV programmes, in newspapers or even in journals dealing with political, military, economic and scientific subjects. So the art of all source collection requires a judicious mix of overt and covert means. The latter process entails penetration of the security barriers erected by the target organisation. Recent publicity in Washington and in London has thrown some light on what covert collection means are available in this day and age. It is not my intention in this paper to pursue collection of intelligence further. As a tailpiece I would only add that all three Armed Services can and do collect intelligence relevant to their needs in peace as well as in war, should this occur. A simple example is that of aerial photography; another is direct observation of naval vessels on the high seas; neither infringe national sovereignties in peace.

The second process in the art of intelligence is to sift and select from the mass of overt and covert material so collected those pieces of the mosaic of intelligence that are relevant to current or potential future requirements. This is the province of the desk officer or analyst who painstakingly, day by day, records new facts as they occur or updates those he has already collected. According to his particular speciality his data base consists of collated material going back for a considerable span of years and cross referenced to other known related facts. The desk officer who keeps it all in his head and poses as the greatest living expert on a given subject, because his memory is good, is in practice a menace. The data base of collated facts needs to be kept in such a way that rapid access to all recorded and related facts is available, irrespective of the length of time in office of any particular individual. Automation has an important part to play, but here, too, lie dangerous reefs. Failure to build in effective data retrieval systems is one. Belief that machines can do much of the thinking is another; nevertheless automatic data processing can and should be developed to manipulate related facts in a number of ways to help in the process of analysis. Such winnowed data from a wide range of collated records should ideally itself be capable of automated storage and feed back in written or pictorial form.

Assessment and Dissemination

The third process of intelligence production is that of assessment. This jargon phrase conceals the fact that assessment is the process by the trained and experienced human mind of reaching a positive judgment based upon the evidence. In a complex problem, many specialist desk officers dealing with various facets will have produced their analyses and be regarded in depth perspective, extruded from their collated data base. These officers need to be cross-examined to see that their ideas stand up to related findings by other analysts handling some other facet of the problem. Finally a positive judgment, on which policy makers can lean, if they are so minded, has to be made. This is the province of the senior hierarchy of the intelligence world. Issues concerning war and peace prospects lie at the summit of the edifice. Below the Olympian heights lie a myriad of other subjects which require judgments before delivery to customers.

Delivery to customers has another jargon name—dissemination. Essentially this means getting the intelligence assessment to the right people at the correct time and in a form which will hopefully answer questions in their minds as they approach decision making. This process is a great deal less simple than it sounds. Much leg work is needed constantly to keep up with how policy thinking is developing and at what level it has now reached. Intelligence assessments need to be injected early in some cases because vested interests may otherwise reject them as being inconvenient. The same story has to be told throughout the whole process, subject only to updating by new facts, otherwise the credibility of the intelligence judgment can be damaged. Then there is the vexed question of how best to inject assessments, orally or in writing. In a complex case much supporting evidence is needed if the judgment is to carry conviction. This evidence may reveal sensitive sources, so its security has to be safeguarded. Dissemination is therefore by no means the least important part of the four stages of the intelligence process—collection, collation and analysis, assessment, and finally getting the product out to those who need to know. On the contrary, it requires as much imagination and drive as any of the other elements that go to make the finished intelligence article. The truth is that intelligence is only as well regarded by those who come in contact with it as its assessments contribute to their own thinking. To sit like a broody hen upon huge masses of acquired knowledge without telling anyone about the end product is the negation of common sense for the intelligence staff officer.

Writers of fiction love to personalise all these processes into the activities of one or a smaller group of heroes whose individual endeavours save the day at the last moment. This, of course, is the warp and wool of thriller writing. Reality is far more diffuse and prosaic and so it rarely if ever receives treatment by the publicity media or the novelists. For example, the President of the United States may make a public statement that certain military and technological developments in Russia are a cause of grave concern both to himself and to NATO. But nobody inquires as to how he has been briefed to speak as he does, other than perhaps to refer to the odd CIA release or NATO spokesman. Opponents of the President's policies then try to denigrate such sources. In practice the devoted work of many thousands of both civilian and Service members of the intelligence community will have gone into fashioning the finished product used by the President to justify new policies to be pursued by his country and by NATO.

At a more specific level, the UK has recently announced its decision to proceed, at least for the moment, unilaterally with the design and production of a new tank. It is betting on a certainty that such a decision did not derive from the activities of a latter day James Bond swinging on the chandeliers of the Kremlin. For real, the painstaking efforts of a myriad of people, collectors, collators, analysts and assessors will have gone to provide a new assessment of intelligence tailored to the needs of decision takers in the specific area of military technology.

* * * * *

Policy Makers

At this point it may be helpful to define policy makers more definitively. A better term might be decision takers, that is those from governmental level downwards who are able to take decisions on which orders for action are framed in detail. But those who are charged with preparing policy options also need intelligence input, often of a more detailed kind. So do those who execute decisions, again in more detail and covering other facets not germane to the formulation of policy. For these reasons I define policy makers as the whole range of those who need to take into account the situation in some way or another outside our own islands. For example, those who negotiate with the Russians at Vienna over Mutual Balanced Force Reductions need a specific kind of input on Russian intentions and capabilities. Those who seek some solution to the baffling problems of Rhodesia are another. Decisions to reinforce the British garrison in Belize in recent times illustrate the range of supporting intelligence input needed from politico-military assessments of Guatemalan intentions to the precise military nature of the local threat to Belize that must govern in large part the scale and nature of the reinforcements required. The commander on the spot needs yet further intelligence of a tactical nature to make his dispositions on the ground.

In these, and a host of other, world-wide problems the collation of the basic facts must necessarily have started much earlier in time. Unless the customer—the policy maker—has indicated his potential interest well in advance, then the intelligence assessor will be faced with trying to make bricks without straw at the last moment. Given unlimited resources, of course, then the ideal for the intelligence staff consists in studying all possible subjects in the greatest possible depth. Would it not be nice to know everything there is to know about the Beagle Channel now under dispute between the Argentine and Chile? But if it was known, what could the policy maker in his wisdom in London do about it? In times of financial stringency, strictly finite limits have to be set on how scarce resources are to be used to conform to national priorities. The policy maker must enter into the labyrinth from a fully informed standpoint because he will pay the penalty if he has denied intelligence the authority to study a particular subject in depth in good time. In the event he, or more probably his successor, will blame intelligence for not providing the answers, denying all knowledge of resources having been cut out long since to collect against a certain target so that the vital collated data base could be built and constantly updated against the day it would be needed.

* * * * *

Defence Expenditure

In Russia, the defence budget as published is a travesty of the truth. Far from being static or even declining in recent years, it has in real terms been increasing steadily each year by about 40 per cent. Public statements in recent Defence White Papers and in Washington have disseminated intelligence assessments on this most important topic. The figure of 11 per cent to 13 per cent is indeed a daunting one, and its detailed breakdown of very great significance to NATO. This is where the economic intelligence analysts come into the picture. It is their task to accumulate a mass of detailed collated material on what it actually costs the Russians to equip a tank division complete with all the latest equipment, to produce ever larger numbers of nuclear submarines, strategic missiles, variable geometry aircraft and so on.

They need to cost the men who man these advanced weapons systems and the base facilities from which they operate.

Given a sufficiently detailed series of "building blocks" on costs of all the multifarious items that go to make the totality of Russian defence spending, then a pretty accurate estimate of what Russian defence expenditure really amounts to can be made. When compared with published Russian figures, gaps in the latter become rapidly apparent. The dismay of the Russians over public statements of this kind is evident from the intensity and direction of their hostile press campaign in reply. No doubt the Politburo are extremely anxious to sustain their propaganda line every bit as much for internal reasons as for consumption by their sympathisers in the world at large.

Perhaps the most worrying aspect of Russian spending is the allocation of resources to military research and development, to the detriment of similar expenditure for civil purposes. This is an important area of effort for scientific and technical analysts. As ever with Russian targets, much reliance has to be placed on covert collection sources if complex programmes in progress are to be read aright. At the top end of the spectrum come future Russian strategic systems, either land, undersea or air launched. In the middle are those nuclear delivery systems that threaten Europe directly. In recent years the ballistic missile system SS20 has had pride of place in analytical studies. Nuclear weapons can of course be delivered by aircraft and by long range artillery and rockets, so all of these figure in scientific and technical analyses. Last, but by no means least, come the very wide range of applications of new applied scientific principles to naval, land and air forces. At the forefront of the thinking of all three Services is the need to grapple with the tactical implications of sensors in the electromagnetic part of the spectrum which are constantly being evolved by the Russians.' Electronic warfare has been a reality for many years. Now the electro-optical part of the spectrum is creating additional problems in terms of new laser sensors. At sea, the quest for better anti-submarine warfare technology is likely to figure as of very high priority in the thinking of Admiral Gorshkov. Acoustic intelligence in the sonar field continues to grow in importance. The list of Russian research projects is huge; the most worrying aspect is the sheer breadth of their programmes. The depth of research possible through allocation of resources larger than those of the NATO Allies including the United States is equally disturbing. The implications for NATO decision takers are evident.

* * * * *

I set out first of all to explain the process of intelligence. Collection is the keystone of the whole arch and tends to be the most expensive item in the intelligence inventory. The more secure the target, the harder to find out the essential facts and the greater the expense of so doing. Future intentions are the hardest targets of all to unravel. Difficulty of access is compounded by the fact that final decisions tend to be taken very late in the day. The elements of chance and of personal whim are often the decisive factors. So pity the poor collector, and the assessor who ultimately depends upon him, in this most opaque of all intelligence tasks. Policy makers rightly place assessments on intentions, as and when needed, at the top of their lists of requirements. But failing the fictional hero source, assessments have to be made the hard way by piecing together those pieces of the jigsaw puzzle that collection has brought up in the trawl. Success will hinge to no small extent upon sufficient resources having been made available far back in time to make sure the mesh of the collector's trawl is sufficiently fine.

Recognition of the all important key pieces of the jigsaw is the province of the desk officer analysts. These, too, require sufficient resources allocated to them by policy makers so that the individual span of speciality is not too wide. Should this become the case, then the collated data base is likely to be too thin to ensure that a particular analyst will be able to see the importance in perspective of one small but vital piece of raw intelligence. In consequence, a warning of ill-tidings of great importance may be missed.

Above the analyst level there must be adequate senior intelligence officers who can survey wide reaches of the spectrum of intelligence coverage authorised by policy makers. In the case of defence intelligence these need to cover the special needs of the Royal Navy, the Army and the Royal Air Force, together with the wider horizons of relevant foreign defence policies and strategic thinking. In addition to politico-military efforts, there is a specific need to oversee scientific and technical, economic and logistic collation and analysis leading to finished assessments in all these fields.

The size of such staffs is directly related to the range of subjects that policy makers place upon them. So ultimate responsibility for the success or failure of intelligence input to the full range of defence intelligence customers rests with policy makers. Given adequate resources, given the right questions to address their minds to, and given high grade people, then intelligence should come up with most if not all the answers. Intelligence, in the widest definition of that term as set out in this paper, will indeed prove to be the handmaiden of policy.

66. 1981

Civil Defence—A View for 1981

C. N. Donnelly

Soviet Studies Research Centre, The Royal Military Academy, Sandhurst.

* * * * *

It cannot be denied that Soviet doctrine does hold that it is possible to fight and win a nuclear war. The USSR not only maintains its Armed Forces for deterrence but also studies and practices how to fight nuclear war if it occurs. War is seen by the Soviets as a tool of policy, and the policy of the Soviet Union is the eradication of capitalism. Furthermore, there can be little doubt that if the USSR actually came to believe that she possessed the ability to destroy the USA whilst avoiding significant retaliatory damage, she might, under certain political conditions, be tempted to use that ability if no practicable alternative existed and *if she were absolutely certain of total success.*

Nuclear war may break out by accident or due to a "third party" conflict. If it does so, it will clearly be the Soviet aim to accomplish the destruction of the USA whilst achieving the survival of her own form of socialist society. In this light, the Russians have never accepted the US doctrine of Mutually Assured Destruction, they regard this attitude with contempt, considering it accurately summarised by its initials.

The Soviet Government, however, in no way underestimates the effects of nuclear war. One of Lenin's most fundamental political principles was that, once the Communist Party came to power, it should retain power at all costs. A large scale nuclear attack on the USSR would present the greatest imaginable threat to Party power. Consequently, the Party takes every possible step to ensure its survival, and to ensure that it retains its grip on the reins of power. Its main means of achieving this in a post-strike period will be through the various military agencies (the troops of the Ministry of Defence, the MVD and the KGB) plus its widespread and disciplined Party membership.

The Soviet Civil Defence organisation, as a Regular branch of the Soviet Armed Froces with perhaps as many as 40,000 regular troops, will provide a most important framework which will enable military control over the civilian structure of society to be more easily enforced. This, they hold, will better enable the fabric of society to survive intact. Through the Party's control of the military, and the military control of the civilian regular and part-time CD organisation, the Party will be able better to re-establish control post-strike, and its chances of survival in power will be greatly enhanced. One must add to this political reasoning two further points in justification of the CD system: the natural Russian tendency to over-insure in military matters, and the supporting natural human instinct to provide some form of protection in case of attack, no matter what. The result, seen through official Soviet eyes, is that Civil Defence is just common sense.

One must also guard against seeing the Soviet's situation through Western eyes. The USSR believes that she has not only the US nuclear threat to contend with. She may also face a nuclear strike from China; or in event of European war in which the US was not involved, from Britain and France. In the foreseeable future other nations, such as India or Israel, may get nuclear weapons. The consequent danger of a limited nuclear attack poses a more manageable problem than even a partially effective Civil Defence system might hope to cope with.

It must not be assumed from this that the Soviet system does not encounter problems similar to those of the West. In fact, the Civil Defence systems of East and West seem to have remarkably similar problems. The most basic problem is a larger degree of scepticism amongst the average Russians, especially in the countryside, as to the efficacy of Civil Defence. This is, if anything, made worse by the Civil Defence publicity and the basic instructional course which frightens people, especially when coupled with the obvious inadequacies of the preparations in some areas. Nevertheless, the USSR has one great advantage—it does not need to heed public opinion, and so it is proceeding with the development of its Civil Defence programme, "consumer resistance" notwithstanding.

Special Features of the Soviet Civil Defence System

(a) Soviet perceptions of national vulnerabilities

There are certain factors of Soviet governmental organisation which, it is normally stated, tend to make centralised control in emergency more effective. However, there are reciprocal factors which may counteract this tendency. For example, the economic centralisation means in practice that in towns all heavy plant machinery and lorries are held in central transport depots. Consequently this material can probably be brought into action more quickly than would be the case in a Western country where similar machinery would be dispersed and under private control. On the other hand, in virtually any Western Country, proportionally more useful industrial and agricultural plant machinery will be available due to the relative affluence of the civilian sector in Western economies. This produces a greater potential, provided that an organisational framework for its use in emergency is established and practised.

The great size of the USSR will mean that it can obviously absorb a great number of nuclear strikes, and still have much of its territory unharmed. The relative isolation of large cities provides vast spaces into which their populace could be evacuated. However, the very centralisation of the Soviet population and economy presents an easy nuclear target. Soviet heavy industry is based on a very small number of extra-large-scale facilities and there are only about 220 towns with over 100,000 inhabitants. As the power base of the Communist Party is in the urban population (62 per cent of the USSR's 255 million) and the Party is very weak in the countryside, this makes Party organisation very vulnerable to counter-value targetting. Moreover, conditions of village life are very primitive when compared to the industrial societies of the West. The ability of Soviet villages to cope adequately with streams of evacuees would certainly be less in both human and material terms than the ability of villages near a Western industrial city to deal with a similar problem. Inadequte though the facilities of the villages of northern Hampshire would be to cope with the refugees from a nuclear strike on Southampton, they are certainly superior to facilities in the Byelorussian villages surrounding Minsk.

Another serious Soviet problem in nuclear war is posed by the difficulties of transport between the centres of industry and population. There is only the sparsest of road networks and the paucity of road transport (at harvest time, even military transport from Soviet forces in East Germany is sent back to the USSR to supplement the domestic fleet) would hinder not only recovery, but also evacuation procedures.

To these geostrategic features of Soviet Civil Defence must be added one feature so basic to Soviet life that it is often overlooked by analysts of Soviet Civil Defence capabilities—the climate. The intense cold of winter makes human survival post-strike far less easy; the snow and ice hinder effective decontamination; the wet spring and autumn render much of the countryside totally impassable. The harsh climate in itself is not an insuperable problem; Russians have successfully coped with it for centuries, but it complicates post-strike recovery enormously. Side effects of living in this climate produce other, normally uncalculated, effects. Centuries of living in village poverty (until 1927, 85 per cent of the Soviet population could accurately be described as peasants), throughout the long winters have made the average Russian citizen of today far less sensitive to personal hygiene and sanitation than his Western European or American counterpart. This, any Western visitor to the USSR can readily affirm. Despite a certain natural immunity that the Russian has acquired, the poor quality of sanitation and lack of sanitary habits would, in war, make for extreme vulnerability to epidemics. The significance of this is reflected in the importance attached to epidemiology by the Soviet Military Medical Services.

The variety of nationalities in the USSR must be considered a further problem in event of nuclear war. In peace-time nowadays, nationalities and minorities problems have largely been assuaged. However, in event of a terribly destructive war, the antipathy of some non-Russian nationalities to Moscow, plus the undoubted antipathy of much of Eastern Europe, would certainly complicate the internal security problem to an enormous degree, just as it did in the early stages of the last war.

Finally, the stringencies of the Soviet economy (of which the Kremlin can hardly but be aware) must be taken into account. The traditional overtasking of factory managers, the constant raising of norms to be met—in short, the Russian tradition of trying to get a quart out of a pint pot—means that the time and materials which should be allocated to Civil Defence training and preparation in factories and local governments are not available. Resources will not be adequate except in those cases where the party intervenes directly to enforce the regulations. The "on paper" Civil Defence capability will, as a result, appear somewhat greater than what can be accomplished in practice.

(b) The Soviet approach to the problems of war survival

Above are listed just some of the special features of the Soviet environment with which any Civil Defence system, to be effective has got to cope. Yet the Soviet plan for war survival does not depend solely on its Civil Defence network. Rather, it depends on a whole range of organisational and control measures, of which the Civil Defence network is just one.

The Communist Party itself, with members in every town and village, every school, factory, and farm in the USSR is the most important of the country's tools for war survival. The Party provides the leadership, direction and discipline that the Russian nature needs, and it controls even in peacetime every social activity in the USSR. The means of control which it employs in peacetime will be those it relies on for post-strike recovery. The prime means of control are: the

Armed Forces, the Police and MVD forces, the KGB, and regular Ministry of Defence troops; the central and local Government system, with its almost total centralised control of food and food distribution, the construction industry, production facilities and the labour force, all national finance, and the national communications network, and the mass media and educational systems, both entirely in Government hands.

The chief functions of the Soviet Civil Defence network are; firstly, to co-ordinate under Party control the relevant Civil Defence functions of all these agencies; secondly, to provide comprehensive studies and plans, and the framework for exercises, so that Civil Defence procedures can be perfected; thirdly, to provide an organisational and personnel structure to enable the mass of the population to be educated in and mobilised for effective Civil Defence work.

* * * * *

The official aims of the Soviet Civil Defence organisation are as follows:
   (a) Protection of the population from weapons of mass destruction.
   (b) To prepare economic facilities (factories, farms, administrative organisations) for stability of operation in the face of attack.
   (c) To perform rescue and emergency restoration work at sites of destruction in order to establish national recovery.

Protection of the population is to be achieved by the following measures:
   1. Effective early warning.
   2. Effective training in early warning drills.
   3. Dispersal of essential work force from large town to surrounding villages.
   4. Evacuation of non-essential populations deep into the countryside, evacuation and concealment of essential R & D establishments.
   5. Providing essential individual means of protection.
   6. Providing blast shelters for key governmental personnel and industrial work force in urban areas, and fall-out shelters for the population in evacuation and dispersal areas.
   7. Creation of food and water reserves.
   8. Effectives training in decontamination and protection.
   9. Providing a contamination monitoring and recce service.
   10. Implementing anti-epidemic measures.
   11. Organising search and rescue operations.

The efficiency of the Soviet Civil Defence system is admitted to be patchy. A simple rule of thumb is that the more important a town, factory or similar site is to the functioning of first, the military and second, the governmental systems, the better will be the level of Civil Defence preparations. For example, in major naval dock areas or towns with important army installations, in factories producing vital defence material, or in areas close to the key Party and governmental control centres, the number of Civil Defence shelters and the level of popular training will be far higher than in a town or installation without any such important facility.

There is available a comprehensive, though not entirely accurate, Civil Defence handbook for all personnel, and over the past few years there have been an increasing number of combined military and civilian exercises involving towns or certain factories in a given area. The fact that these have to date been on a limited scale should not be taken to indicate a lack of serious effort in this regard. The programme of integrated exercises is in its infancy, and what we are seeing is a very gradual development of effective Civil Defence capability within a limited budget. There can be little doubt that the system has been much improved over the last ten years, and that continued improvement can be expected.

Little progress has apparently been made in the field of planning mass evacuation, and no really large-scale exercises have been reported. In the provision of shelters, a recent construction programme has gone some way to providing protection facilities for key personnel in important areas, but there is little evidence that any mass shelters are in existence, and even obvious shelters such as underground railways do not yet appear to have been adequately equipped for the task. Whilst stockpiles of food and fuel reserves exist, and stocks of protective masks and clothing may also exist, there is little evidence of exercises in the distribution of these, such as were common in European countries before 1939.

Consequently, all things considered, one is forced to conclude that some of the enthusiastic claims made for the Soviet Civil Defence system, particularly in some US circles, are unsubstantiated. We see no evidence that the USSR could at the moment hope to provide adequate protection to ensure beyond doubt the survival of its social system with minimal damage. However, this is not to say that the Soviet Civil Defence effort is to be derided, or that it is not worthwhile. The following points must be considered.

Despite a long history and tradition of Civil Defence, the achievements of the present Civil Defence system should really be measured from 1971, when it was rejuvenated. Since then it has made great strides and will probably continue to do so in the foreseeable future.

The system has achieved an increased level of awareness amongst the public and appears to be overcoming public apathy, particularly in those key areas where a lot of time and money is put into Civil Defence effort.

There is now an established shelter network for key personnel that would ensure at least the survival of a governmental command structure.

There is in existence a comprehensive framework of a Civil Defence system which could be activated fairly quickly, were the Party to make it a priority.

The Civil Defence system is part of an overall system of social control which has proven its ability to function in adversity.

The high level of Armed Services in the USSR provides a powerful disciplinary force in the country, to which the Civil Defence system could harness much of the civilian organisation in emergency.

The Russian's natural character is to respond very positively to coercion. Force has always been a most effective means of rule, and in an emergency, these last three points would combine to counteract the post-strike chaos.

Most importantly, when studying the Soviet efforts at Civil Defence, there are certain points that are instructive to study. We believe these to be as follows.

Firstly, the USSR has severe problems when trying to ensure its own war survival, but the Government recognises these problems and attempts to adapt its systems to cope with them. It does not take the line that "because we cannot do everything to ensure survival, we should do nothing".

Secondly, the Soviet Government, with its complete control of the media, had in the past little success in overcoming popular scepticism, although it could afford to ignore this. However, in those areas where recently real interest and effort have been displayed, and adequate funds have been made available, there the popular enthusiasm for Civil Defence has visibly increased, and exercises appear to be quite successful.

Thirdly, the Soviets do not appear to assume that any attack on their country is certain to be an all out one. Limited attack is possible, and in this event a Civil Defence organisation might well expect to cope. It would certainly help in reducing local casualties.

Fourthly, despite popular cynicism and apathy, mass movements do even so provide a common purpose for a society, and act as an outlet for human desire to "do something" to ensure society's (and individual) survival.

* * * * *

The Lives of a Soldier

(*Crown copyright*)

THE BRITISH ARMY IN GERMANY—ROYAL ARTILLERY
The detachment prepares a 175-mm self-propelled gun for action

(Picture Post Library)

MALAYA. A man has been wounded by a terrorist and his comrades blow up trees, and clear the undergrowth, so that a helicopter can land to take the man to hospital.

The jungle shown in the picture is typical of many parts of Malaya

SOUTHERN ARABIA

Two night-fighting patrols leave their base camp in the Radfan mountains

(Crown Copyright)

THE NORTHERN FLANK

(*Crown copyright*)

Snow-shoe patrol—1st Battalion Coldstream Guards in Norway, March 1969

(*Crown copyright*)

Evacuation of casualty by ski—Men of 45 Commando training in Norway

(*Crown copyright*)

Belfast 1970

Russia Goes to Sea

Z CLASS: A typical Z Class long range torpedo-firing submarine, from which the first generation of ballistic missile submarines, the Z V Class SSB, was developed

Z V CLASS: The Z V Class, carries two surface-launched missiles stowed vertically in the fin, and the Russians claim that the first firing of a submarine-launched missile took place in 1955

CONFRONTATION IN THE AIR AND AT SEA

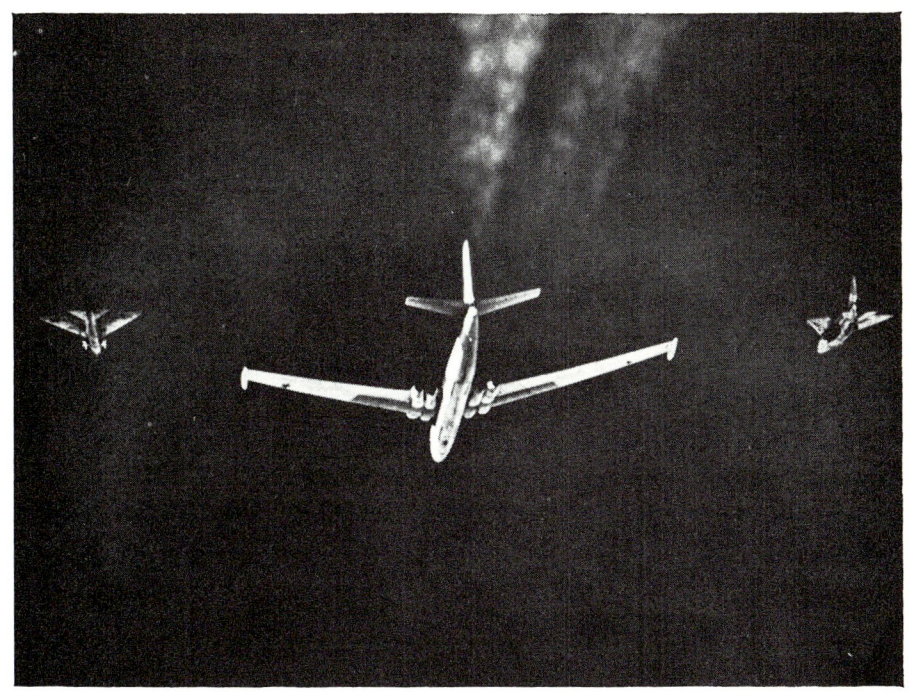

(Crown copyright)

Lightnings of Strike Command escort Soviet Badger

(Crown copyright)

Russian *Mayak*-class trawler observes HMS *Ark Royal* on Exercise Northern Wedding, September 1970

(*Photo: MOD*)

The new Soviet warship *Kiev*, photographed in the Mediterranean by an RAF Nimrod of 203 Squadron based in Malta, after passing through the Bosphorus from the Black Sea. Note the Soviet V/STOL fighters on deck and four Hormone helicopters.

PART X: 1982–1986
Full Circle

Introduction	363
67. 1982 *Trident*—A Candidate for Cancellation? IAN BELLANY	365
68. 1982 Merchant Shipping and the Maritime Threat H. G. DAVY	373
69. 1983 Strategic Weapons LAWRENCE FREEDMAN	380
70. 1985 Space: The Military Applications Today and Tomorrow GROUP CAPTAIN T. GARDEN	388
71. 1986 British Defence Issues ADMIRAL SIR JOHN FIELDHOUSE	396

Introduction

In 1886 the dilemmas facing defence policy makers were those of relating changing military technology to an unstable international system. How were national security to be maintained and national interests furthered? Could war be avoided? If not, what would be the effect of the new technologies on strategy and tactics? Could a nation meet the increasing burden of defence costs without running the risk of an overstrained economy and social collapse?

Expressed in such categories, 1986 will present striking similarities. The mere existence of two radically differing political and ideological systems among the advanced nations will make continuing international instability inevitable. Economic failure, political and sectarian frictions in the Third World will lead to local conflicts with the possibility of Great Power intervention. Technological advance, in the form of more destructive and accurate weapons, and the wider issue of anti-ballistic missile defence and the militarisation of space, make the likelihood and nature of future wars impossible to predict. All that can be certain is that if they do occur they will be infinitely more destructive (perhaps to the extent of destroying civilisation) than the gloomiest prophet of 1886 could have envisaged.

For Britain, now a second rank rather than a dominant world power, the dilemmas of defence are far more challenging than they were a century ago. The problems of relating resources and costs to commitments are so much more acute that radical adjustments will have to be considered. These involve not only the impact of the acquisition of *Trident* on other procurement programmes but also strains on manpower which once again are producing problems of retention of skilled and experienced personnel. Underlying specific difficulties is the great uncertainty arising from the declared policy of the Labour party not only to give up Britain's own nuclear capabilities but also to close all United States nuclear bases and installations. All these difficulties, coming at a time when the reach and flexibility of Soviet strategy have considerably increased, can only be resolved by sound decision making by politicians and military leaders. In such a situation, the accurate information and balanced discussion, which has characterised Brassey's from its beginning will be even more necessary as it enters its second century.

Recommended Reading

Lawrence Freedman, *The Evolution of Nuclear Strategy* (1981).

67. 1982

Trident—A Candidate for Cancellation?
Ian Bellany

Professor Bellany is Head of the Centre for the Study of Arms Control and International Security, University of Lancaster.

The decision to replace the Polaris flotilla with a new flotilla of similar size based upon the Trident C4 missile will almost certainly have to face reconsideration sometime in the next few years on grounds of economy. The reason is that British defence expenditure chronically outstrips British resources, and British resources are shrinking, relatively, and even (at the latest count) absolutely.

The diagram in the Appendix shows that the British share of (or contribution to) Western European defence expenditure far exceeds the British share of the Western European GNP. Put another way, and slightly more tendentiously perhaps, the British contribution to the defence of Europe substantially exceeds its share of that which is to be defended.

It is not within the scope of this article to account for the "excessive" British contribution to Western European defence, but the fact of its existence is strong circumstantial evidence that the protection afforded to defence expenditure in the early period of the Thatcher government will not be indefinitely extended.

Once the search for cuts in defence expenditure is begun, a prime target will be the Polaris replacement programme. Not because it is particularly expensive compared to other, conventional, defence programmes and activities, but because it will be at the dangerous stage of sucking in large amounts of money while still some years away from entering service. And, unlike the run of defence programmes, which can always rely upon fierce protective lobbying from their parent branch of the Armed Services, the strategic nuclear force is a comparative orphan—not without supporters, certainly, but they are widely dispersed. Nor, as we shall see, can the Polaris replacement programme, by its nature, equip itself with the various anti-cancellation protective devices that military aerospace projects, for instance, have evolved in recent years.

A clearer picture of what is being said here may be obtained by use of an analogy. Suppose that instead of defence policy we were dealing with energy policy. Suppose the replacement issue in question concerned the new generation of nuclear power stations. Argument would then rage over whether new power stations were needed at all; what the balance should be as between nuclear and conventionally fired stations; and if there were to be any nuclear, what type they should be.

How would the choice of new power station be settled? It would be settled in two parts. The first part would be "rational". Careful assessment would be made of the costs and benefits of the various choices available and the decision made according to what maximised the latter and minimised the former. But implementing the rational choice always takes time, and the rational choice then has to enter the political arena and survive changes of government, unexpected shifts in the external environment which seem to call into question the correctness of the original choice (shifts the relative prices of fuel, in this example), and technical and managerial failures in the construction phase, all of which give new heart to the supporters of the alternative possibilities that have been discarded at the rational stage.

The survival rate of the rational choice amongst British publicly funded capital projects is not markedly good, whether we are speaking of power-station choices or new sites for the third London airport. Those that do survive are those that by luck or design have about them qualities that make the choice politically robust in the buffeting phase between the end of the rational choice-making process and the coming into operation of the project in question.

The Rational Choice

There is little doubt that the decision to replace Polaris is "rational" enough.

The British defence budget pays for a deterrent, defensive strategy of bare but calculated sufficiency. In theory, NATO declares that it pursues what the United States has for a long time wished it would pursue in practice, namely a flexible response strategy which is supposed to be about being able to deter Soviet aggression by offering resistance in depth to Soviet force at whatever level of violence it is pitched at. In fact NATO's conventional forces are too small and its reinforcement capabilities too arthritic to stop a full-scale push by Soviet conventional forces, and the balance is maintained by leaving the Soviets in doubt as to the exact point at which their conventional successes (or semi-conventional, if battlefield nuclear or chemical weapons had already been used) would force the West to escalate the war to the highest possible level of attacking Soviet cities with strategic nuclear weapons.

The exact size of the conventional armies NATO needs in order for this strategy to work cannot be calculated very precisely. Very large armies would probably be read abroad as a signal that the introduction of strategic nuclear weapons would be long delayed, perhaps indefinitely so; by contrast very small armies might be read as signifying that defence was simply not being taken seriously and that an application of Soviet pressure would actually meet with no resistance. But between these extremes a wide range of conventional or semi-conventional capabilities, provided this was coupled with an unquestioned ability to take the war to the extreme pitch of violence, will do.

This method of keeping the peace was conceived in the days of Western nuclear superiority and it was officially disbanded for flexible response when this superiority began to dissolve as the Soviet Union equipped itself with a full range of nuclear and thermonuclear weapons and delivery systems to match. In practice it has remained undisbanded since the original expectation that the strategy would become untenable because the Soviets would not be daunted by what amounted to a Western threat to commit suicide, i.e. to trigger a nuclear exchange at the highest pitch of violence, has been modified. A threat to commit suicide gains in credibility (as aircraft hijackers remind us almost daily) the more desperate the threatener's position seems to be. And desperation will be the last thing the West will be short of when its comparatively feeble armies are on the point of being crushed by a Soviet offensive.

NATO practice is British practice writ large. If NATO were to disappear tomorrow, the balance between Britain's conventional and strategic nuclear capabilities would not need to be altered at all. Moreover, Britain's strategic nuclear capacity helps to maintain in working order the mechanism that connects NATO reverses in the field to the introduction of US strategic nuclear weapons. The Soviets could scarcely offer to leave US territory out of the war once its own chief cities had ben razed by a British strike. Or to put it another way, provided the British strategic nuclear force can hit the most cherished Soviet targets, Britain has a veto upon any wartime arrangement being reached between the superpowers, tacitly or otherwise, to limit the European war to levels of violence acceptable to them.

If savings are to be forced on British defence expenditure, what the defence budget buys will be least affected if the savings were made on the conventional side, since British defence policy is in fact, whatever may be declared to be the case, relatively insensitive to the size of its conventional component. And, of course, there is far more scope for cuts in the conventional side since it nowadays amounts to 95 per cent of the budget. The main danger of making large conventional cuts is not that they would shorten an already fairly short fuse between the outbreak of war and the point where strategic nuclear weapons were used, but the psychological one that the Soviets might read into such cuts that the will in Britain (and, by implication, in NATO) to resist Soviet pressure was evaporating. To avoid that risk the best point at which to cut conventional capabilities is at the same time as improvements are being made to the nuclear force.

The rational choice may of course be disputed as not so much incorrect as fundamentally misconceived. Just as a rational national energy policy choice may dictate that a certain proportion of nuclear-fired stations should be built, and this may be objected to on the principle that nuclear stations are (allegedly) peculiarly unsafe and not to be built at any cost, so obviously can there be root-and-branch objections to a defence policy with such a prominent nuclear element. The defence policy being pursued certainly does not include much of a margin of safety in the event that deterrence fails. It may also very well be judged morally inferior to possible alternatives, since it rests upon a willingness not only to use nuclear weapons but also to use them first and, moreover, to use them deliberately to kill civilians in their many hundred of thousands. But any orthodox strategy which placed more weight on conventional forces would be extremely expensive. To push nuclear weapons firmly into the defence background, to become solely a means of deterring nuclear attack, NATO would need a conventional capability in Europe at about twice the current level, with the British contribution up in proportion.* Since this article is not unreasonably predicted upon economic stringency portending cuts in British defence expenditure, of as much as 20 per cent (see Appendix), an orthodox conventional strategy calling for a substantial increase in defence expenditure is out of the question.

It may be retorted that this is no answer to radical objections and is merely a defence of the rational choice. Defending a large nuclear energy programme by showing that relying solely on conventionally fired stations would work out very expensive, says almost nothing about the viability of a programme based upon a mass of windmills and solar panels. Equally it is true that this procedure for making a rational defence policy choice is not capable of considering the merits of a British or West European defence policy modelled, say, upon the Swiss, revolving around light armaments and a citizen army.

---

*This is based upon what would be needed to match, on paper, unit for unit, the Warsaw Pact offensive capacity in Central Europe. This probably overstates somewhat the extra expenditure that would be required, but it takes no account of the cost of meeting subsequent Warsaw Pact force increases stimulated by the NATO move.

The Robustness of the Rational Choice

The rational choice made, in this instance to replace Polaris and to replace it by Trident C4†, we can now turn to assessing its chances of surviving cancellation and reaching the point of coming into service some 10 or 15 years hence.

The sign of a public capital expenditure project in distress is either cost over-run or failure to keep to schedule or both, and a prime cause of both is the unexpected technological hitch. In the case of Trident, technological problems have been guarded against virtually entirely by the purchase of as tried and tested a system as is feasible—short of buying something already on the verge of obsolescence. And buying off the shelf obviously also helps. Trying to develop a similar missile within Britain or even a greater part of the missile than simply the warheads, as is the plan, would have been exceptionally risky. And this is known, as near as can be, for certain, because the evidence of the until recently secret, homemade, Chevaline project—to make the British Polaris A3 missiles better at penetrating Soviet anti-missile defences—is before our eyes.

The Chevaline development phase alone took about 8 years (1972–80) and cost at least £1,000 million (1980 prices). The object seems to have been to retro-fit into the Polaris A3 missile a British-made front end to produce a hybrid system with most of the features of the American Poseidon missile, except that no front-end can be used to hit more than one area target.* In other words, Chevaline is a two-dimensional MIRV capable of subjecting Moscow (or any other Soviet city) to a spatially well-separated multiple warhead attack, with some warheads plunging in almost vertically and with others arriving at a flatter angle.

But Chevaline cannot be deemed a success. It appears to work, but it is coming into service very late, much nearer the end of the useful life of the Polaris system than can originally have been planned. Moreover, its cost does seem astounding, especially when set against the

---

†With the rational decision to replace Polaris taken, it is not altogether accurate to assume that the specific replacement choice, Trident C4, emerged purely on the basis of cost/benefit calculations. It was no doubt preferred over its nearest rivals—sea-based or air-based cruise missile systems—partly for reasons of this kind, but it was also preferred (in the same way the pressurised water reaction is now being preferred over the steam-generating heavy-water reactor) because it is more of a known quantity, not certain to be free of problems in the development and construction phases but more likely to escape the sort of embarrassingly large problem a relatively untried system might encounter. Students of game theory will recognise they are in the presence of a "maxi-min" decision—aiming not at obtaining the best possible outcome but at avoiding the worst possible.

*The name "Chevaline", assuming that the Ministry of Defence is only human, is itself a slight give-away. Chevaline means "horsy" or perhaps "horse-like". Poseidon, as well as being god of the sea, was also the god of horses.

£5,000 to £6,000 million that will apparently buy the entire new Trident system, submarines included. And, besides, Chevaline was begun after the 1972 Moscow Agreements, which strictly limited the number of Soviet anti-missile interceptors and which inaugurated an era of low Soviet interest in this area of military activity; in other words, Chevaline, even had it come in at a much lower cost, and more quickly, was probably unnecessary anyway.

It might be wondered why Chevaline survived cancellation when doubts are in order about the robustness of the Trident project. The answer is that Chevaline was smuggled through behind the Official Secrets Act and was thus never exposed to the buffeting open projects have to endure. If it had been conducted openly it might have been cancelled in mid-stream; alternatively, openness might have imposed better discipline on the project managers and got the thing completed sooner and more cheaply.

The Trident project cannot be conducted wholly secretly; it is too prominent politically, looms too large in the defence budget to be overlooked, and, anyway, nothing can be kept secret which involves £1,500–£2,000 million worth of key components being bought from the United States.

Two other classic project defence mechanisms are also out of reach. Military (and civilian) aerospace projects find a safeguard against cancellation in international collaboration. Involving other governments as joint participants in major aerospace projects may or may not reduce unit costs below what could have been achieved nationally; equally it may or may not improve equipment standardisation amongst allies; but it does give a kind of guarantee against cancellation. If it had been possible to build the Polaris replacement in collaboration with a foreign partner (and France would have been the only feasible contender), some new freedom from cancellation would have been won. But to set against this there would have been greater technological risk, problems about how far Britain could be open with France with respect to American technical information Britain already had, and possibly too a sense that the collaboration would have been unequal since such an arrangement would have given France a theoretical veto on the future of the British strategic nuclear force without giving Britain a reciprocal veto (since France apparently has a national capability, with or without foreign assistance, to stay in the nuclear weapons business indefinitely).

The other defence mechanism denied the Polaris replacement project is the export market: it is not inconceivable, on the other hand, that should the Trident C4 have encountered development problems in the United States, one or two voices might have been heard saying that pressing on would permit some to be sold to the British.

Of course no project can be made cancellation proof, because while it is well enough understood that in the United Kingdom almost every medium or high technology capital project with a unit cost of about £300 million upwards has a tendency to outwit its managers and to cost much more and take far longer to build than had been planned, no-one quite understands why. Ordinary inflation may play a part and so may also the absence of economies of scale: possibly if Britain required fifty nuclear power stations, or SSBNs, rather than (typically) five, the average unit cost and completion time would be acceptable.

What the Trident planners seem to have done to protect the project is to choose a route, which will be as free as possible of technical and managerial surprises, by "cloning" the earlier Polaris project, which was even for those less inflationary days a triumphant exception to the rule of budget overshooting and production delays. It is not exactly clear why the original Polaris project behaved as well as it did, but it is probably connected with the fact that its managers resisted the temptation to tamper with the original American designs.

But there is one important obstacle to a successful cloning. The marriage of the Trident C4 missile to a purpose-built sixteen-tube SSBN, which is apparently what is being proposed, has never actually been made before, even in the United States. There, the sixteen-tube Trident C4 boats now in existence were originally built for the Poseidon missile but with Trident in mind in that the boats were deliberately built over-size so that when the time came they would be roomy enough to take Trident C4 (just as the later Polaris A3 boats were built a little over-size so that they could subsequently be fitted out with Poseidon). Thus there is no complete sixteen-boat American blueprint for the British to follow. There are complete blueprints for a twenty-four-tube boat (the *Ohio* class), purpose-built to carry Trident C4, and built over-size so that it can later also take the successor missile Trident D5. But such boats will be more costly than the sixteen-tube models, and, moreover, themselves a managerial risk in that British yards have no experience of building submarines of this size.* The smallest managerial risk would seem to be associated with a sixteen-tube Poseidon-style boat with tubes reamed out to their full diameter to take Trident C4 from the beginning. The flaw in this choice would be a vulnerability to a shift in the external environment. The Trident C4 will be rather old by the time it enters service with the Royal Navy: older than the SSBNs carrying it. Soviet anti-submarine warfare techniques may have improved so much and Soviet anti-missile defences been strengthened so far that the greater range and payload of a retro-fitted D5 might be seen as necessary to maintain the British capability to wage war at the furthest extremity of violence.

*The lack of American experience at building boats of this size has been reflected in cost over-runs and completion delays: and it is too early to be certain that the *Ohio* class will not in fact itself be cancelled.

The other change since the days of the original British Polaris project is that the depressing knowledge that the British capacity to manage large-scale projects is very limited, has sunk in everywhere. The responsible planner, then, will be torn between doing what he can to protect the project from cancellation—and if this means getting on with the job cheaply and quickly so much the better—and taking out some insurance against cancellation being a complete disaster, by arranging from the beginning for the project to be capable, if necessary, of taking on a cheaper and more acceptable form as an alternative to outright cancellation. The responsible planner in this case might want to be able to offer as an alternative the conversion of the SSBN boats to SSN boats (by removing, or not building, the middle, missile tube, section) and the sale back to the United States of the missiles. The responsible planner will also know that planner's logic is not always political logic; for projects have sometimes been carried through to completion because there was no escape and for no better reason. A politically more astute form of insurance might simply be an eventual unspoken willingness to see the four-boat flotilla cut down to three.

Summary

What makes the Polaris replacement project special is not its cost (conventional weapons programmes are at least as expensive as nuclear these days) but its cancellability. Unlike Chevaline it cannot be smuggled through its secret, and unlike, say, the Stingray torpedo project, it has high political visibility and cannot expect simply to be overlooked; unlike the MRCA-Tornado aircraft there is no foreign partner, for fear of offending which cancellation becomes unthinkable; and, unlike the through-deck cruiser, export markets do not even exist in theory. The rationality of the original decision is no safeguard; the long-term cost advantages of a defence policy with a prominent strategic nuclear component will mean very little to a government anxious for immediate savings.

In the absence of a miraculous transformation of the national economy, pressure for a substantial reduction in defence expenditure cannot be long delayed. What is cancellable will be cancelled. If the Polaris replacement project is perfectly managed, perhaps a 2 in 3 chance of escaping the axe would be about right. But if, as seems more likely, the project will encounter turbulence over the choice of the appropriate size of the SSBN and its manufacture, since there is not today as there was in 1962, an obvious and proven American design to be replicated, a 1 in 2 chance may be more accurate.

* * * * *

68. 1982

Merchant Shipping and the Maritime Threat

H. G. Davy

Mr. H. G. Davy, M.B.E., is the Director responsible for Defence affairs at the General Council of British Shipping.

The phrase "We are an island nation" once immediately evoked patriotic outbursts about its implications, not least in defence terms. Nowadays it is treated as a statement of the obvious, of a fact of life with which we must live and to which no special considerations need apply. A few moments' though will, however, bring a somewhat changed reaction.

There is the fact that we still import some 93 per cent (by weight) of our requirements by sea, that the EEC imports about 60 per cent of its requirements by sea and that 98.2 mn tons of goods move between the United Kingdom and the Community annually by sea. There is, of course, also our trade with other NATO partners.

A conflict in Europe would necessitate massive supply and resupply of men and military equipment from the United States and there would be the need to sustain fortress Europe under siege with the food and essentials of life for civilians as well as troops, quite apart from home production of military supplies. In world defence terms one must add the consideration that major hostilities centred elsewhere than in Europe will surely necessitate extended lines of communication in which merchant shipping will play a major part.

Then there is our experience during two world wars. Looking at the United Kingdom alone, between 1914 and 1918 we lost 7·7 mn grt. Between 1939 and 1945 we lost 11·4 mn grt (2,570 ships) and over 30,000 merchant seamen. In 1942 the Allies lost over 6·0 mn grt of shipping— 1,160 ships all by submarine attack.

These factors surely underline the implications of the position of the United Kingdom and the dependence which it and, indeed, Europe must place on adequate maritime communications. They demonstrate also the vulnerability of these lines of communication to attack at sea, especially from submarines.

The Size of our Merchant Fleet

The United Kingdom, and for that matter most of its Allies, cannot afford to sustain a merchant fleet of a size and composition entirely suitable to military or other wartime requirements. Allied merchant fleets have to live in an increasingly competitive international environment, and thus the size of merchant fleets and their composition must reflect the basic economic considerations and ship types. We Allies are competing with each other and the United Kingdom must certainly keep in the lead in developing specialist ships necessary to achieve the maximum economic potential which, in turn, can require specialist port facilities. The growing trend towards specialist ships (e.g. containers) and specialist port facilities involves extreme vulnerability in war. So a very relevant factor is the size and composition of our merchant fleet and that of the other members of NATO.

In 1939 the United Kingdom owned 27 per cent of the world tonnage. By 1950 that figure was 21 per cent; now it is about 6 per cent. But the UK merchant fleet remains comparatively large—we are still fourth in the world league table. Much of it is normally engaged in cross-trading, which is, of course, of tremendous value to the nation in terms of invisible exports. In 1979 UK-owned merchant ships contributed £1,139 mn (net) to the UK balance of payments.

It is not only the British merchant fleet which has changed in size, so too have most of our NATO partners. The tables at the end of this article indicate what has happened in general terms. The following tables summarise developments regarding some specific types of ship.

GENERAL CARGO

	1971		1975		1980		Change	
	Ships	dwt/000	Ships	dwt/000	Ships	dwt/000	Ships	dwt/000
UK	1356	8889	1086	6508	830	4374	−526	−4515
Greece	1330	7955	1514	9629	2041	15,521	+711	+7566
USA	1042	10,393	558	5850	446	5057	−596	−5336
Balance of NATO	6062	20,676	4756	16,398	4043	14,630	−2019	−6046
Total NATO	9790	47,913	7914	38,385	7360	39,582	−2430	−8331
USSR	1507	7900	1757	9494	1793	9929	+286	+2092

CELLULAR CONTAINER

	1971		1975		1980		Change	
	Ships	dwt/000	Ships	dwt/000	Ships	dwt/000	Ships	dwt/000
UK	51	647	91	1242	74	1529	+23	+882
Greece	—	—	4	44	7	45	+7	+45
USA	75	1054	103	1628	87	1572	+12	+518
Balance of NATO	60	615	78	1211	116	2757	+56	+2144
Total NATO	182	2314	276	4125	284	5901	+98	+3587
USSR	—	—	11	69	37	274	+37	+274

It is true that the Greek fleet has increased, much of the superior tonnage registered under flags of convenience is under the control of the United States or other NATO countries and could be expected to become available to NATO if required. But so far as the United Kingdom is concerned the trend in overall fleet size and certainly numbers of ships has been downward and must be kept under continual scrutiny. The size and composition of Free World merchant fleets is a matter of general concern because undesirable dependency on the fleet of others must be avoided.

The Soviet Merchant Fleet

Western shipowners have become increasingly concerned about the build-up of the Eastern Bloc, particularly the Soviet, fleet in the past few years. The Soviet fleet is now the sixth largest in the world with over 23 mn grt. It is not, however, the total size of the Soviet fleet which causes anxiety but that of the Soviet cargo liner fleet, which is now the largest in the world.

Already in the bilateral trades, Soviet shipping carries 78 per cent of imports and exports between the USSR and the United Kingdom and 75 per cent between the USSR and Western Germany and the USSR and Japan.

This is achieved by the manipulation of cargo to their own ships, by the use of cif (carriage–insurance–freight) sale/fob (free on board) purchase terms and by enforcing artificially low rates for such balance of general cargo as the ships of their trading partners and of third flags are permitted to carry.

British and other Western shipowners recognise the desire of COMECON countries to engage in international trades. On the liner side they are prepared to encourage Soviet lines to become conference members under normal conference criteria (e.g. taking account of traffic generated by the applicant's country, the number of cross-traders already in the conference, access by other members to the trade of the applicant's country, whether or not the whole conference trade is over-tonnaged and whether it is growing or diminishing). But in recent discussions with most conferences Soviet lines have made demands which are unreasonable by these criteria, having first established themselves as "outsiders" by charging rates which are uneconomic by free market economy standards.

The Trans-Siberian Railway (TSR) is a special problem, not perhaps obvious from the viewpoint of maritime defence in a narrow sense. However, it carries as a transit operation 10 per cent of the high-quality containerised general cargo moving in both directions between Western Europe and the Far East: further TSR expansion, even up to a hypothetical 50 per cent of the traffic, is envisaged. More is said about this later.

Functions and Organisation of Soviet Merchant Shipping

In the Soviet Union the division of functions between the merchant fleet and the Navy is less sharp than in the West. The Soviet commercial fleet appears to have at least five main functions and at any moment any particular ship may be serving more than one of these:

(i) the transport of Soviet internal and external trade, the most important single function;

(ii) the improvement of the Soviet Union's hard currency position, both by earning foreign currency in cross trades and also by reducing the need to import foreign shipping services in the Soviet Union's direct trades;

(iii) direct auxiliary support for the Soviet Navy, principally involving logistic support but also possibly involving mine laying and sabotage within enemy harbours during times of war;

(iv) as an intelligence-gathering organisation;

(v) the support of Soviet political objectives by facilitating trade, including the arms trade with Soviet allies.

* * * * *

NATO Merchant Shipping and an Emergency

What would the NATO merchant fleet be called upon to do in the event of rising tension or hostilities?

It is well known that in a period of rising tension a factor of deterrence, though also of final preparation, would be the movement of men and military equipment across the Atlantic to reinforce Europe. Equipment particularly would move by sea and plans are laid for it. On the basis of military planners' requirements, sufficient allied merchant ships of various types would at present be available to carry these cargoes within the envisaged time scale and on the assumption that actual hostilities have not broken out.

The plans as to the military requirements and the availability of shipping are kept under review. But as time passes these plans need particular scrutiny to ensure that changes in the size of the allied merchant fleet, its composition, the time scale and the requirements for protection are all taken into account.

During such a period of rising tension merchant shipping would expect to be harassed in various ways. They may find strategically placed and exceptionally large exercise areas being used by unfriendly countries; their ships may be brought under close surveillance at sea— some may even be boarded under some pretext or other—they could experience problems at unfriendly ports and so on. In all such situations close liaison with governments and with NATO is vital, including, on the one hand, the availability of some form of physical protection and, on the other, complete understanding as to the commercial considerations which may face an owner in avoiding what may or may not be a hazard to his ship and cargo. This is another area where existing NATO planning must keep up to date.

If attempts to deter aggression fail, merchant shipping must expect to face merciless attack, either at long range or by close encounter, especially by submarines.

The permanent and semi-permanent installations in the North Sea would be particularly vulnerable to long-range attack and to sabotage, as would the ports generally and the highly sophisticated terminal installations (e.g. for containers) in particular. Our planning must surely rely on very restricted use of highly specialised ships requiring comparably specialised facilities, but it would not be realistic or feasible to equip the entire dry-cargo fleet with ro-ros. However, it should be noted that the handy-sized general-purpose fleet is the one which is being driven out by economic progress.

Those responsible for sustaining civilian life and production of essential goods as well as military equipment need to be clearer as to their requirements, both in commodity terms and in their transportation needs.

It may be that hostilities would not focus wholly on Europe and/or that other areas of conflict would develop. Relatively few sophisticated ports would be readily available under friendly control and, over extended lines of communication, there would probably be a great responsibility on sea transport for troop movements by comparison with air. One should reflect also on the limitations of planning between all NATO countries because of the limitations of the NATO area and the constraints on mutual assistance between such countries elsewhere. Other alliances are needed.

Our capacity to defend ourselves in the modern world, just as much as in the past, depends on economic strength as well as the amount of hardware and the number of effectives that we can put into battle. Not only military strength but political influence and intelligence, in the widest sense, are dependent on trade and shipping.

The tremendous increase in Soviet naval power in the last two decades has been matched by rapid expansion of the Soviet Merchant Navy which not only serves the requirements of Soviet foreign trade but increasingly participates in cross-trading and acts as logistic support in intervention abroad.

Operating in the Western world market economy, shipowners have sought efficiency, and in most countries they wish to remain free of government support and direction. They are now equipped with specialist ships of much greater carrying capability and lower unit cost than we were operating 10 years ago, or are operated by the Russians.

In comparison the Russians have a very large fleet of multideck conventional liners, with a growing proportion of roll-on/roll-off ships; so-called Combo ships, which carry containers and have deep capacity for break-bulk cargo with their own discharging gear aboard, and a few LASH—lighter aboard ships vessels—which are not commerically viable but are ideal for military support.

Historically, most bulk cargoes were carried by handy-sized tramp ships. Their modern equivalents are of a much larger size and more efficient economically. But their numbers are relatively fewer, certainly in the British fleet, and the rate of attrition which can be contemplated raises serious questions.

The tanker fleet has developed in much the same way but to even larger sizes. The supply of oil is obviously crucial and thus raises serious questions.

A fleet of 18 large container ships can do the work of 120 conventional ships: only 18 missiles are needed to do the work of 120 torpedoes in the last war. The implications are obvious.

Shipping and National Security

Those nations which have sought to exercise imperial power without proper application of the influence of sea power have in the end been humbled by it. Sea power, commonly referred to as maritime power to embrace all forms of air-power at sea, consists not only of fighting units but also of the strength of a strong mercantile fleet developed and maintained in peacetime. This element is the safeguard against dangerous dependency on others, the means of support of national existence and the support of the armed services in time of war.

The failure of Allied governments to consider the role of merchant shipping in terms of national security may result in the Alliance finding itself without adequate tonnage under Allied control for diversion to wartime military use and maintenance of civilian supply.

To have reached such a point in a time of tension, together with the failure to provide enough fighting ships to defend the sea lanes, means that it is too late to seek a remedy.

* * * * *

69. 1983

Strategic Weapons

Lawrence Freedman

Dr. Freedman is Professor of War Studies, King's College, London.

In the early 1970s it seems that the future of strategic nuclear weapons was one of declining significance. A common view was that the offensive nuclear forces of both superpowers neutralised each other, and could therefore be of slight relevance in everyday international affairs. Given this, there seemed little point in adding to nuclear arsenals when the increments were so expensive. It was widely assumed that successive rounds of the Strategic Arms Limitation Talks (SALT), which got under way in the last months of the 1960s, would confirm the stability of the strategic relationship and diminish the stockpiles of nuclear weapons and their international role.

By the early 1980s events had moved in quite the opposite direction to the one that had been expected. Nuclear weapons appeared at the centre of domestic and international rows in both North America and Europe. New systems were being introduced by both superpowers and yet more were under development. SALT, and arms control in general, had reached a crisis point with its value seriously questioned in the United States and the key negotiations stalled. The Reagan Administration which came into power in January 1981, contained many hawkish critics of SALT and advocates of greater US exertions in the strategic nuclear field. However, this view gradually has had to be qualified. The Reagan Administration has been forced to recognise the grave anxieties that a stress on a nuclear strategy raises within its own population as well as among its allies. From its first days in office it has found pressure from the allies not to abandon arms control negotiations. These various pressures have led to the opening of negotiations with the USSR on the intermediate nuclear forces (INF) based in and around Europe, preparations for new strategic arms talks orientated towards substantial reductions and even tentative plans for a Reagan–Brezhnev summit. Meanwhile, there have been a variety of financial, technical, political and environmental objections to specific programmes—most importantly the M-X ICBM. Lastly, there has even been a revival of questioning of NATO's policy of refusing to discount the first use of nuclear weapons in a European war.

What seems to have happened is that those pushing for nuclear rearmament, a more credible nuclear strategy for NATO, and a sceptical approach towards arms control have been unable to overcome the crude, grisly but compelling logic of the "balance of terror". This balance represents the most fundamental feature of the strategic relationship. It means that one superpower cannot launch a nuclear attack against the other without risking retaliation in kind. It has been like this since the 1950s and despite the current uncertainties it is likely to be so until the next century. Most developments over the past couple of decades have served to reinforce rather than undermine the stalemate. There has been a persistent belief in the possibility that a dramatic technical breakthrough might cause some decisive shift in the balance of advantage, allowing the successful power in rescue its nuclear strength from the constraints on its use imposed by the nuclear strength of the adversary.

If one side did manage to achieve such a breakthrough, creating an ability either to destroy enemy forces in a surprise first strike or to shoot down an incoming attack, then the strategic relationship would be completely transformed. It would be possible to contemplate a victor in a nuclear war. For this reason, many surveys of developments in strategic weaponry are informed, implicitly or explicitly, by this prospect of a quantum jump, something equivalent to the first atomic bomb, or the movement from fission to fusion weapons, or from aircraft to ballistic missiles. One of the consequences of this preoccupation is a tendency to attribute an excessive significance to technological innovations. A developing asymmetry in one particular capability, say the capacity to attack hardened targets or to protect a portion of the population from fallout, may get presented as an historic shift in the international balance of power.

*　*　*　*　*

ICBM Vulnerability

The growth in the number of Soviet MIRVed missiles could be seen as no more than an effort to catch up with the United States. It has been taken to be more ominous because of a basic feature of Soviet ICBMs as against US ICBMs: they are much larger. Throw-weight is the weight of the post-boost vehicle (warheads, guidance systems, penetration aids) that can be delivered over a given range. The Soviet SS-17 and SS-19 ICBMs have 3 to 4 times the throw-weight of the

Minuteman III and the SS-18 has up to 10 times the throw-weight. The total throw-weight of all Soviet ICBMs and SLBMs comes to some 10·0 million lb as against the comparable US figure of 4·2 million lb.

Aggregate throw-weight by itself does not mean very much. It certainly does not confer by itself any significant superiority on the Soviet Union. When the throw-weight equivalent for bombers is considered, that available to the United States is some 3 million lb, compared with 1·8 million lb in the Soviet Union. Throw-weight is best seen as an indicator of potential. Practical consequences depend on the efficiency with which the available space is used. The Americans have put a lot of effort into increasing the explosive yield for a given amount of nuclear material, and into the miniaturisation of the electronics contained in the missile nose-cone.

The most immediate consequence of high throw-weight is a high explosive yield. The Soviet Union can detonate explosions of some 25 megatons (Mt), that is equivalent to 25 million tons of TNT. This is a frightening prospect, but it is not a lot more frightening than being attacked by weapons of 1 Mt as nuclear explosions are subject to diminishing marginal returns in terms of destructive effect. For attacks on populations, the concentration of the populations of both superpowers in a limited number of large cities, means that extra megatonnage confers no serious advantage.* If, however, the warhead can be split up through MIRVing, then a different sort of target structure can be attacked.

During the 1960s it was felt that placing missiles in submarines or hardened silos would keep them invulnerable. To be sure of destroying individual ICBM silos at least two attacking ICBMs would be needed, so the exchange ratio would be unfavourable. MIRVing initially did not help because the process involved a loss of accuracy and yield. With improvements in guidance systems, the most significant and persistent technological phenomenon of the 1970s, the capacity of even small warheads to destroy hardened targets was greatly enhanced.* The large size of the Soviet warheads means that they can each accommodate a number of re-entry vehicles of high yields that can individually threaten US ICBM silos without placing excessive de-

---

*As destructive power does not grow proportionately with yield, a better measure is equivalent megatonnage which is expressed as the two-thirds power of yields below 1 Mt and the square root of yields above 1 Mt. If the US bomber force was equipped with gravity bombs to the exclusion of Short-Range Attack Missiles (SRAM), then it would ensure that US equivalent megatonnage was greater than that of the USSR.

*The capacity to destroy hardened targets is known as lethality. This is directly proportional to yield 2/3 and inversely proportional to CEP^2. Circular error probable (CEP) is a measure of accuracy. This formula has the unfortunate mathematical property of creating an infinite lethality when CEP reaches zero. However, it does indicate the great sensitivity of lethality to improvements in accuracy.

mands on accuracy. Although the United States has been reinforcing its missile silos, this is more than compensated for by the gradual improvements in Soviet accuracy which now approach American standards.†

There is general agreement that the Soviet Union is approaching a capacity to destroy virtually all of the US ICBM force in a surprise attack using a relatively small portion of its own ICBMs. The United States will not have a similar capability. Even when the Mark 12A front-end is in service, the match of warheads to targets will still not provide a complete counter-ICBM capability.

* * * * *

A Military Break-Through?

This question of comparative counterforce capabilities has become the cause of great anxiety. In a much-reported speech in Brussels in September 1979, Dr. Henry Kissinger suggested it represented a fundamental change in the strategic situation of the United States and threatened the credibility of the nuclear guarantee. It is not self-evident that this need be so, particularly without complementary improvements in other Soviet capabilities.

There have been rumours of Soviet break-throughs in particle-beam technology that would permit an effective ballistic missile defence or at least serious interference with space-based support systems relevant to reconnaissance, navigation, command, control and communications. The possibilities for directing high-energy beams to targets instantaneously and with a 100 per cent destructive efficiency will command attention in the research laboratories of both superpowers. However, whatever the potential for the future, it is extremely difficult to see how energy could be generated in sufficient quantities, particularly if the system were space-based, or how reliability could be ensured, or how the problems of covering large areas of space, tracking and identifying individual objects and overcoming a variety of possible (and cheaper) countermeasures could be overcome. Typically, in the enthusiasm for the concept, the promise is underlined while the obstacles are played down.

Nor does it seem likely that any Soviet civil defence programme will be able to reduce the impact of a nuclear attack to tolerable levels. The Soviet Union has taken civil defence more seriously than the United

†Despite its critical importance, little official information is released on accuracies of either Soviet or American ICBMs. Best public estimates suggest that the best Soviet missiles presently attain accuracies just about 1,000 ft while the best US accuracies are measured in hundreds of feet and may soon be measured in tens of feet.

States, where the consensus opinion has been that protection against nuclear explosions can only be limited. Little could be done to spare most people the consequences of an all-out nuclear attack. The possibility of mass evacuation from cities has been mooted. There is no evidence of actual exercises in the USSR to assess the feasibility of this sort of operation. It is doubtful that the organisational and social problems created by mass evacuation to unprotected camps would be a sensible move in a nuclear crisis. The only form of mass civilian defence which might make some sense would be fall-out shelters, but only in the case of a limited attack not directed at cities.

Bombers and SLBMs are not expected to face the same vulnerability problems of ICBMs. Alert bombers can take off on warning and be recalled if the warning turns out to be false. ICBMs can also be launched on warning. This prospect might well weigh heavily with a Soviet planner contemplating an attack, especially as there is some evidence that they themselves are drawn towards this. As it would involve supreme dependence on warning systems to trigger a nuclear war this is not a policy to be encouraged.*

Bombers, unlike missiles, face the problem of dense active defences. With defence suppression missiles, such as SRAM, and air-launched cruise missiles which, though individually vulnerable because of their slow speed, can saturate air defences or take detours, bombers will still represent a potent attacking force. Anti-submarine warfare has still a long way to go before it will reach a state permitting high-confidence attacks on large numbers of submarines. Those at port, however, are extremely vulnerable and this is an area of significant Soviet vulnerability because of the low percentage of its force on patrol at any time.

* * * * *

A New Arms Race?

Nevertheless, it is the Americans who have become most preoccupied with the vulnerability issue and the question of whether comparative counter-force capabilities now provide the key to the strategic balance. The ideal weapon is now seen to be one which is both capable of attacking protected targets yet on a launch platform itself relatively invulnerable to attack. Obviously both sides cannot achieve both objectives at the same time. The attempt to do so could well be a recipe for an arms race.

*Using a depressed trajectory missile from a submarine close to the United States it might be possible to attack bomber bases without adequate warning time for the bombers to escape. It has been suggested that this hypothetical capability would be an appropriate topic for strategic arms control.

Up to now it has been assumed that ICBMs offered high accuracy and high vulnerability, whereas SLBMs were exactly the opposite. Current US plans envisage improving the accuracy of SLBMs and the invulnerability of ICBMs.

The most difficult of these objectives is to improve the invulnerability of ICBMs. The design of a new missile—the M-X—with an impressive counter-force package of 10 MK 12A warheads has not caused serious problems. The major difficulty lies with the basing mode which, in addition to providing survivability, must also be cost-effective, environmentally tolerable and capable of verification in an arms control regime. By the end of the Carter Administration some 30 different proposals for basing had been evaluated, before settling on a system that would have 200 missiles on launchers moving between 4,600 shelters, along interconnecting roadways. However, the Reagan Administration continued the tradition of an annual summer review of M-X basing. It became worried at the prospective delays caused by political and environmental objections as well as the cost. A number of ideas, such as putting M-X to sea on a large number of small vessels, were revived. In October 1981 President Reagan announced that the first batch of M-X ICBMs would be fitted in reinforced Titan and Minuteman silos, pending a more survivable long-term solution. Congress baulked at the expense of the reinforcement. The latest scheme involves packing silos close together so that through the "fractricide" effect, the attacking missiles will neutralise each other rather than the targets.

Meanwhile the Administration is accelerating the pace of development of submarine-based force. New Trident I missiles (with 8 warheads of 100 kt each) are being fitted onto old Poseidon submarines. A completely new type of submarine, the Ohio class, is being produced to carry 24 Trident I missiles at greater speed and less sound. By the end of the decade, the next generation SLBM, the Trident II, will be in service with 14 warheads and, unlike Trident I, a hard target capability.* It was also decided to convert large numbers of general purpose submarines, and even battleships, to take sea-launched cruise missiles (SLCMs). Although some 3,000 of these may be deployed by the 1990s, most of these will have conventional warheads and be used as anti-ship weapons. Up to a quarter will be able to launch nuclear attacks with great accuracy against land targets.

Finally, the Administration has reversed President Carter's 1977 decision not to proceed with a new manned bomber, the B-1. It was then felt that the extra lease of life provided to the B-52 force by air-

---

*The Trident II missile is to be purchased by Britain as the successor to its Polaris force.

launched cruise missiles would be sufficient. The new bomber will probably be a revised version of the B-1.

The likely Soviet response to all these moves is uncertain. In the past the USSR has on occasion developed unique systems, but these reflect distinct requirements rather than technological superiority (a good example of this is an anti-satellite satellite, active testing of which resumed in early 1981, which is comparable to a type rejected by the United States in the early 1960s; the United States has been developing superior systems). Any sense of overall superiority results from the greater number and larger size of individual systems. Its main technical effort over the past decade has been devoted to MIRVing, following a US example which it has been able to exploit to greater effect.

Little effort has been devoted to new forms of launch platform or basing. Harold Brown's last report as Secretary of Defence noted: "We have been expecting the Soviets to develop a new long-range bomber for several years", but there is yet to be conclusive evidence. There has been a long-standing interest in mobile ICBMs. The SS-16 was developed in the mid-1970s but it does not appear to have been particularly impressive and deployment would be specifically prohibited under SALT II (though the two-stage version, the mobile SS-20 is now the backbone of the Soviet theatre force).

The first of a new class of large submarine, the *Typhoon*, has been launched with a new long-range missile under development for its 20 tubes, making it comparable to the Trident system. Reports suggest a displacement of 30,000 tons (compared with 19,000 for the Ohio class). The reason for this enormous size may be to accommodate a large propulsion unit to provide for greater speed and depth, which would suggest a continuing preoccupation with the quality of Western anti-submarine warfare capabilities.

All these uncertainties compound the political problems surrounding strategic arms control. None of the possible technical developments discussed above would actually have been halted by the prompt ratification of SALT II, which deals with quantity more than quality (except that it would only allow for one new type of ICBM each). Even opponents of SALT seemed more concerned that the Treaty might lull the public into a false sense of security than over particular inhibitions on weapon development.

The Reagan Administration has taken time to sort out its own arms control. It took care not to take irreversible steps which would undermine the Treaty. The predilection of President Reagan appears to be for "deep cuts" rather than ceilings. In November 1981 he proposed to the USSR a "zero-option" of no intermediate range nuclear forces (such as ground-launched cruise missiles and SS-20s). His intention for strategic arms has been indicated by his desire to rechristen them the

Strategic Arms Reduction Talks (START). Whether his intentions will at all correspond with those of President Brezhnev remains to be seen.

All this uncertainty is creating much anxiety over the future. The continued difficulties with arms control, the regular consideration of new systems and technical tricks on both sides and the argument over doctrine and theories of limited nuclear war, have pushed the nuclear issue well to the fore of international politics in an unusually disturbing manner. Yet is is important to remember that despite all the expense and ingenuity devoted to nuclear weapons over the past decades, convincing answers have yet to be found to the problem of developing a nuclear strategy that offers the prospect of a decisive advantage and removes the risk of devastating retaliation. Given this, there is no reason to believe that the stalemate of the balance of terror is close to being broken.

70. 1985

Space: The Military Applications Today and Tomorrow
Group Captain T. Garden

Group Captain Garden is currently Director of Defence Studies for the Royal Air Force. The views expressed in this article are those of the author, and do not necessarily reflect official opinion.

On October 4, 1957, the successful launch into orbit of Sputnik I startled the West, by demonstrating that the Soviet Union had an advanced capability to operate in space. The repercussions for deterrence, and NATO strategy were profound. This was not as a result of concern about future war in space, but because of the potential use of ballistic missiles to threaten the continental United States with nuclear weapons. More than a quarter of a century on, it might have been expected that the military exploitation of space would be both comprehensive and perhaps of over-riding importance. It is of interest to compare the developments of military aviation over a similar period, in the much less rapidly advancing technological period of the early part of this century. In 27 years, aviation advanced from the Wright Brothers first steps into the air, to a worldwide use of air power for offensive, defensive, reconnaissance and transport operations. The use of space has advanced in a much less dramatic way. Military satellites have been developed and deployed, often as a more economical way of carrying out a particular role. On the civil side, manned and unmanned systems have extended scientific knowledge about the earth, the solar system and deep space, but even in this field the level of activity has been patchy. The exploitation of space is undoubtedly an expensive and difficult operation. Is it so expensive and difficult that it will only have marginal strategic importance for the foreseeable future, or is it the new strategic high ground as some would suggest? To answer these questions, we need to look at the military applications of space today, and how they may be developed, and also at the proposals and possibilities for novel systems in the future.

Military uses of Space in 1985

Satellites are used extensively by the Soviet Union and the United States for military purposes. They can be launched at short notice for specific missions, or can be placed in medium or long term orbiting positions. With the exception of the Soviet anti-satellite devices, there is no evidence to suggest that any of the satellites currently deployed are armed. Satellites may be used for reconnaissance, communications, navigation, meteorological purposes, geodetic survey and anti-satellite operations.

Reconnaissance satellites

Military reconnaissance satellites can be used for photographic reconnaissance, electronic information gathering, ocean surveillance or for early warning purposes. Around 40 per cent of all military satellites launched are used for photographic reconnaissance. High resolution is obtained by returning the film to earth for processing. Rapid, but less detailed reconnaissance can be achieved by in-flight processing and data transmission to earth. Unclassified estimates of the likely resolution achievable by such systems are of the order of 15–30 cm, which would be more than enough for identifying individual pieces of military equipment deployed. The USA, the USSR and China all operate reconnaissance satellites. France and Japan are developing a capability in this area.

The detection and analysis of electronic signals by satellites can be achieved by adding appropriate sensors to photographic reconnaissance satellites, or specialist vehicles may be used. Little is published about the capabilities of either the American or Soviet electronic surveillance systems. The large area covered by direct line-of-sight gives such satellites a capability to detect signals across the electromagnetic spectrum, with the appropriate receivers. These can then be relayed by downlink to earth stations.

Another specialist reconnaissance role is that of ocean surveillance. Satellites to detect surface ships and submarines are of interest to both superpowers. Detection can be made by integrating a number of different sensor methods: visual light, electronic radiation and radar.

Finally, reconnaissance satellites are of strategic importance in the early warning role. The use of satellites can extend early warning time to around 30 minutes. The sensors detect an enemy missile almost immediately after launch, through the infra-red radiation of the rocket exhaust.

Reconnaissance satellites operate by detecting a particular part of the electromagnetic spectrum from radio waves, through radar to infra-red and visible light. The frequency band chosen, and the sensor material available, will condition the resolution, and the degradation from atmospheric conditions. Micro-processor developments will enhance the sensor capabilities significantly in the near term. Atmospheric degradation, and ground and sea clutter remain limitations.

Communication satellites

Satellite communications are becoming a vital part of the worldwide civil telecommunications network. The vastly increased line-of-sight range from a space system makes for a much reduced re-broadcasting network. For military uses, satellite communications are allowing very long range control of the battlefield at the lowest level. This may not necessarily be an advantageous development. The Falklands conflict of 1982 demonstrated the importance of satellite communications. Satellites are deployed for command and control of strategic forces, worldwide communications, data relay from other satellites, and for tactical communications. Signals can be carried on frequencies throughout the electromagnetic spectrum, including laser light.

Navigation satellites

Satellite navigation systems have been in operation since 1960. Initially, the principal purpose of the satellites was to act as an in-flight up-date for inertial navigation systems on strategic missiles. More recent developments allow worldwide three-dimensional position fixing for all military applications. The Soviet Union will be deploying GLONASS, and the United States NAVSTAR GPS in the immediate future, to give a universal navigation facility to their forces.

Meteorology

The widescale usage of civilian meteorology satellites has obvious military applications. Current satellites can provide prompt information of weather parameters on a worldwide basis for any military operation. Such information has a bearing on military planning, reconnaissance priorities and navigation of ICBMs.

Geodetic survey

Satellites specifically designed for geodetic survey can provide the essential mapping for the guidance of both ballistic missiles and cruise missiles. The difficulties in realistic testing of missile systems over their wartime orientations has heightened the importance of such survey systems.

Anti-satellite weapons

Both the Soviet Union and the United States have been carrying out research and development on anti-satellite weapon systems. Their approaches have been significantly different. The Soviet Union began testing satellite interceptors in 1967, and has continued a programme since then. The anti-satellite weapon is launched in the same way as a satellite, in order to achieve a close pass to the target. When within range of the target satellite, the anti-satellite interceptor is exploded. While the United States has been conducting research along similar lines, more interest has been generated by its programme to develop an anti-satellite missile which could be launched from an F-15 aircraft at high altitude. Other possible avenues which are being explored include the use of the electro-magnetic pulse (EMP) of an exo-atmospheric nuclear explosion. Particular sensors of reconnaissance systems may also be significantly vulnerable to ground or space-based laser systems.

Current Strategic Importance of Space

Such a review of the developments in the military use of space so far, suggest a relatively restrained use of the new high ground. Those systems deployed are extremely vulnerable to anti-satellite technologies which are either available now, or are soon to be available. The developments have tended to reflect space's ability to provide a better and more cost-effective service in peacetime, rather than a more dependable wartime facility. When considering the strategic importance of space today, the interaction of the vulnerability of the systems against their importance must be examined. There is no doubt that, in the West there is a tendency to increasing reliance. This dependence may not be as significant for the Soviet Union. Reconnaissance systems can be complimented by their use of manned aircraft and drones; loss of navigation facilities need not be critical; geodetic survey is essentially a peacetime activity; and meteorological information need not depend on space-based systems. The weak link is the increasing dependence on communication, command and control, and strategic warning through satellite systems. If commanders in peace come to rely on the assured ability to control operations through communication and data transfer via space, they will be offering a vulnerability to be exploited by the opposition. For this reason, research on offensive anti-satellite systems is bound to continue, as must development of defensive measures for irreplaceable systems.

Space and Strategic Defence

If, after 27 years in space, man has not made it another battlefield, it is not for the lack of imagination. During the 1960s, Robert McNamara looked for ways to limit the damage to the American homeland in a future superpower conflict, so as to enhance deterrence. Among the measures considered was the development of anti-ballistic missile (ABM) systems. Yet after significant expenditure, he concluded in 1965 that the enemy could always frustrate any attempt at strategic defence at considerably less cost than the measures taken by the United States. Although we were not privy to the debates which undoubtedly took place in the Soviet Union, it is likely that they also concluded that effective ballistic missile defence (BMD) was likely to founder at that time on financial and technical grounds. Deployment of systems on both sides was likely only to lead to increased deployment of ICBMs.

It was, therefore, in the interests of both the superpowers to prevent wide-scale deployment of ABM systems, and as a result the ABM treaty was negotiated as part of the strategic arms limitations talks in 1972. The SALT 1 ABM Treaty prohibited each country from deploying ABM defensive systems at more than two sites. The 1974 protocol reduced this limit to one site on each side. In the event, the United States has no system deployed, and the Soviet Union has a light ABM ring around Moscow which consists of 32 Galosh missiles.

Ballistic missile defence foundered in the 1960s because the capability which the technology could provide was of dubious effectiveness either in protecting cities or even in a limited aim of protecting missile sites. The cost of deployment would have been very high, given its limited effectiveness, the enemy could counter at much less cost by deploying more ICBMs or more warheads. The effect on the strategic balance, and on stability, was uncertain. In 1983, the debate was re-opened:

> Let me share with you a vision of the future which offers hope. It is, that we embark on a programme to counter the awesome, Soviet missile threat with measures that are defensive. ... What if free people could live secure in the knowledge that their security did not rest upon the threat of instant US retaliation to deter Soviet attack; we could intercept and destroy strategic ballistic missiles before they reached our own soil or that of our allies?

President Reagan: Speech to the Nation, March 23, 1983.

> Should this conception be converted into reality, this would actually open the floodgates of a runaway race of all types of strategic arms, both offensive and defensive. Such is the real purport of the seamy side, so to say, of Washington's "defence conception".

General Secretary Andropov: Pravda, March 26, 1983.

The reason for the re-opening of the debate was the promise of the new BMD technologies. New technologies for space-based systems appeared to be offering the prospect of a real defence against nuclear weapons. But can these new technologies promise a strategy, which enhances stability, and can be afforded, or are we merely seeing history repeating itself? Is the slow development of the military uses of space a reflection of its inherent irrelevance to future conflict? To seek the answer to this question, we can examine the well documented US proposals for a ballistic missile defensive system of the future.

The High Frontier

The main elements of a next generation BMD system are summarised in the High Frontier study. A layered defensive system, which seeks to destroy most hostile ICBMs in their early boost phase, is proposed. Those remaining would be re-attacked during their mid-trajectory phase. Those warheads penetrating the first two layers would be attacked during the terminal phase by ground-based ABM systems. Civil defence measures complement this defensive programme. The study claims that:

> We can deploy in space a purely defensive system of satellites using non-nuclear weapons which will deny any hostile power a rational option for attacking our current and future space vehicles or for delivering a military effective first strike with its strategic ballistic missiles on our country or on the territory of our allies. Such a global ballistic missile defence system is well within our present technological capabilities and can be deployed in space in this decade at less than other options that might be available to us to redress the strategic balance.

Not everyone shares this degree of confidence in such a global BMD deployment. In early 1983, President Reagan set up a commission to review the strategic modernisation programme of the United States. Among other topics, the Scowcroft Commission looked at the question of BMD. They found that:

> Applications of current technology offer no real promise of being able to defend the United States against massive nuclear attack in this century. An easier task is to provide ABM defence for fixed hardened targets, such as ICBM silos. However, even this will be a difficult feat if an attacker can use a large number of warheads against each defended target.

* * * * *

Strategic Implications

Even if it is assumed that all the technological problems can be overcome, at a price which is affordable, what are the implications for the security of the West? None of the proponents of BMD, make the claim that it can ever produce a 100 per cent effective defence. The exact per centage of "leakage" is a matter of some debate. At the upper end of effectiveness, a full deployment would protect adequately against all contingencies except a massive retaliatory counter-city attack. At the lower end of the assessment, the system would merely reduce the vulnerability of the fixed land-based ICBMs from a pre-emptive first strike. Any assessment of the effect of BMD deployment on stability must depend on the prior assessment of effectiveness.

If a high "leakage" rate, say greater than 10 per cent of incoming missiles, were found to be the likely value, then deployment could not be cost effective. The enemy could always counter more cheaply by increasing his missile numbers. The vulnerability would only be that of land-based systems, and the continuing invulnerability of submarine-launched ballistic missiles would ensure that deterrence continued on the same basis as before.

If the BMD deployment appeared to hold in prospect a real change in the strategic balance, what would the effect be? A simultaneous deployment by both sides of an effective BMD system, might make the possibility of fighting a conventional war both more likely, and more acceptable, in that the risks of unacceptable nuclear destruction of the homeland had been reduced. Should one superpower consider that the other might achieve deployment first, then the prospects for stability could be of concern. The nation losing the race for deployment might feel so threatened, that it needed to take pre-emptive action against the BMD system. Certainly, both sides will feel the need to investigate the technologies, to have the capability to deploy counter-measures, should the other side produce an operational BMD system.

The deployment of only a moderately effective system by either of the superpowers would have significant implications for lesser nuclear powers. In as much as the independent decision centres of Britain and France enhance NATO deterrence, the reduction of the credibility of these systems would have a negative influence on stability. It can be argued that the advantages of acquiring small nuclear weapon capabilities would be reduced to an extent whereby other nations would not wish to proliferate. However, this is by no means necessarily the case, as the pressures on proliferating states come from other lesser powers, who would not be deploying their own BMD systems. Only if the superpowers took on the role of world policeman could their BMD deployments have a positive effect on countering nuclear proliferation.

In summary, because perfect defence is unattainable, nuclear deterrence, through the threat of massive retaliation, should remain effective. Depending on the level of BMD effectiveness, it might be that the prospect of limited war between the superpowers became possible again. In any event both would need to deploy appropriate counter-measures, which could include significant increases in their nuclear systems.

* * * * *

The Future

The ABM Treaty came out of a mutual realisation that neither superpower could benefit from the deployment of a BMD system in the 1960s. Technology may or may not be able to provide a reasonable degree of protection against ballistic missiles. It does not appear that it will be able to do it cheaply. The strategic implications of deployment of a BMD system are not clear. It may be good politically to be able to assure populations that money is being spent to reduce the risk of their nuclear annihilation, but if this makes war more possible, it is counter-productive. The abrogation of the ABM Treaty could signal another twist in the downward spiral of superpower relations, and thus add to the possibility for conflict. Nevertheless, given that both sides are carrying out research in this area, each must develop appropriate counter-measures against the deployment of a BMD system by the other.

It seems that the train of events is not greatly dissimilar from those in the 1960s. Technology promises some capability, but at a price which may not be affordable. The implications for stability are again ambiguous. The result may then be once again that both sides decide eventually, having conducted sufficient research, that it is in their mutual best interests to abide by the ABM Treaty.

Space offers no easy option; it offers no sanctuary and no automatic control of the earthbound battlefield. Unlike the development of air power, it has not yet shown signs of conferring advantage on the side which can control it. Indeed the history of the military applications of space so far, suggests that it has more pitfalls than advantages. Over reliance on vulnerable space-based systems, which may be available in peacetime, can reduce effectiveness when they are destroyed in the early minutes of any war. Development of future systems will involve financial undertakings, which given finite resources, may threaten far more urgent military needs. Finally, were success to be achievable in space-based defensive systems, it is not clear that the prospect of stability would be enhanced.

71. 1986

British Defence Issues
Admiral Sir John Fieldhouse
Chief of Naval Staff and First Sea Lord

This article originated in lectures on The Year Ahead given at the RUSI from October 1984 to June 1985.

This article is set against the background of a very uncertain world: uncertain and in many ways exciting too. Defence is but part of many complex measures that any country needs to take to ensure that its way of life is sustained and its fortunes improved. The world community of which Great Britain is a member is in the course of rapid change in many fields, and for some time we seem to have been approaching the limits of our planet's capacity. Burgeoning population, food production, availability of resources and effluent control and disposal, the continuing recurrence of national disasters, some self-inflicted, all underline the demands upon the self-control and co-operation of the world community. Yet "community" is too comfortable a word in some ways and, despite the very great progress which has been made in areas such as India's green revolution or Third World birth control, and despite the vital opportunities which Emerging Technology (ET) offers, the scene is set for continuing unease, and inevitably, with it some sharp conflicts of interest. Tension, spanning war, insurrection and terrorism will threaten peace in many places. Compounding this is the ideological clash apparent in so many political and religious areas. There is a climate of instability and uncertainty and it is the need to contribute to containing and counteracting this which must be one of the driving forces underlying British defence issues. Effective defence is about providing stability and reassurance in a troubled world: and to achieve that does, I suggest, require the professionalism, flexibility and readiness which our defence effort, uniformed and civilian, can muster.

I intend to discuss the relationship between defence and other aspects of government, and to bring out the time factor in defence planning, particularly affecting men and hardware. I will touch on some aspects of strategy and features of technology, before turning to the ever present challenge of resources: not least the need to harness wider national resources in some measure to support our defence requirements.

Defence and Government

It is clear that increasingly defence must be closely linked with other aspects of government. There are the obvious relationships: internal security and the direct safeguarding of the country's infrastructure; there are close ties with ministries concerned with the safeguarding of offshore resources, and further afield, all three services in carrying out their many roles, work in concert with the Foreign and Commonwealth Office. Next is the relationship in terms of resources: not only priorities between various calls on the nation's wealth, but also ensuring that national resources are used to best effect. There are important interfaces *within* defence: the links between the three services, for instance, have never been closer and there are few situations in which a military operation would not now be tri–service. This was abundantly proved during The Falklands Campaign. It was reassuring after such a long period of comparative peace to find that our plans and expectations came to fruition quickly and effectively when tested.

We have subsequently been examining and learning the lessons of The Falklands and building on them: one particular lesson is to ensure that this ability to act jointly is formally recognised in both exercises and in our command arrangments. It is a witness of this close inter-service relationship that I, as the First Sea Lord, feel able to write on defence related issues in this way. Indeed, I have no option, since I must be able to relate sea power to defence in the widest sense and to look upon sea power as being totally complementary to, and interactive with, air power and land power. All three prime areas of military capability are essential to effective defence and thus in generating stability. Victory will continue to be won through land power supported by air and sea power—and the maintenance of continuing peace and stability is, in itself, victory. Defeat can, however, be attracted, possibly over a period of decades, by not giving enough weight to air and sea power. Sometimes there is too much loose talk about destructive inter-service rivalry. That is not my experience in practice: we work extremely closely and effectively together and we will ensure that this continues, for it is essential.

Future Defence Issues

The next 12 to 18 months will see the continuing and relentless evolution of technology. Nearly every technological trend needs to be taken into account when considering the best shape of our defence programme. And here is a defence issue which is a continuing dilemma. Against the moving backdrop of world events and the fast developing technological scene it is, necessarily, extremely difficult to plan hardware for all three services. Not only does such hardware take several years to come to fruition through research, development and production, but it must also then remain in service *and effective* for a considerable number of years. I *mean* a considerable number too because, for instance, from the first gleam in the eye of the naval designer a class of say twelve ships will probably not finally conclude their active service for half a century. We must also remember the time factor relating to trained manpower; it takes at least ten years to instil the training and experience in senior ranks or rates which is essential to professionalism.

However, with these facts, two truths emerge, which merit special mention. First, in defence procurement for all three services we have to achieve maximum flexibility. By this I mean, achieve a range of features not only in basic design but also in concept. For example, the tactical flexibility of air power can be increased significantly by in-flight refuelling and this is a feature of our defence planning which is getting reinforced emphasis. On the naval side, it remains the policy to ensure that our ship designs incorporate a good measure of all round capability so that any ship can carry out the many basic tasks required at sea as part of our defence requirements. The second point concerns our men. The Falklands Campaign reaffirmed how essential is the high quality of service manpower if we are to be effective. For the future I see no change in the need to recruit very high quality people and to give them the requisite training and experience. This must instil in them the ability to think flexibly and use their initiative—facets of the military vocation on which we pride ourselves but which I believe sometimes come under threat in these days of computer aids, good rearward communication and automation. We must ensure job satisfaction to retain men in our volunteer services if we are to get the best out of the hardware we have. However much we try, on the day the hardware will not be precisely what we would have wanted; some will be obsolescent, some will be only newly tried and it will be the ability and determination with which it is handled in all three areas of warfare which will mean the difference between success and failure.

Strategic Concerns

Now for one or two aspects of strategy which I have chosen because they are current. In general, I see a reassuring continuity in our strategic thinking: it has after all been successful for a significant period during which it has certainly been tested. Part of that success has been our independent strategic deterrent, and this underlines the importance of the firm decision that there should be a replacement for the Polaris force. Because of the success of nuclear deterrence and the part we have played in developing international acknowledgement that this ultimate weapon plays a vital part in peacekeeping through the acceptance that it must never be used, I remain in no doubt that this country should not give up its independent strategic deterrent, I believe that as a nation with our unique experience and reputation for responsibility we have a long-term role to play in this increasingly complicated world of balanced deterrence in which, for better or for worse, we live. I believe our Western Allies and, in honesty, many other countries would wish that we kept this independent voice at the highest level of military capability; a capability whose sole purpose is to provide that balance which prevents it ever being used; and I would remind you that several, including some less than responsible, nations are likely soon to have nuclear weapons. Our independent strategic deterrent not only contributes to world stability and reassurance but protects us from nuclear blackmail whatever the source.

Staying with strategy, the North Atlantic Treaty Organisation must continue to play the crucial part in providing peace in the NATO area to which we have grown accustomed over the last 40 years. It is by no means a full peace, but it does provide an essential balance at the European pressure point and thus helps to avoid the dangers of a world conflict. It is very firmly in this country's interests to be a member of NATO and to give full support to that organisation, I see no change to that. Equally I see no change in, or need for NATO to be concerned about, its boundaries. Although these are not recognised by the Soviet Union, who will most certainly take action beyond them if necessary, nevertheless they are clear cut and there is no doubt in anybody's mind within the alliance or outside it about NATO's formal commitments. I do, however, welcome the growing sensibility of the Western Allies, and indeed of other countries too, such as Japan, to the wider interests that we all have far beyond NATO's borders. This has been well recognised in recent Defence White Papers and indeed Mr. Heseltine spoke of it in his address to the RUSI last October. Another important strategic point is that within NATO we have a classic military posture, a front line in Europe supported by an Atlantic Axis and with maritime/land flanks to the North and in the Mediterranean. There must be no weak link in relation to the others in this interlocking structure.

A further aspect: there is a growing and inevitable interplay between political and military considerations in defence and operational planning. The political requirement for military power can be at odds with the military requirement. To be more precise: in the Central Region there is a need for the Allies to stand together as far forward as possible, and this is clearly understandable in political terms, but inevitably it raises problems militarily, where the need for the commander on the ground to be able to act flexibly is of great importance. Another example is the use of maritime power, on the flanks of NATO. In the Norwegian Sea in the early days of tension, amphibious capability provides the means of deploying marines towards the Northern Flank and to have ships well forward in the Norwegian Sea not only as an outward sign of resolve but also carrying out important surveillance and warning tasks. Equally important will be the flexible use of air power in that area. These are measures which contribute significantly to deterrence in a period of rising tension and they have substantial political attraction. Militarily, however, such measures require risk judgments. For example, sea and air control in the area would be vital to ensure that the forces involved had a sensible level of supporting capability.

Political imperatives are facts which the defence planner has to recognise. Our business is about deterrence and avoiding the calamity of its failure. Deterrence is very much a meld of the military and political. The effectiveness of deterrence is measured through the perception of the opposition. This is a complex mix of certainty and uncertainty. Polaris, for instance, is a strategic certainty in terms of capability and readiness, but an uncertainty with regard to its whereabouts and, in the mind of the opponent, the precise circumstances of its use. Equally, conventional deterrent moves need to send clear strategic s gn. ls whilst retaining a wide range of tactical options. This approach to deterrence may have to be, to a degree, at the expense of pure military considerations, but this must be a recognised and calculated factor within the wider issue of keeping the peace. A difficult part of assuming an effective deterrent posture lies in assessment of risk. Military units must be put at risk to fulfil their role and to show resolve. There will be danger of damage and loss, but the aim is to ensure that the application of force contributes to deterrence or, if necessary ultimately, achieves victory. In the context of deterrence there is great advantage in being able to deploy a capability without commitment. Both the high seas and international air space provide this opportunity and enable us to deploy forces in the shape of aircraft, surface ships and submarines into areas which are totally international and which thus facilitate the sending of clear signals without the shackles of commitment sometimes involved in land based deployments.

A final strategic point. A long-term planning guideline was recently agreed by NATO relating to the concept of the Follow On Forces Attack (FOFA). It seems to me to be sensible to use technological developments to enable us to improve our capability to strike at the supporting lines of any hostile attack. Indeed, this decision highlights the attraction to the Warsaw Pact of attacking the Atlantic Link. This is where the bulk of NATO's follow-up forces comes from, and we have seen for years the dangers of a threat to that link. Here technology has at last given us the opportunity to talk strategically in terms of "tit for tat". It is a readjustment of balance of a major nature and I welcome it. In the broad strategic scene I welcome also the emphasis which is now being placed on sustainability. History teaches us that more often than not a conflict becomes longer and more drawn out than expected, indeed it is surely part of deterrence to ensure that one's defence capability can be sustained over a period of time. This is tangible evidence of will to fight and fight hard. It lends added weight to the concept of effective reinforcement as well as to the sensible and measured use of stockpiling and reserve forces as complementary parts of Alliance sustainability.

Emerging Technology

Turning to Emerging Technology (ET), a relatively simple platform can now launch clever weapons which makes the two an extremely practical and formidable combination. We saw some of this in the Falklands: for instance, the Harriers which so notably won their spurs. However, we must not forget either simple capabilities or reliability, each of which play an important part in our business. Some years ago during the confrontation of Malaysia we found ourselves embarrassed with sophisticated frigates which did not have small calibre guns to deal with the sampans and other small craft which the opposition were using to some effect. It is often very much part of flexible response for all three services to be able to deploy a basic, reliable capability which can be seen clearly and is adequate to the threat. There is a subtle and elusive balance between the quantity and the quality of our defence forces. I know my fellow Chiefs of Staff have equally difficult debates, but on the naval front, for instance, I am investigating the evolution of the patrol vessel and am supporting a policy to move towards simpler mine counter-measures vessels and thus increase the numbers we can purchase. Quantity and quality might increasingly become a factor in the manpower field where a tendency in all three services to rely more heavily on reserves represents the pressure to sustain numbers even at the expense of training and experience. I am concerned lest technology should lead us and shape concepts and policy rather than vice versa. This is sometimes very difficult to avoid, but it is a major step at least to remain aware of the problem, and to strive to keep a firm grip on what technology generates and ensure that it does not lead us by the nose. Potentially it can blur defence issues dangerously in areas such as the increasingly complex conventional/nuclear and theatre/strategic issues, and could undermine the very stability and reassurance we seek to provide. Finally, defence must help to ensure that this country builds on its particular areas of technological expertise. To have viable and internationally competitive defence industries is extremely important and wherever possible we should nurture and promote the proven areas which offer future promise. Examples are VSTOL technology and many facets of anti-submarine warfare. We should not be tempted away lightly.

Resource Allocation

Which brings me to resources and their allocation, a central issue of defence. The Government has to regulate public spending and the Defence Budget is only one feature, albeit a substantial one, in those considerations. This is no new situation: the problem is how do we face up to a projected plateau in defence spending after 1985/6. We have a whole range of opportunity—a fitting subject for an article itself—in the fields of specialisation, rationalisation, collaboration and interoperability, despite all the undoubted but entirely understandable problems which they pose. Suffice to say, that while recognising the real difficulties, we will persevere.

Historically it is national resources as a whole which underpin defence when a particular crisis arises. In the Falklands Campaign we were able to put to the test the taking up of ships from trade and this was not only outstandingly successful but crucial to our success. This philosophy can and must be extended. The services already have flourishing and expanding reserve and volunteer forces who contribute a great deal, not least in peacetime and during exercise periods but in support of our overall aim. We must do everything we can to prevent war, including using civilian and national resources sensibly to complement the defence forces in peacetime. Moving in a measured and sensible way, we can learn a great deal from commercial practice. For instance, in the area of support it is likely that we can move more confidently away from the traditional intramural approach firmly under the control of the defence organisation towards one in which we rely more on commerical firms. This is a particular development which must continue during the next few years, for without it the size and shape of all three services are likely to come under increasing pressure. We have to make the best use of our national resource base to ensure that the cutting edge of all three services is maintained. Inevitably priorities within the overall defence scene will be considered. I am quite sure that our strategy is well established and that broadly the priorities fall out of that. Clearly, we must be cautious not to get drawn too closely into set scenarios. If anything, the NATO Area tends to channel thought towards the set piece. We must remember the uncertainty of the world scene with which I opened and accept that it does not lend itself to set pieces. In addition, we must not be drawn to situations and assumptions simply because they are quantifiable. This is a great temptation in our computer age, and computers, though they can be invaluable servants, must not become our masters. Deterrence *is* to do with capabilities, but *also* with perception, political will and demonstrable confidence to deal with the unexpected factors which cannot be over-constrained by the quantifiable but arguably simplistic scenarios which have sometimes dominated our considerations in the past. The lessons of the Falklands, embodied in the subsequent White Paper stated: "In allocating defence resources we shall be taking measures which will strengthen our general defence capability by increasing the flexibility, mobility and readiness of all three services for operations in support of NATO and *elsewhere.*" I cannot express it more clearly than that.

The MOD Reorganisation

Before I conclude I will mention, if only briefly, the reorganisation within the Ministry of Defence. Early days these are for substantive comment, and I can say no more than that as we said we would, we are making it work.

One dilemma which I *hope* will remain with us is simply this: the success of defence planning and indeed of the military vocation today is that peace is kept. In doing so, however, military expertise puts itself under threat because by keeping the peace successfully its ability to prove that it is correct in its judgements is to a degree undermined. I would not have it otherwise, although I fear that in practice this uncertain world will from time to time give us the yardstick to prove our planning. Nevertheless, with an increasing number of people moving into positions of influence in government, the Civil Service and elsewhere, who have had no experience of the realities of major war, it becomes more difficult to carry the military viewpoint in the defence debate. It is a truism that defence is a national insurance policy and that, paradoxically, the more satisfactory the policy proves, the greater is the temptation to erode the premium in order to ensure that, for instance, health, education and social security within our particular and cherished way of life are given due investment. It is not for me to make a judgement on the balance of national investment. The military man, by doing his job and keeping the peace, runs the risk of losing, to a degree, his authority. Therefore it behoves those who are concerned with defence decisions to remember this and give due weight to military judgement when developing defence policies which will ensure continued success in peace keeping.

In summary from my perspective the main points are that in this uncertain world of ours it will be important, whilst not departing from the well-proven strategy of NATO, to ensure that we build into all three services flexibility and resilience to deal with that uncertainty. With regard to resources, our main way ahead is to ensure that we get the best value from the capabilities that we possess as an Alliance by promoting inter-operability. As a nation we must continue to draw increasingly on our national resources base as a whole in order to maximise the use of defence resources towards the front line. Finally, I have been writing about defence issues in my capacity as First Sea Lord, which underlines the fact that the three services work very closely today and will do so in the future. Whilst the three principal disciplines of warfare remain quite distinctive, it will be important to meld them to the optimum as a joint concern to ensure success in the future as peacekeepers. Such success will further our national way of life and continue to provide reassurance and stability in a world where these will assuredly be needed, and the defence forces are capable of continuing to meet this challenge.

Index

Abbreviations used: IGN, Imperial German Navy; IJN, Imperial Japanese Navy; USAF, United States Air Force; USN, United States Navy.

Adalbert, Prince of Prussia 67
Ailleret, General 290
Aircraft
  Britain: Buccaneer 276;
    Canberra 242, 243; F.111A 283;
    Gannet 76; Gnat 246;
    Hunter 293; Javelin 248;
    Meteor 242; P-1 248; P-1154 272;
    Phantom 276; Sea Vixen 276;
    Swift 243; Tornado 372;
    TSR-2 272, 283; Valiant 243, 248;
    Venom 243; Victor 246, 248;
    Vulcan 245, 248; Wessex
    (helicopter) 276
  Germany: Gotha 136; Zeppelin 136
  Japan: Short, KF-1 143
  Russia: Backfire "B" 338, 339;
    Mig-15 242
  United States: B-1 386; B-47 224;
    B-52 224; Sabre 242; S-2F
    (helicopter) 276
Alfred Holt Steamship Co. 187
Altham, Captain E. 139–143
Andropov, Y. 392
Armoured Fighting Vehicles
  Russia: BMD 340; BMP 340;
    T-72 340
Armstrong Whitworth 31
Asquith, H. 67, 68
Attlee, C. 258

Bacon, Rear-Admiral R. H. 24–27, 121–125
Baddeley, Sir Vincent xvi
Badger, Rear Admiral, USN 118
Baldwin, S. 153
Banks, M. 326–332
Barclay, C. N. xix
Beconsfield, Earl 9
Beatty, Earl, Admiral of Fleet 164
Bellany, Ian 365–371
Beresford, Admiral Lord Charles 13, 77, 78
Bevan, A. 255
Brancker, Air Marshal Sir S. 132–138

Brassey, T. xv
Brassey, T. (1st Earl) xv–xvii, 36, 37
Brassey, T. (2nd Earl) xvi–xvii, 30–31, 32–37
Brown, George 255
Brown, Harold 386
Buchan, Alastair 263–271
Bülow, Prince 69
Bushnell, D. 49
Buzzard, Rear-Admiral Sir Anthony 256

Cahill, Joe 330
Caprivi, General 28
Carter, President J. 344, 385
Charmes, Gabriel 13
Che Guevara 293
Churchill, Winston 59, 79, 89, 100–101, 335
Clark, Rear-Admiral B. 46
Clausewitz, K. 319
Crepin, General 285
Cunard Steamship Co. 60, 61
Cunningham, Admiral Sir Andrew 191

Dabezies, P. 330–337
Darling, General Sir Kenneth 299
Davy, H. G. 373–379
Dawson, Sir Trevor 90
Desoutter, D. M. 242–250
d'Estaing, Giscard, President 335
Donnelly, C. 354–360
Dulles, J. Foster 221
Dunnell, G. R. 60–61

Edward VII, King 69
Edward VIII, King 170
Eisenhower, General D. 212, 215, 320
Erickson, J. 338–341

Fisher, Admiral Sir John 59, 79
Franklyn, Major-General Sir Harold 217–219

405

406 INDEX

Freedman, Lawrence 380–387
Furness, Shipping Co. 187

Gaitskell, Hugh 255
Garden, Group-Captain T. 388–395
de Gaulle, General Charles 264–268, 287–291, 308, 333–336
Giap, Vo Nguyen 293
Gladstone, W. xvi, 5
Goold-Adams, R. 251–257
Gorshkov, Admiral S. 352
Graham, Sir James 77
Gretton, Vice-Admiral Sir Peter 272–277
Gwynn, Major-General Sir Charles 227

Hall, Rear-Admiral S. 126–131
Hamilton, Lord George 15–17
Hanning, H. 326
Hardy, Admiral Sir Thomas 77
Hawke, Admiral Lord 76
Healey, Denis 266, 269, 281–286, 309
Heinl, Colonel R. D. 306–311
Hesseltine, Michael 399
Hollman, Admiral, IGN 29, 30
Holmes, Admiral E., USN 312–317
Horden, Commander L. H., RN 107–111
Howard, Michael 318–325
Howe, Admiral Lord 197
Hughes, Harrison 188
Hunt, Fletcher 189
Hurd, A. xviii 186–189
Hurford, G. H. 163–165

Inskip, Sir Thomas 174

Jellicoe, Earl, Admiral of Fleet 104
Johnson, President Lyndon 263, 267, 270, 307

Kato, Admiral, IJN 119
Keisuke, Okada, Admiral, IJN 143
Kennedy, President J. F. 258, 289, 307
King, Admiral E. K., USN 166–169
Khruschev, N. 264, 231
King-Hall, Rear-Admiral H. 78
King-Hall, Commander Sir Stephen 252
Kissinger, Henry 383
Koren, G. M. 299–305
Krupps 73

Lennox, Lord Henry 5, 8
Le May, General Curtis, USAF 321

Leyland, J. 41–42, 45–48, 67–71, 76–79, 104–106, 117–119
Londonderry, Lord 164
Luns, J. M. A. H. 342–346

Macmillan, Harold 253, 259
Mahan, Admiral A. T., USN 3, 31, 59, 63, 65
McKenna, Reginald 72–75
McNamara, Robert 258–259, 265, 268–269, 289, 306, 322, 324
Mao Tse-tung 293
Minco, Osumi, Admiral, IJN 143
Missiles
  Britain: Bloodhound 249; Chevaline 369, 370, 372; Fireflash 249; Firestreak 248; Polaris 261, 337, 365, 366, 369–372; Skybolt 235, 261; Trident 365–376; Thunderbird 249
  Russia: ABM 338, 341; ICBM 338; MIRV 338, 341, 351–352; SLBM 338; SSN-7 339; SSN-8 338; SS-16 386; SS-17 381; SS-18 382; SS-19 381; SS-20 352, 386
  United States: ABM 391–395; Minuteman 310, 382, 383; M-X 385; Polaris 310; Poseidon 310, 371; Trident 365–376, 385
Mitterand, President François 335
Moffett, Admiral W., USN 167
Moltke, General von 77, 79
Mondale, Vice President W. 344
Moulton, Major-General J. L. xix, 283–286

Naval Vessels
  Britain: A.1 (S/M) 50; Aboukir 95, 99; Achilles 154; Argus 154; Ark Royal 176; Asturias (hospital ship) 102; Benbow 7; Camperdown 7; Collingwood 7; Courageous 154, 158; Cressy 95, 99; CVA-01 283; Dreadnought 59, 62–65, 74, 75, 80–83; Duncan 74; Eagle 154; Exeter 154; Furious 154; Formidable 74; Glorious 154; Hero 7; Hogue 95, 99; Hood 117; Howe 7; Impérieuse 7; Indomitable 7; Inflexible 60; Invincible 60, 74, 75, 81, 82; King Edward 74; Lord Nelson 74; Medea 15; Neptune 83; Orion 83; Orlando 35; Pandora 15; Pathfinder 97; Polyphemus 49–50;

INDEX

Queen Elizabeth 117; Renown 7; Rodney 7, Royal Sovereign 117; Sans Pareil 7; Sharpshooter 7; Vesuvius 49–50; Warspite 7, 35
France: Gymnote (S/M) 50; Hoche 7; Magenta 7; Marceau 7; Neptune 7; Triton (S/M) 50; Zédé (S/M) 50
Germany: Admiral Graf Spee 165; Brandenburg 42; Deutschland 42, 165; Falke 29; Fürst Bismarck 42; Gefion 29; Hertha 42; Iltis 29; König Grosse Kürfunst 104; König Wilhelm 29, 42; Lübeck 61; Markgraf 104; Ostfriedland 104; Sachsen 42; Sedylitz 104; U-9 97; U-29 97; U-B 107; U-C 107
Japan: Ariuke 141; Chidori 141; Hatsuharu 141; Hatsushimo 141; Jiji Shimpo 119; Manazuru 141; Mikuma 141; Mogami 141; Nenoi 141; Ryujo 141; Tomazuru 141; Wakaba 141; Yogure 141
Russia: Charley (S/M) 339; Foxtrot (S/M) 339; Kara 339; Kiev 339; Krests 339; Krivak 339; Ropucha 339; Tango (S/M) 339; Typhoon (S/M) 386
United States: California 118; Holland (S/M) 50–51; Lake (S/M) 50–51; Ohio (S/M) 371, 386; Tennessee 118
Nelson, Viscount Horatio 14, 76, 104
Niven, Sir John 188
Nimitz, Fleet Admiral Chester, USN 204–207
Norstadt, General L., USAF 291
Northbrook, Earl 5–9, 15

Parker, Lord Justice 329–330
P & O Steamship Co. 186
Parsons, Sir Charles 80
Persius, Captain, IGN 102
Pohl, Admiral von, IGN 102
Pompidou, President George 334
Pratt, Admiral W., USN 149

Reagan, President Ronald 380, 385–386, 392
Richardson, A. xviii
Richardson, J. xviii

Richmond, Admiral Sir Herbert 192
Robinson, Commander C. N. xviii, xix, 38–40, 49, 89–91, 96–99
Ross, H. M. xviii
Runciman, Philip 188
Russell, Bertrand 252, 253

Sandys, Duncan 235, 254
Sato, Naotake 139
Scott, Admiral Sir Percy 64
"Securus" 178–181
Sidmouth, Viscount 5
Sims, Admiral W. S., USN 59, 62–65
Sixsmith, Major-General E. K. G. 226–232
Stewart, Oliver 199–203
Stosch, Albrecht von 67

Tagkaki, Commander, IJN 144–151
Templer, General Sir Gerald 227
Thompson, Sir Robert 293–298
Thomson, Sir Vernon 188
Thursfield, Rear-Admiral H. G. xix, 174–177, 190–198, 237–241
Thursfield, J. R. 18–22, 23, 43–44
Tirpitz, Grand Admiral von, IGN 107
Tryon, Admiral Sir George 19

"Volage" 153–158

Watts, E. H. 170–173
Wavell, Field Marshal Lord 191
Weddigen, Lieutenant Commander, IGN 97
Weyl, E. 28–29
White, Sir William 80–83
William II, Emperor 80–83
Wilkinson, Spenser 78
Willison, Lieutenant-General Sir David 347–353
Wilson, Admiral Sir Arthur 79
Wilson, President Woodrow 146
Wyndham, Colonel E. H. 212–216

Yool, Air Vice-Marshal H. M. 221–225, 258–262

Zhukov, Marshal 341